19th Russian Youth Conference on Physics of Semiconductors and Nanostructures, Opto- and Nanoelectronics (RYCPS 2017)

Journal of Physics: Conference Series Volume 993

St. Petersburg, Russia
27 November - 1 December 2017

Editors:

R.A. Suris
L.E. Vorobjev
D.A. Firsov
V.A. Shalygin

ISBN: 978-1-5108-6164-0
ISSN: 1742-6588

Printed from e-media with permission by:

Curran Associates, Inc.
57 Morehouse Lane
Red Hook, NY 12571

Some format issues inherent in the e-media version may also appear in this print version.

Printed by Curran Associates, Inc. (2018)

For permission requests, please contact the Institute of Physics
at the address below.

Institute of Physics
Dirac House, Temple Back
Bristol BS1 6BE UK

Phone: 44 1 17 929 7481
Fax: 44 1 17 920 0979

techtracking@iop.org

Additional copies of this publication are available from:

Curran Associates, Inc.
57 Morehouse Lane
Red Hook, NY 12571 USA
Phone: 845-758-0400
Fax: 845-758-2633
Email: curran@proceedings.com
Web: www.proceedings.com

Table of contents

Volume 993

19th Russian Youth Conference on Physics of Semiconductors and Nanostructures, Opto- and Nanoelectronics

27 November to 1 December 2017, St. Petersburg, Russian Federation

Accepted papers received: 19 March 2018
Published online: 10 April 2018

Preface

Papers

Bulk properties of semiconductors

Structure growth, surface and interfaces

Heterostructures, superlattices and quantum wells

Quantum dots, quantum wires and other low-dimensional systems

Optoelectronics and nanoelectronics devices

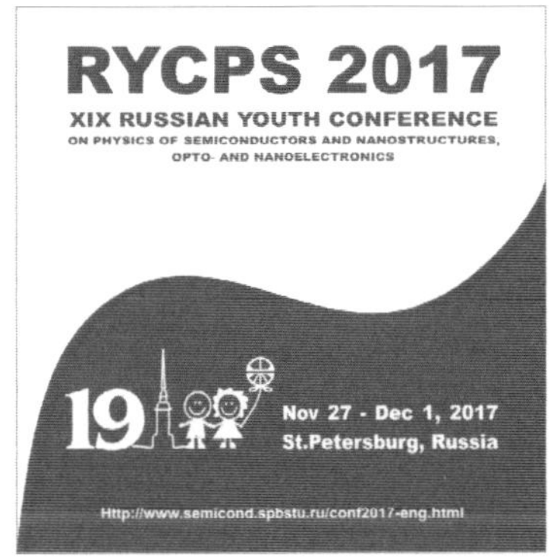

Editorial Preface

The 19th Russian Young Conference on Physics of Semiconductors and Nanostructures, Opto- and Nanoelectronics was held on November 27 – December 1, 2017 at Peter the Great St. Petersburg Polytechnic University.

The program of the Conference included semiconductor technology, heterostructures with quantum wells and quantum dots, opt- and nanoelectronic devices and new materials. A large number of participants with more than 200 attendees from many regions of Russia provided a perfect platform for the valuable discussions between students and experienced scientists.

Conference included three invited talks given by Doctor of Science G.F. Glinskii ("Empirical pseudopotential method in the theory of semiconductor", Doctor of Science M.Z. Schwartz ("Highly efficient cascade photovoltaic solar radiation converters") and Doctor of Science I.V. Rozhansky ("Anomalous and topological Hall effect in nanostructures"). Students, graduate and postgraduate students presented their results on plenary and poster sessions. The total number of accepted papers published in Russian (the official conference language) was 121. Here we publish 40 of them in English.

Like previous years, the participants were involved in the competition for the best report. Certificates and cash prizes were awarded to a number of participants for the presentations selected by the Program Committee. The special E.F. Gross Prize was given for the best presentation in semiconductor optics. Works with potential applications were recommended for participation in the following competition for support from the Russian Foundation for Assistance to Small Innovative Enterprises in Science and Technology.

The conference was supported by the Russian Foundation for Basic Research and the innovation company "ATC – Semiconductor Devices", St. Petersburg.

The official conference website is http://www.semicond.spbstu.ru/conf2017.

Editors

R.A. Suris
L.E. Vorobjev
D.A. Firsov
V.A. Shalygin

IOP Conf. Series: Journal of Physics: Conf. Series **993** (2018) 011001 doi:10.1088/1742-6596/993/1/011001

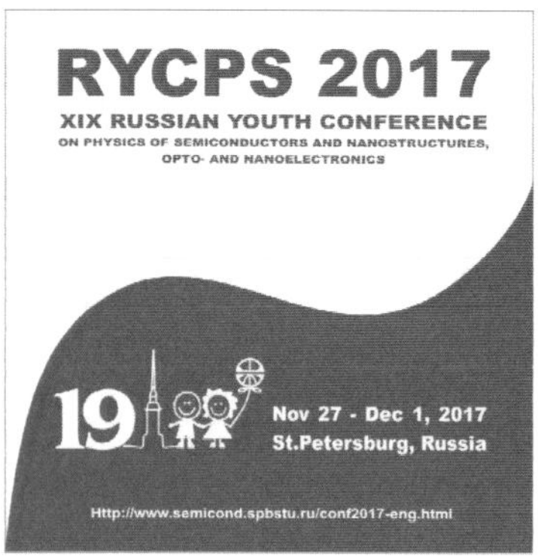

The winner of the E.F. Gross Prize for the best presentation in semiconductor optics

Kozyrev Nikolay

A postgraduate student at Ioffe Institute
(Saint Petersburg)

Presentation:

" Spin-flip Raman scattering in CdMnSe quantum dots"

IOP Conf. Series: Journal of Physics: Conf. Series **993** (2018) 011001 doi:10.1088/1742-6596/993/1/011001

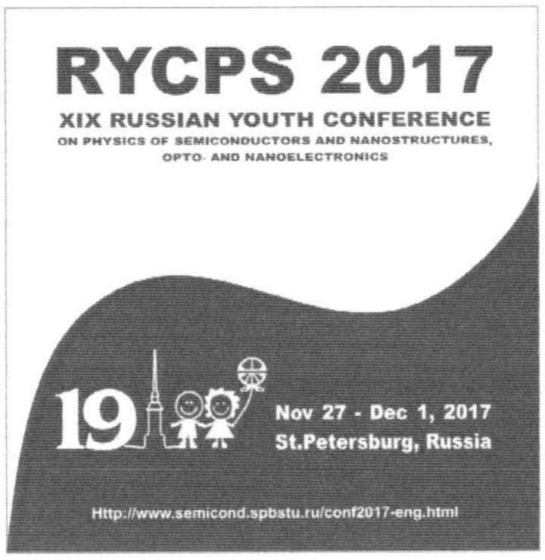

Winners of the competition for the best report

First Prize

Makhov Ivan

a postgraduate student at Peter the Great St. Petersburg Polytechnic University

Presentation:

"Impurity-assisted terahertz photoluminescence in quantum wells under conditions of interband stimulated emission"

First Prize

Muretova Maria

a student at Academic University of the Russian Academy of Sciences (Saint Petersburg)

Presentation:

"Design of Aluminium-free laser heterostructures with asymmetric barrier layers"

IOP Conf. Series: Journal of Physics: Conf. Series **993** (2018) 011002 doi:10.1088/1742-6596/993/1/011002

Peer review statement

All papers published in this volume of *Journal of Physics: Conference Series* have been peer reviewed through processes administered by the proceedings Editors. Reviews were conducted by expert referees to the professional and scientific standards expected of a proceedings journal published by IOP Publishing.

Polarization anisotropy of electroluminescence in indium antimonide

I S Makhov, V Yu Panevin, D A Firsov, L E Vorobjev and M Ya Vinnichenko

Peter the Great St. Petersburg Polytechnic University, 195251 St. Petersburg, Russia

E-mail: MakhovIS@spbstu.ru

Abstract. Mid-infrared electroluminescence spectra under impact ionization conditions in strong electric fields are studied experimentally in monocrystalline *n*-InSb. The emission band observed in electroluminescence spectra is associated with a radiative recombination of electrons and holes between the conduction and valence bands. Spectral measurements of the electroluminescence intensity for different light polarizations relative to the direction of the applied electric field reveal the polarization anisotropy of the electroluminescence intensity in InSb, which is caused by the anisotropy of the distribution function of nonequilibrium electrons as well as by the anisotropy of the angular dependence of the interband optical matrix element in momentum space.

1. Introduction

The interest in narrowband semiconductor structures is caused by their potential application to the elaboration of mid-infrared sources and detectors. A lot of recent publications are devoted to studying the luminescence of narrowband A3B5-based [1-3], as well as HgCdTe-based [4-6] alloys and superlattices using classic techniques for luminescence excitation – interband photoexcitation or current injection in *p-i-n*-diode structures. In the current paper we are paying the attention to an example of an alternative way for nonequilibrium carrier excitation using an impact ionization process. On the one hand, it does not require a complicated *p-i-n* design of the sample structure, but on the other hand, it uses a more technological technique than photoluminescence study. Additionally, this approach provides a way to study new electro-optical properties of semiconductors. A small energy gap and high electron mobility of InSb allows reaching the impact ionization at low enough electric fields that enables observation of luminescence. Significant electron heating accompanying a further increase of the electric field should result in the anisotropy of the distribution function in ***k***-space. Taking into account the anisotropy of the interband optical matrix element [7], the polarization anisotropy of interband radiation relative to the direction of the electric field should be observed. The polarization anisotropy of integrated interband electroluminescence was observed earlier in bulk InSb in Ref. [8], but spectral dependencies of such polarization anisotropy have not been studied yet.

2. Sample and experimental setups

As a sample for experimental studies, we used monocrystalline *n*-InSb with a low donor impurity concentration of about $6 \cdot 10^{13}$ cm^{-3} and an electron mobility of about $6 \cdot 10^{5}$ cm^{2}/(V·s) at liquid nitrogen temperature. The sample dimensions were $0.17 \times 3 \times 4.3$ mm^{3}. Two indium contacts were made in the

IOP Conf. Series: Journal of Physics: Conf. Series **993** (2018) 012001 doi:10.1088/1742-6596/993/1/012001

form of two parallel stripes along the 3 mm sample edge. The contacts were annealed in a dry nitrogen atmosphere at 300ºC. In order to reach the impact ionization conditions, but to reduce an overheating of the sample the electric field was applied in the short pulse regime (250 ns / 10 Hz). For the electroluminescence study, the sample was mounted to a cold finger of a liquid nitrogen bath optical cryostat. The cold finger temperature was about 85 K. The electroluminescence spectra were studied using a Fourier transform infrared vacuum spectrometer operating in a step-scan mode supplemented with a KBr beamsplitter and a HgCdTe photodetector. The photodetector signal was measured with a gated integrator SR250.

3. Experimental results and discussion

The current-voltage characteristic of the *n*-InSb sample placed directly in liquid nitrogen at the temperature $T = 77$ K is presented in figure 1. At the electric field values smaller than 200 V/cm, the current density is approximately linearly dependent on the electric field. At the electric field higher than 200 V/cm, the current density starts to increase sharply, that is associated with the impact ionization processes (this field is marked with the arrow labeled "impact ionization threshold" in figure 1). At strong electric fields, the electron mobility decreases because of the polar optical phonon scattering rise, which results in slowing down the current density rise at electric fields higher than 600 V/cm.

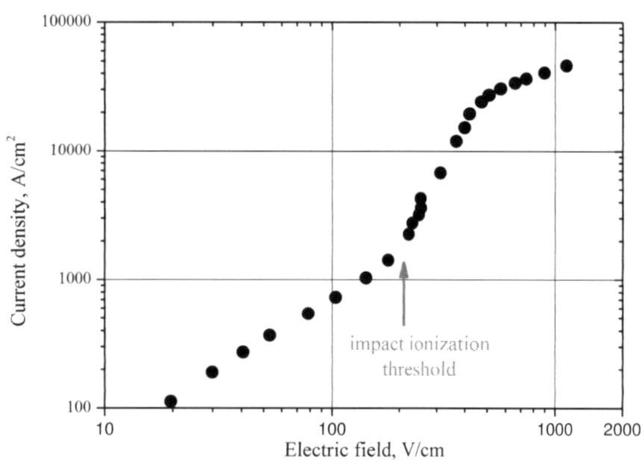

Figure 1. Current-voltage characteristic of the *n*-InSb sample at the temperature $T = 77$ K.

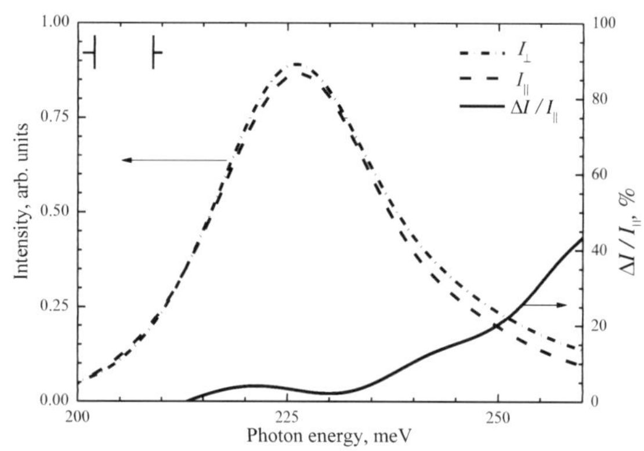

Figure 2. Electroluminescence spectra (dashed and dash-dot curves) of the *n*-InSb sample for different light polarizations (perpendicular $I_{\perp}$ and along $I_{\parallel}$, respectively) relative to the direction of the applied electric field (1 kV/cm) and the spectral dependence of the relative difference between electroluminescence intensities of different light polarizations (solid curve). Spectral resolution is shown on the plot.

Electroluminescence spectra of different light polarizations with respect to the direction of the applied electric field measured in a cold finger liquid nitrogen bath cryostat at the strong electric field (1 kV/cm) are presented in figure 2 (curves marked in the legend of figure 2 as $I_{\parallel}$ and $I_{\perp}$ correspond, respectively, to a light polarization along and transverse to the direction of the electric field). The broad emission band observed in electroluminescence spectra is associated with interband electron-hole radiative recombination because its spectral position is located near the InSb bandgap value (0.23 eV at the $T = 85$ K). A slight difference in spectral positions of the low-energy part of the luminescence band and the bandgap value of InSb could be connected with the rise of self-absorption due to the pinch effect [9], as well as with sample heating in strong electric fields.

One can see that electroluminescence spectra for different light polarizations (see figure 2) differ. In order to better demonstrate the anisotropy of the electroluminescence intensity we plot in figure 2 the relative intensity difference for two light polarizations $\frac{\Delta I}{I_{||}}$, where $\Delta I = (I_{\perp} - I_{||})$.

To compare the obtained spectral dependence of the relative intensity difference with the theory, let us consider the following approach. At the equilibrium, at low carrier concentration and energies, we can use the parabolic band structure model and the Maxwell-Boltzmann distribution function. In the case of nonequilibrium, the distribution function of hot electrons becomes the shifted Maxwell-Boltzmann one:

$$f(\boldsymbol{p}) \sim \exp \frac{(\boldsymbol{p}-\boldsymbol{p}_{dr})^2}{2 \cdot m_e \cdot k_B \cdot T_e}, \tag{1}$$

where $\boldsymbol{p}$ is the electron momentum, $\boldsymbol{p}_{dr}$ is the electron drift momentum in the direction of the applied electric field, m_e is the electron effective mass, k_B is the Boltzmann constant, and T_e is the hot electron temperature. The electroluminescence anisotropy degree $(\frac{I_{\perp}-I_{||}}{I_{||}})$ was estimated in Ref. [8] using the angular dependence of the interband optical matrix element provided that $\boldsymbol{p}_{dr}$ is small relative to $\boldsymbol{p}$:

$$\frac{I_{\perp}-I_{||}}{I_{||}} = 0.1 \cdot (\frac{\boldsymbol{p} \cdot \boldsymbol{p}_{dr}}{m_e \cdot k_B \cdot T_e})^2, \tag{2}$$

where $\boldsymbol{p}$ is the electron momentum of an electron participating in interband radiative recombination.

Regardless of the type of the energy band structure of a semiconductor, the electron momentum $\boldsymbol{p}$ rises with the photon energy. For the accepted parabolic band structure model, the electron momentum $\boldsymbol{p}$ is related to the photon energy $h\nu$ by the following expression:

$$p = \sqrt{2 \cdot m_{eh} \cdot (h\nu - E_g)}, \tag{3}$$

where E_g is the InSb bandgap and m_{eh} is the electron-hole reduced mass. According to the equations (2) and (3), the anisotropy degree should increase with a photon energy increase:

$$\frac{I_{\perp}-I_{||}}{I_{||}} = 0.2 \cdot (\frac{\boldsymbol{p}_{dr}}{m_e \cdot k_B \cdot T_e})^2 \cdot m_{eh} \cdot (h\nu - E_g). \tag{4}$$

From the spectral dependence of the relative difference between electroluminescence intensities for different light polarizations presented with a solid curve in figure 2, the experimentally determined anisotropy increases with the photon energy increase, which corresponds to the theoretical dependence (see Eq. (4)). Thus, we can connect the observed polarization anisotropy of electroluminescence with the polarization anisotropy arising in a strong heating electric field and is related to the anisotropy of the distribution function of nonequilibrium electrons and the angular anisotropy of the interband optical matrix element [7] in momentum space.

Conclusion

Luminescence spectra in *n*-InSb in strong heating electric fields have been studied. The spectral dependence of the polarization anisotropy of electroluminescence intensity in a strong electric field in InSb is revealed. The obtained results are in good agreement with theoretical data.

Acknowledgments

This work was supported by the Russian Foundation for Basic Research (grant 16-02-00863), Russian Federation President grant for young scientists No-4616.2016.2 and the Ministry of Education and Science of the Russian Federation (state assignments, projects 3.933.2017/4.6 and 3.5518.2017/7.8).

References

[1] Belenky G, Wang D, Lin Y, Donetsky D, Kipshidze G, Shterengas L, Westerfeld D, Sarney W L and Svensson S P 2013 *Applied Physics Letters* **102** 111108

IOP Conf. Series: Journal of Physics: Conf. Series **993** (2018) 012001 doi:10.1088/1742-6596/993/1/012001

[2] Selivanov A V, Makhov I S, Panevin V Yu, Sofronov A N, Firsov D A, Vorobjev L E and Shterengas L 2015 *Journal of Physics: Conference Series* **643** 012078

[3] Rogalski A, Martyniuk P and Kopytko M 2017 *Applied Physics Review* **4** 031304

[4] Rhiger D R 2011 *Journal of Electronic Materials* **40** 1815

[5] Morozov S V, Rumyantsev V V, Antonov A V, Kadykov A M, Maremyanin K V, Kudryavtsev K E, Mikhailov N N, Svoretskii S A and Gavrilenko V I 2014 *Applied Physics Letters* **105(2)** 022102

[6] Rumyantsev V V, Fadeev M A, Morozov S V, Dubinov A A, Kudryavtsev K E, Kadykov A M, Tuzov I V, Dvoretskii S A, Mikhailov N N, Gavrilenko V I and Teppe F 2016 *Semiconductors* **50(12)** 1651

[7] Baumgardner C A and Woodruff T O 1969 *Physical Review* **173(30)** 746

[8] Vasil'eva M A, Vorob'ev L E and Stafeev V I 1969 *Soviet Physics Semiconductors* **3(9)** 1374

[9] Shotov A P, Grishechkina S P and Muminov R A 1966 *Soviet Physics JETP* **23(6)** 1017

Incoherent interlayer transport in single-crystal films of $Nd_{2-x}Ce_xCuO_4/SrTiO_3$

A S Klepikova[1], D S Petukhov[1], O E Petukhova[1], T B Charikova[1,2], N G Shelushinina[1] and A A Ivanov[3]

[1] M.N.Mikheev Institute of Metal Physics Ural Branch of RAS, Ekaterinburg, Russia

[2] Ural Federal University, Ekaterinburg, Russia

[3] National Research Nuclear University MEPhI, Moscow, Russia

E-mail: klepikova@imp.uran.ru

Abstract. The conductivity in $Nd_{2-x}Ce_xCuO_4/SrTiO_3$ single-crystal films with the c-axis perpendicular or parallel to the substrate plane was studied. A comparison of the results for two types of films reveals the quasi-two-dimensional character of transport processes in stoichiometric (optimally annealed) single crystal films with $0.135 \le x \le 0.15$ at the antiferromagnetic - superconducting phase transition.

1. Introduction

The problem of the resistivity anisotropy in the normal state of copper oxide systems has long attracted the attention of researchers. The resistivity in different directions differs not only in the magnitude but also in the character of its temperature dependence. The difference between a metallic behavior in the conducting planes ($d\rho_{ab}/dT > 0$) and an insulating behavior in the *c*- direction ($d\rho_c/dT < 0$) is a subject paid close attention by theorists [1-3], as it can give indications of the models appropriate for analyzing the normal-state resistivity of cuprates.

A strong anisotropy of the conducting properties ($\rho_c/\rho_{ab} >> 1$) when a nonmetallic conductivity along the *c* axis is combined with a metallic conductivity in the *ab* plane was repeatedly observed in underdoped and optimally doped *hole-type* HTSCs [4, 5]. This is evidence of the quasi-two-dimensionality of oxide systems that consist of highly mobile CuO_2 layers separated by buffer layers [3]. The nonmetallic character of ρ_c in most superconducting high-T_c compounds suggests an unconventional conduction mechanism between CuO_2 planes.

The cerium-doped cuprate of $Nd_{2-x}Ce_xCuO_{4+\delta}$ has a layered quasi-two-dimensional perovskite-like crystal structure [6]. As compared to other cuprate superconductors, $Nd_{2-x}Ce_xCuO_{4+\delta}$ has many unique properties that make it an attractive subject for investigations. This is a superconductor with an *electron-type* conductivity whose structure contains a single CuO_2 plane per unit cell. In optimally annealed crystals, there are no apical oxygen atoms between neighboring conducting CuO_2 planes. Therefore, $Nd_{2-x}Ce_xCuO_4$ crystals have clearly pronounced two-dimensional properties.

The $Nd_{2-x}Ce_xCuO_{4+\delta}$ compound is characterized by an ability to reversibly absorb and desorb oxygen, and its properties are very sensitive to the oxygen content. Pure Nd_2CuO_4 is dielectric, and superconductivity appears only in related solid solutions of $Nd_{2-x}Ce_xCuO_{4+\delta}$ ($x > 0.13$) after additional annealing in an oxygen-free atmosphere [7]. The main goal of such annealing is to remove excess nonstoichiometric oxygen (predominantly from the apical positions). This treatment reduces disorder, so electrons become delocalized and exhibit superconducting properties. After optimum doping and annealing, an $Nd_{2-x}Ce_xCuO_4$ crystal consists of a set of isolated conducting CuO_2 planes spaced 6 Å apart and is strongly anisotropic.

In bulk $Nd_{2-x}Ce_xCuO_{4+\delta}$ single crystals, a very strong anisotropy of resistivity is observed ($\rho_c / \rho_{ab} \sim 10^4$) [8-10]. However, the nonmetallic temperature dependence of $\rho_c(T)$ is quite rare. This is apparently due to the enhanced sensitivity of transport properties of the Nd system to the content of non-stoichiometric oxygen (δ) and difficulties in achieving an optimal annealing regime ($\delta \to 0$) for bulk samples. On the other hand, single-crystal $Nd_{2-x}Ce_xCuO_4/SrTiO_3$ films (up to 5000 Å thick) are well suited for different annealing procedures.

Earlier we obtained and investigated high quality $Nd_{2-x}Ce_xCuO_{4+\delta}/SrTiO_3$ single-crystal films with the c axis perpendicular [11] and parallel [12] to the substrate plane ((001) and (1 $\bar{1}$ 0) orientations, respectively). In [13], $\rho_{ab}(T)$ and $\rho_c(T)$ dependences for $Nd_{2-x}Ce_xCuO_{4+\delta}/SrTiO_3$ single-crystal films with $x = 0.12$ (nonsuperconducting underdoped films), $x = 0.15$ (optimally doped films with a maximal T_c) and $x = 0$ 17; 0.20 (overdoped films) at (001) and (1 $\bar{1}$ 0) orientations have been comparatively analyzed.

Emphasis has been placed on the overdoped region, that is, in comparison with the region of optimal doping, the "damping" region of superconductivity with a gradual decrease in T_c down to $T_c \to 0$ for $x \approx 0.22$. In our Nd-system (an n-type HTSC), we have found a transition from a quasi-2D anisotropic system ($d\rho_c / dT < 0, d\rho_{ab} / dT > 0$) to a 3D system with metallic conductivity both in the ab- plane and in the c-axis direction ($d\rho_c / dT > 0, d\rho_{ab} / dT > 0$) with increased doping, x, and at higher temperatures, which is similar to that previously found in the La- system (a p-type HTSC) [5]. A conclusion about the correlation of quasi two-dimensionality of the system with the emergence of superconductivity in copper oxide compounds can be made from [5, 13].

The region of the appearance of superconductivity with increasing cerium doping (at $x \approx 0.13 \div 0.14$) thus remained unexplored. Presently, the advances in technology have allowed us to grow high-quality $Nd_{2-x}Ce_xCuO_{4+\delta}/SrTiO_3$ single-crystal films with $x = 0.135$ and $x = 0.145$ in which the c- axis was both normal ((100) films) and parallel to the substrate plane ((1 $\bar{1}$ 0) films) in order to study the processes of charge carrier transfer in the region of the antiferromagnetic (AFM) – superconducting (SC) quantum phase transition.

The study of the properties of these samples under optimal annealing in conjunction with the optimally doped ($x = 0.15$) samples is the subject of this work. We are the first to experimentally observe the $\rho_c(T)$ dependences with nonmetallic behavior for samples with $x = 0.135$ and 0.145 that is near the threshold x values for the AFM – SC phase transition. A comparison of the results obtained for the two types of films allowed us to demonstrate the quasi-two-dimensional character of carrier transfer in them.

2. Experimental results and discussion

In table 1, we present the characteristic parameters (film thickness and the onset and completion temperatures of the superconducting transition) for both types of the investigated $Nd_{2-x}Ce_xCuO_4/SrTiO_3$ films. The results for in-plane or out-of-plane resistivities as functions of the temperature in the samples optimally annealed in vacuum with $x = 0.135$, 0.145 and 0.15 are shown in figure 1 (a, b).

It is seen that the normal state conductivity in the *ab*- plane is metallic with the manifestation of 2D weak-localization effects at $T \leq 70K$ in a sample with $x = 0.135$. The normal state out-of-plane resistivity ρ_c across the blocking layers is large with respect to the in-plane resistivity ρ_{ab} and demonstrates a non-metallic temperature dependence ($d\rho_c / dT < 0$) for all the investigated films up to 300 K.

The resistivity anisotropy coefficient is large even at room temperature ($\rho_c / \rho_{ab} \cong 10 - 10^2$) for all the investigated films (figure 2). This parameter dramatically increases with decreasing temperature and reaches the values of $\rho_c / \rho_{ab} \cong 10^3$ for the compounds with x =0.145 and 0.135 due to resistivity divergence $\rho_c(T) \sim 1/T$ at low temperatures (see figure 1).

Table 1. The characteristic parameters for both types of investigated $Nd_{2-x}Ce_xCuO_4/SrTiO_3$ films.

Type of film	x	T_c^{onset}, K	T_c, K	d, Å
	0.135	13.7	9.6	3300
(001)	0.145	15.7	10.7	2500
	0.15	23.5	22	2000
	0.135	12.8	-	4200
(1$\bar{1}$0)	0.145	15.8	6.3	4000
	0.15	21.2	7.8	1400

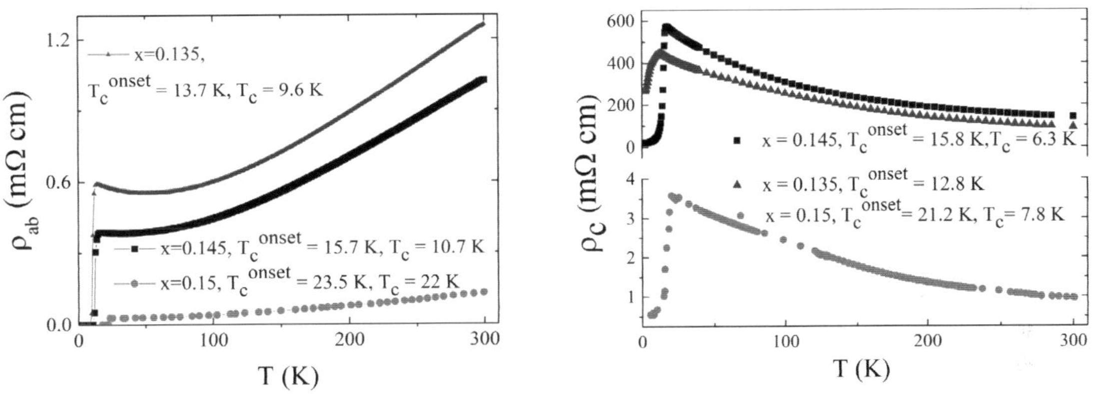

Figure 1. Temperature dependencies of in-plane (left), ρ_{ab}, and out-of-plane (right), ρ_c, and resistivity of $Nd_{2-x}Ce_xCuO_4/SrTiO_3$ films at different doping levels and optimal annealing for x = 0.135; 0.145 and 0.15.

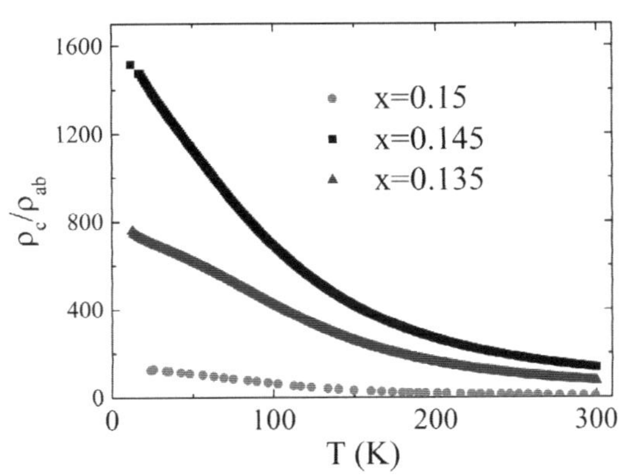

Figure 2. Temperature dependence of the resistivity anisotropy coefficient for optimally annealed $Nd_{2-x}Ce_xCuO_4/SrTiO_3$ films with various cerium contents.

A number of microscopic models for describing the deviation from a phase coherence in c-axis transport have been proposed [1-3, 14-17]. However, only the simplest one-dimensional Kronig-Penny model with its ideal periodicity (and thus the coherence) can explain the metallic nature of the interlayer conductivity.

In [14-17], the non-metallic behavior of $\rho_c(T)$ in layered oxides was attributed to *incoherent* tunneling of charge carriers in the *c*-axis direction. Incoherent transport between CuO_2 layers occurs when the probability of carrier scattering in the plane ($\hbar/\tau$) is much higher than the interlayer hopping integral t_c ($\equiv \hbar/\tau_{esc}$) between the planes. Here, τ is the carrier relaxation time in the plane, and τ_{esc} is the escape time from the given plane to the neighboring one.

If an electron experiences many collisions before moving to another plane, then the subsequent tunneling processes between the planes are uncorrelated. The interlayer conductivity is then proportional to the rate of tunneling between just two adjacent layers, and the diffusion coefficient along the layers ($D_{\|}$) and across them ($D_{\perp}$) reads (see [15], [16] and references therein):

$$D_{\|} = l^2/2\tau, \quad D_{\perp} = c^2\tau/2\tau_{esc}^2, \tag{1}$$

where l is the mean free path in the ab-plane, and c = 6 Å is the distance between neighboring CuO_2layers. Thus, we can empirically estimate the ratio of the characteristic times as:

$$\tau_{esc}/\tau = (c/l)\sqrt{\rho_c/\rho_{ab}} \tag{2}$$

Using the estimates of the mean free path and the anisotropy coefficient (ρ_c/ρ_{ab}) at $T < 50$K, we get $\tau_{esc}/\tau \cong 10$ for x = 0.135; 0.145 and $\tau_{esc}/\tau \cong 2$ for x = 0.15. Thus, at low temperatures, the transport in the c-direction is incoherent ($\tau_{esc} >> \tau$) for underdoped Nd systems and approaches a coherent transfer ($\tau_{esc} \approx \tau$) for the optimally doped one.

We emphasize that the low-temperature anisotropy coefficient is maximal for x = 0.145 ($\rho_c/\rho_{ab} \sim 2\cdot 10^3$) and x = 0.135 ($\rho_c/\rho_{ab} \sim 10^3$) and much less for x = 0.15 ($\rho_c/\rho_{ab} \sim 10^2$) (see figure 2), in contrast to the situation in the overdoped region where the values of ρ_c/ρ_{ab} and T_c directly correlate (see figure 4 in [13]).

Using the model of a natural superlattice (when CuO_2 layers are quantum wells and Nd(Ce)O blocks serve as barriers of the effective height Δ) [17-19], we can consider the disorder that is undoubtedly inherent in this system (the chaotic impurity potential) as a cause of the incoherence in the c-axis transport.

Indeed, if the wave function of the electron is localized in the c-direction with a characteristic radius of localization r_0, which is less than the distance between adjacent CuO_2 planes, then, according to [20]:

$$t_c \sim \exp(-c/r_0)\exp(-\varepsilon_a/kT), \tag{3}$$

where $r_0^{-1} = \sqrt{2m\Delta/\hbar^2}$, and ε_a is the spread of electron energy in the wells due to the fluctuations of Δ values, the same as in the one-dimensional Anderson model.

The first factor in (3) (overlap integral) determines the dependence of the transition probability between the layers on the barrier height, and the second one leads to a nonmetallic temperature dependence of the conductivity at low temperatures (analogously to the conductivity within the impurity band of semiconductors [21]).

With increasing temperature, the contribution to the conductivity associated with the thermal activation of carriers across each barrier begins to play an increasingly important role [17]:

$$\sigma_{term} \sim \exp(-\Delta / kT) \tag{3}$$

and for $kT > \Delta$, there should be a transition to "metallic" conductivity. Based on the form of the ρ_c *(T)* temperature dependence, we can estimate the effective barrier height as Δ > 300 K for all the investigated samples.

3. Conclusions

We have grown high-quality (1 $\bar{1}$ 0) single-crystal $Nd_{2-x}Ce_xCuO_4/SrTiO_3$ films in which the *c* axis is parallel to the substrate plane, in order to exclude the effect of carrier transfer in the CuO_2 planes on the transport properties along the *c* direction. As a result, we were able to experimentally observe nonmetallic behavior of the normal state $\rho_c(T)$ dependence ($d\rho_c/dT < 0$) in $Nd_{2-x}Ce_xCuO_4$ samples both with $x = 0.135$; 0.145 near the AFM - SC quantum phase transition and at optimal doping with $x = 0.15$ in the SC phase.

A combination of metallic behavior of the $\rho_{ab}(T)$ dependence and nonmetallic character of the $\rho_c(T)$ dependence for stoichiometric (optimally annealed) underdoped and optimally doped NdCeCuO samples is an indication that the investigated system is quasi-two-dimensional. Specifically, two-dimensional conduction occurs along delocalized states in CuO_2 planes, and incoherent tunneling (hopping) takes place across the blocking Nd(Ce)O layers along the *c* direction.

In the present work, we found that the transport in the *c*-direction is sufficiently incoherent in the region of coexisting antiferromagnetic and superconducting ordering at $x = (0.13 \div 0.14)$ and becomes more coherent in the region of SC ordering at $x = 0.15$. We assume that the fluctuations due to competition between the AFM and SC types of ordering, as well as the impurity disorder-induced spread of electron energy levels in CuO_2 quantum wells, promote the incoherent character of *c*-axis transport in the investigated structures.

Acknowledgments
This study was performed under the state assignment of the Federal Agency for Scientific Organizations of Russia (theme Electron, no. 01201463326) and in part supported by the Russian Foundation for Basic Research (project no. 18-02-00192) and by the Ministry of Education and Science (contract no. 14 Z 50.31.0025).

References

[1] Anderson P W and Zou Z 1988 *Phys. Rev. Lett.* **60** 132
[2] Kotliar G, Abrahams E, Ruckenstein A E *et al* 1991 *Europhys. Lett.* **15** 655
[3] Sadovskii M V *Superconductivity and Localization* 2000 (World Scientific).
[4] Ito T, Takagi H, Ishibashi S *et al* 1991 *Nature* **350** 596
[5] Nakamura Y, Uchida S 1993 *Phys.Rev.B* **47** 8369
[6] Plakida N M 2012 *High-Temperature Superconductivity: Experiment and Theory* (Springer Science & Business Media)
[7] Tokura Y, Takagi H, and Uchida S 1989 *Nature* **337** 345
[8] Wang Z Z, Chian T R and Ong N R 1991 *Phys.Rev.B* **43** 3020
[9] Charikova T B, Ponomarev A I, Ignatenkov A N *et al* 2001 *Phys.Met. Metallogr.* **91** 150
[10] Charikova T, Ignatenkov A, Ponomarev A *et al* 2003 *Physica C (Amsterdam)* **388–389** 323
[11] Harus G I, Ponomarev A I, Charikova T B *et al* 2002 *Physica C (Amsterdam)* **383** 207
[12] Charikova T B, Ponomarev A I, Kharus G I *et al* 2007 *JETP* **105**, 626
[13] Ponomarev A I, Charikova T B, Kharus G I *et al* 2007 *Phys.Met. Metallogr.* **104** 72
[14] Kumar N and Jayannavar A M 1992 *Phys. Rev. B* **45** 5001
[15] Cassam-Chenai A and Maily D 1995 *Phys.Rev.B* **52** 1984
[16] McKenzie R H and Moses P 1998 *Phys.Rev.Lett.* **81** 4492
[17] Giura M, Fastampa R, Sarti S and Silva E 2003 *Phys.Rev.B* **68** 134505
[18] Kapaev V V and Kopaev Yu V 1998 *Jetp Letters* **68** 223

[19] Ponomarev A I, Ignatenkov A N, Sabirzyanova L D *et al* 1998 *Proc. Int. Conf. Phys. Semicond. (Ierusalem)* CD-ROM, Sec.5, Subsec.A, N 36
[20] Landsberg P 1974 *Zadachi po termodinamike i statisticheskoi fizike* (Moscow : "Mir") ch.19
[21] Shklovskii B L and Efros A L 1979 *Elektronnye svoistva legirovannych poluprovodnikov,* (Moscow : "Nauka")

$MgFe_2O_4$ phase effect on magnetic properties of $BiFeO_3$ in ceramic composites of (1-*x*) $BiFeO_3$–*x* $MgFe_2O_4$

A A Kamynin, M V Khakhlenkov, N A Tolstykh, A I Bocharov and S A Gridnev

Voronezh State Technical University, Moskovsky prospect, 14, 394026, Voronezh, Russia

E-mail: silentcurve@gmail.com

Abstract. A two-stage ceramic technology was used for producing a family of (1-*x*) $BiFeO_3$ - *x* $MgFe_2O_4$ heterophase systems. The study of the magnetic hysteresis loop revealed the presence of a ferrimagnetic response. The experimentally obtained concentration and temperature dependences of the magnetic permeability made it possible to consider the influence of the $BiFeO_3$ phase on the magnetic response of the studied samples.

1. Introduction

In recent years, there has been an increased need for technologies related to the creation and improvement of elements that combine the advantages of non-volatile magnetic memory, low power consumption and high-speed electrical information processing systems [1]. To solve this problem, one could use magnetoelectric multiferroics in which the magnetic state is varied by applying an electric field [2], which could significantly improve energy efficiency.

One of the promising multiferroics now is the bismuth ferrite $BiFeO_3$ (BF). The main advantage of this material is high Curie and Neel temperatures (T_C = 1083 K, T_N = 643 K). In bismuth ferrite, all types of magnetoelectric interactions described in [3] take place. The main BF problem is the absence of the bulk magnetic moment. This is caused by spatial modulation of the magnetization with a period of 62 nm [4, 5], which leads to a "zero" magnetoelectric response. Attempts to eliminate this disadvantage used various methods to destroy the spin cycloid: applying high magnetic fields (~ 200 kOe) [6], doping of BF with rare-earth and transition metals [7], preparing solid solutions based on BF [8-10], and so on.

In this paper, the magnetoelectric ordering of the BF was attempted by creating a heterophase system, where the ferrimagnetic compound $MgFe_2O_4$ was chosen as the second phase. Thus, the aim of this work is to study the magnetic properties of the ceramic heterophase system of (1-*x*) $BiFeO_3$-*x* $MgFe_2O_4$ at various magnesium ferrite contents (x = 0.01; 0.05; 0.1; 0.15; 0.2).

2. Experiment

The initial components of the (1-*x*) $BiFeO_3$ – *x* $MgFe_2O_4$ system were obtained by ceramic technology from simple oxides of Bi_2O_3, Fe_2O_3 and MgO (99.5%) previously dried and then mixed in a stoichiometric ratio. Synthesis of $BiFeO_3$ was carried out in two stages with intermediate grinding. The first stage included annealing at T_1 = 800 °C for 10 hours. After that, the sample was ground and re-pressed to complete the synthesis. The second stage of synthesis was carried out at T_2 = 800 °C for

5 hours. $MgFe_2O_4$ was synthesized at 1100 °C for 48 hours. The obtained samples of the initial components ($BiFeO_3$, $MgFe_2O_4$) were crushed and ground, after which their powders were mixed in a ratio calculated by the mole fraction x, and pressed in the form of tablets. Samples of (1-x) BiFeO3 - x $MgFe_2O_4$ (x = 0.01, 0.05, 0.1, 0.15 and 0.2) were sintered at 850 °C for 4 hours. The samples were coated with silver electrodes by firing a silver paste.

X-ray diffraction analysis was performed on a Bruker D2 Phaser diffractometer using Cu Kα radiation. Micrographs of the samples of the investigated magnetoelectric composites were obtained using a scanning electron microscope (JEOL 6510 4V).

The magnetic hysteresis loops were obtained using a vibrating magnetometer. For temperature measurements of magnetic permeability, a copper wire with high-temperature insulation was wound on a ceramic (Al_2O_3) tube. The samples were placed inside the tube. The inductance of the coil was detected with an E7-20 immittance meter. The magnetic permeability was calculated using the formula for a solenoid.

3. Results and discussion

X-ray diffraction analysis of the synthesized samples (figure 1) revealed that apart from the cases of $x = 1$ (magnesium ferrous spinel) and $x = 0$ (bismuth ferrite), the system is a two-phase structure. This proves the presence of the reflexes inherent in both $MgFe_2O_4$ and BF on the X-ray patterns for all the compositions. As the content of one of the components of the heterophase structure increases, the intensity of its reflexes increases, while the intensity of the reflexes of the other component decreases. It is established that the cubic phase corresponding to $MgFe_2O_4$ has a symmetry group $Fd\overline{3}m$ and a lattice parameter $a = 8{,}369$ Å. In the samples with $x = 0$ (BF), three phases were detected: the main rhombohedral phase (with lattice parameters $a = 5{,}563$ Å; $c = 13{,}829$ Å,) corresponding to BF with the symmetry group R3c; the orthorhombic $Bi_2Fe_4O_9$ phase with lattice parameters $a = 7{,}936$ Å, $b = 8{,}418$ Å, $c = 5{,}985$ Å and the spatial symmetry group Pbam; and the cubic Bi_2O_3 phase with a lattice parameter $a = 10{,}119$ Å and the symmetry group I23. According to the literature data, the presence of secondary phases in bismuth ferrites is associated with the instability of the main phase at the synthesis temperature and the incongruent melting temperature near it. In our case, a ceramic material containing 71% of the BF phase was obtained.

As seen in figure 1, no other phases appear in the system under study with the increase in the $MgFe_2O_4$ content. The content of the magnesium ferrite phase increases, while the content of secondary phases in $BiFeO_3$ decreases. Thus, the system under study is not a solid solution but a two-phase system consisting of $BiFeO_3$ and $MgFe_2O_4$ – a magnetoelectric composite.

It is known from the literature that doping of $BiFeO_3$ ceramics leads to a variation in the average grain size that affects its physical properties. Therefore, to reveal the effect of the composition on the average grain size, the grain structure on the cleavage of the samples of the (1-x) $BiFeO_3$-x $MgFe_2O_4$ composite was determined using a scanning electron microscope (figure 2).

One can see that in all cases the grain size does not exceed 10 μm. The concentration dependence of the average grain size is presented in figure 3. As can be seen, even a small content of the $MgFe_2O_4$ phase leads to a significant decrease in the grain size. It is possible that the addition of the synthesized $MgFe_2O_4$ powder alters the sintering mode of the composite samples. The number of defects pinning the interphase boundaries and hindering the growth of grains in the process of high-temperature recrystallization increases. Therefore, the average grain size decreases with increasing magnesium ferrite content.

The dependence of the magnetization J on the amplitude of the magnetic field intensity H (figure 4) shows that the magnetization J in the "pure" BF is almost independent of H, since bismuth ferrite has an antiferromagnetic ordering. However, even a small addition of $MgFe_2O_4$ leads to a nonlinear response, which increases with increasing x.

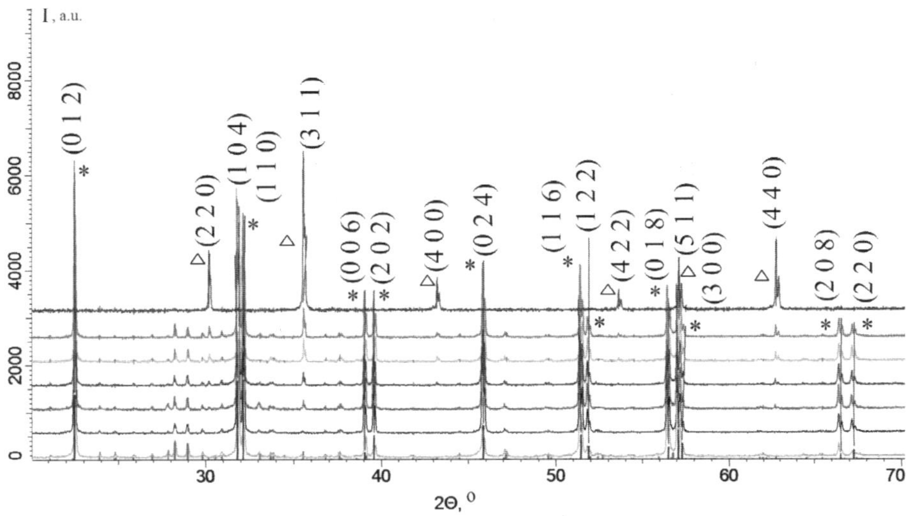

Figure 1. XRD patterns for the composite system of (1-x)$BiFeO_3$ – x $MgFe_2O_4$ at x = 0; 0.01; 0.05; 0.1; 0.15; 0.2 and 1.0 (bottom to top) at room temperature.
* - $BiFeO_3$ phase, Δ – $MgFe_2O_4$ phase

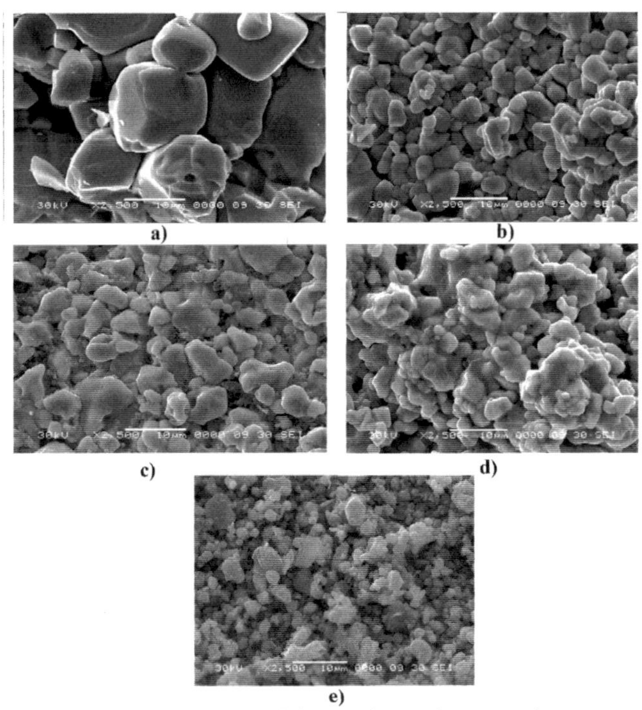

Figure 2. Morphology of the surface of composite samples of (1-x) $BiFeO_3$ - $x$$MgFe_2O_4$, at x = 0 (a); 0.01(b); 0.1 (c); 0.15 (d) and 1.0 (e).

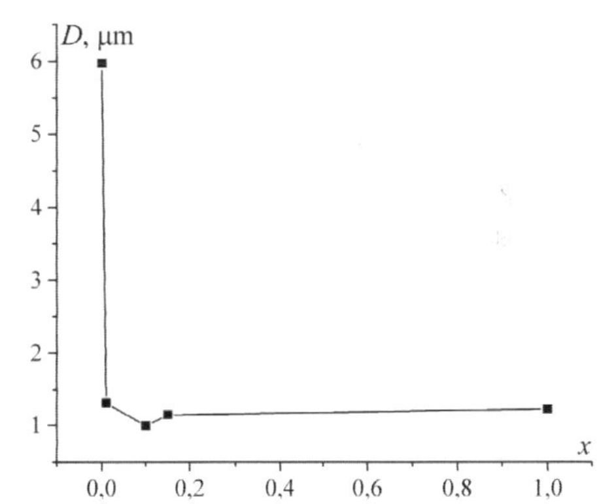

Figure 3. Average grain size dependence on the $MgFe_2O_4$ content, calculated from microphotographs.

We note that magnesium ferrite itself, being a ferrimagnet, has a similar nonlinear behavior J (H). In order to understand whether the magnetization of the BF phase makes any contribution to the magnetic response of the entire composite, we have studied the concentration dependence of the magnetic permeability of the composite. This dependence was obtained using two different methods: magnetic loops (the initial part of the magnetization curve) and the coil inductance measurement (figure 5).

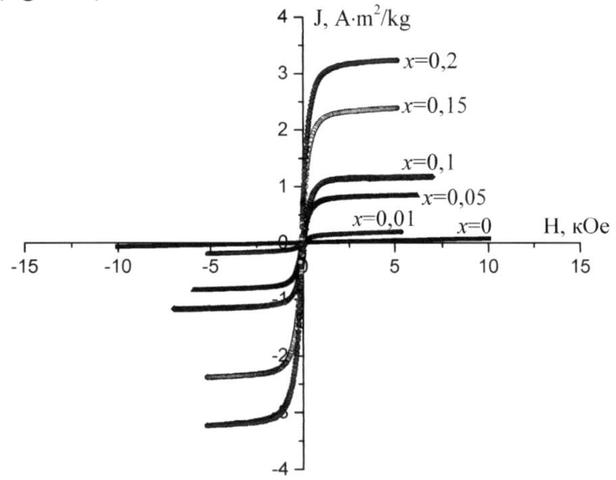

Figure 4. Dependences of the magnetization on the magnetic field strength for the composite of (1-x) $BiFeO_3 - x\ MgFe_2O_4$.

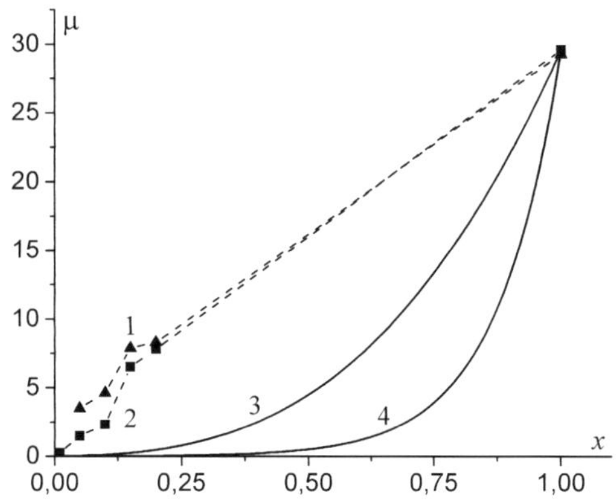

Figure 5. Dependences of the magnetic permeability of (1-x) $BiFeO_3 - x\ MgFe_2O_4$ on the content of the $MgFe_2O_4$ component: 1 – experimental data obtained from measurements of inductance; 2 – experimental data obtained from magnetic loops; 3 – dependence calculated according to the Luang formula; 4 – dependence calculated using the K. Lichtenecker formula.

Figure 5 shows that the experimental dependences μ(x) obtained using two different methods coincide with each other quantitatively and qualitatively, which indicates the reliability of the experimental data. As the content of the magnesium ferrite phase increases, the magnetic permeability of the composite increases. The analysis of the obtained dependence was carried out as follows. The studied composite is a particulate one (figure 2), where BF plays the role of a matrix, and statistically distributed magnetic granules of $MgFe_2O_4$ serve as filler. Then, if the phases of BF and $MgFe_2O_4$ do not interact with each other, the effective permeability of the composite can be calculated using the Luang formula:

$$\mu_{eff}^{1/3} = (1-x)\mu_{BF}^{1/3} + x\mu_{Mg}^{1/3}, \qquad (1)$$

or the K. Lichtenecker formula:

$$\mu_{eff} = \mu_{BF}^{1-x}\mu_{Mg}^{x}, \qquad (2)$$

where μ_{eff} is the magnetic permeability of the composite, μ_{BF} is the magnetic permeability of $BiFeO_3$, and μ_{Mg} is the magnetic permeability of $MgFe_2O_4$.

It follows from formulas (1) and (2) that the increase in the content of the ferrimagnetic $MgFe_2O_4$ phase should lead to the growth of the magnetic permeability of the composite (figure 5, curves 3 and 4). However, as can be seen in figure 5, the experimental data lie much higher than the curves calculated by mixing formulas, which indicates the effect of the magnetic subsystem of magnesium ferrite on bismuth ferrite, leading to a partial suppression of the spin-modulated ordering of the magnetic moment in BF. Such influence can be explained as follows. During high-temperature sintering, the magnesium ions migrate from the $MgFe_2O_4$ phase to the BF phase, and the bismuth ions (r_{Bi}^{3+} = 0.096 nm) are replaced by magnesium ions (r_{Mg}^{2+} = 0.066 nm) in the $BiFeO_3$ lattice, which leads to a variation in the exchange interaction determining the magnetic ordering in $BiFeO_3$. To test this assumption, the temperature dependences of the magnetic permeability were studied (figure 6), which made it possible to observe a shift of the temperature of the magnetic transition in BF as a function of the $MgFe_2O_4$ phase content.

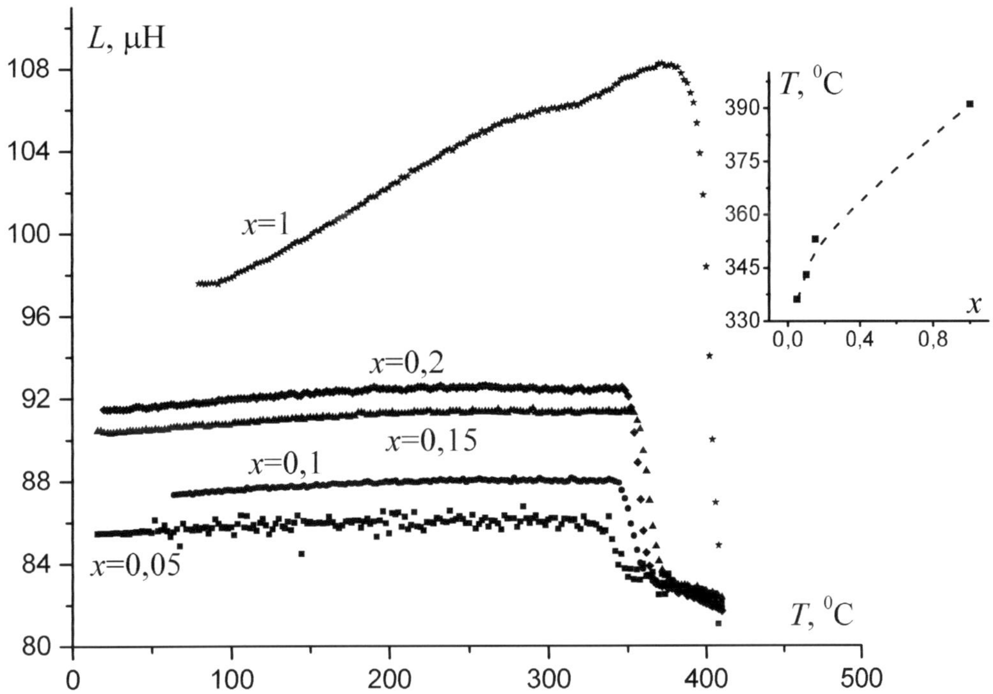

Figure 6. Temperature dependences of the magnetic permeability for the composites at different $MgFe_2O_4$ contents. On the inset: Dependence of the magnetic transition temperature in BF on the $MgFe_2O_4$ content.

A sharp decrease in the magnetic permeability in the temperature range from 350 °C to 400 °C can be observed in figure 6, which is associated with a magnetic phase transition. We note that the form of the dependences obtained for $x < 1$ is characteristic of an antiferromagnetic substance and can be associated only with $BiFeO_3$. Thus, the Neel temperature in $BiFeO_3$ increases with the addition of magnesium ferrite, as shown in the inset of figure 6.

The increase in the Neel temperature can be explained in the following way: as the ions of Bi^{3+} are replaced by smaller Mg^{2+} ions, the volume of the unit cell decreases. This leads to a change in the energy of the exchange interaction and requires a greater thermal energy for magnetic disordering. Therefore, the Neel point shifts toward higher temperatures.

4. Conclusions

An attempt was made to suppress the spin-modulated structure of $BiFeO_3$ by creating a heterophase system of (1-x) $BiFeO_3$-x $MgFe_2O_4$ at x = 0.01; 0.05; 0.1; 0.15 and 0.2. X-ray diffraction analysis made it possible to establish the presence of the main phases and a small number of secondary phases, the appearance of which is associated with the synthesis of $BiFeO_3$. A study of the morphology of the surface showed that the average grain size of the samples studied decreases with the increase of

$MgFe_2O_4$ content, which leads to a decrease in the content of secondary phases. Measurements of magnetic properties allow us to conclude that a ferrimagnetic response in $BiFeO_3$ appears. An analysis of the concentration and temperature dependences of the magnetic permeability indicates a partial suppression of the spin-modulated structure in $BiFeO_3$.

Acknowledgments
This work was supported by the Russian Foundation for Basic Research (Grant #16-32-00697).

References

[1] Belashchenko K D, Tchernyshyov O, Kovalev A A and Tretiakov A 2016 *Appl. Phys. Lett.* **108** 132403
[2] Liu H and Yang X 2017 *Ferroelectrics* **507** 69
[3] Pyatakov A P and Zvezdin A K 2012 *Phys. Usp.* **55** 557
[4] Sosnowska I, Peterlin-Neumaier T and Steichele E J 1982 *J. Phys. C.* **15** 4835
[5] Sosnowska I and Zvezdin A K 1995 *J. of Magn. Magn. Mater.* **140-144** 167
[6] Popov Yu F, Kadomtseva A M, Krotov S S, Belov D V, Vorob'ev G P, Makhov P N and Zvezdin A K 2001 *Low Temp. Phys.* **27** 478
[7] Cheng Z X, Wang X L, Du Y and Dou S X 2010 *J. Phys. D. Appl. Phys.* **43** 242001
[8] Pokatilov V S, Rusakov V S, Sigov A S and Belik A A 2017 *Physics of the Solid State* **59** 1558
[9] Makoed I I, Ravinski A F, Lazenka V V, Galyas A I, Demidenko O F, Zhivul'ko A M, Yanushkevich K I and Moshchalkov V V 2017 *Physics of the Solid State* **59** 1536
[10] Uniyal and Yadav K L 2009 *J. Phys.: Condens. Matter* **21** 012205

IOP Conf. Series: Journal of Physics: Conf. Series **993** (2018) 012004 doi:10.1088/1742-6596/993/1/012004

Effect of bismuth substitution for lead in $CsPbBr_3$ perovskite

M V Elizarov, O V Lozkina, D I Yeletz, A V Emeline , V K Ryabchuk and A A Murashkina

Photoactive Nanocomposite Materials Laboratory, St. Petersburg State University, Ulianovskaia Str. 1 St. Petersburg, 198504, Russia

E-mail: st040426@student.spbu.ru; murashkinaaa@mail.ru

Abstract. The effect of bismuth substitution for lead at concentrations of 0.5 – 5 % on absorption and luminescence of dispersed $CsPbBr_3$ perovskites has been studied by means of diffuse reflectance spectroscopy and luminescence techniques, respectively. The considerable red shift of the absorption edge of $CsPbBr_3$ has been observed, as well as a decrease in exciton luminescence with increasing dopant concentration.

1. Introduction

Organic-inorganic hybrid lead halide perovskites attracted the attention of researchers during the 1970s and again in the 1990s [1], particularly for their unique semiconducting and conducting properties [2]. Perovskite-based materials of the general formula ABX_nY_{3-n}, where A = $CH_3NH_3^+$ or another monovalent organic cation, B = Pb (or Sn), and X / Y = Cl, Br, I anions, have promising optical and electrophysical properties. In particular, they are characterized by low Urbach energy of ~ 15 meV; a Stokes shift of less than 20 meV; a high mobility of charge carriers of more than 10 $cm^2/V{\cdot}s$; a significant diffusion length of up to 1 μm [3]; and high absorption in the visible range of wavelength (with an absorption coefficient of more than 104 cm^{-1}) [4]. These unique properties of the perovskite family come to naught due to their instability under the action of moisture, oxygen, UV radiation, heat (especially in the case of Sn-containing perovskites) [5] and lead toxicity [6]. Therefore, an intensive search for new perovskite (perovskite-like) lead-free materials with the useful properties displayed by lead trihalides is ongoing. Two possible search directions were offered in 2015 - 16: 1) chalcogenide perovskites [7], 2) halide double perovskites of the general formula $A_2BB^{/}X_6$ where A = Cs, organic cation (e.g., $CH_3NH_3^+$), other monovalent cations, B = Sb, Bi; $B^{/}$ = Cu, Ag, Au, K; and X = Cl, Br, I [8,9].

The first direction is represented by just a few publications. There are serious difficulties in the synthesis due to the formation of several phases even at an insignificantly elevated temperature. The second search direction gave encouraging results due to Density functional theory (DFT) ab initio calculations [10] and the possibility of varying the perovskite composition [11]. For example, when examining $Cs_2AgBiBr_6$, the authors of [12] found a long-lived photoluminescence (about 660 ns) at room temperature and offered this material as promising for photoelectrochemical applications. In addition, the authors showed that, in contrast to the halogenide perovskites (ABX_nY_{3-n}), the halide double perovskites ($A_2BB^{/}X_6$) provide an opportunity for variation of the perovskite composition. In the present work, we investigated the influence of Bi-concentration on optical properties of $CsPbBr_3$ perovskite.

IOP Conf. Series: Journal of Physics: Conf. Series **993** (2018) 012004 doi:10.1088/1742-6596/993/1/012004

2. Experiment

The incorporation of dopants was achieved in situ by adding a mixture of trivalent-ion (Bi^{3+}) bromide salt and CsBr into a crystallization solution containing a mixture of $PbBr_2$ (99%, Sigma) in hydrobromic acid, and then heating to the required temperature for solvent removal and annealing at 150°C in N_2-atmosphere. Then, the obtained powders were dissolved in Dimethylformamide (DMF) (0.2 g/5 ml) and applied on glass substrates with subsequent heating at 120°C. We observed a color change from orange to deep red in Bi-doped powders.

The phase formation was analyzed by the X-ray diffraction method using a Rigaku Miniflex II diffractometer with Cu Kα radiation (with an anode current of 15 mA and an accelerating voltage of 30 kV) in the angle range of $10° \leq 2\theta \leq 60°$ with a scanning speed of 5.0°/min. Structural data for the perovskite phase was taken from the ICSD database. The phase analysis by the Rietveld method was carried out using the TOPAS (Bruker AXS) software.

The diffuse reflectance spectra $R(\lambda)$ and absorbance spectra ($A(\lambda) = 1 - R(\lambda)$) were recorded in the 380 - 800 nm spectral range at ambient conditions using a Cary 5000 spectrophotometer equipped with a DRA 2500 external diffuse reflectance accessory. The optical-grade Spectralon was used as a reference standard.

Photoluminescence spectra of perovskites studied have been obtained in the invisible range using a Fluorolog-3 spectrafluorimeter (Horiba Jobin Yvon).

The surface morphology was explored by scanning electron microscopy (using a Zeiss Supra 40 VP system) for all synthesized thin films.

3. Results and discussion

Before carrying out any experiments, we evaluated the perovskite tolerance factor, which should be calculated to estimate the possibility of a heterovalent incorporation (isoelectronic with Pb^{2+}) and must be within the range 0.80 – 1.00 to allow the formation of a perovskite structure. The tolerance factors for the samples of $CsPb_{(1-x)}Bi_xBr_3$ (x = 0.0, 0.5, 0.1, 0.2, 0.5 wt) were calculated to be 0.861, 0.862, 0.862, 0.863 and 0.864, respectively.

The phase composition and structure of the films were identified as the orthorhombic perovskite phase of $CsPbBr_3$ at room temperature. The X-ray diffraction (XRD) pattern of $CsPbBr_3$ shows strong diffraction peaks, which can be assigned to the orthorhombic crystal structure (space group Pbnm) (figure 1(a)).

Note that the growth of Bi concentration results in a decrease in the crystal lattice parameter (figure 1(b)). This is probably due to the size factor: the ionic radius of Bi^{3+} at the VI-coordination position (by Shannon) is 1.03 Å, and the ionic radius of Pb^{2+} at the same position is 1.19 Å. Therefore, the bismuth doping with substitution of Pb positions decreases the crystal lattice parameters.

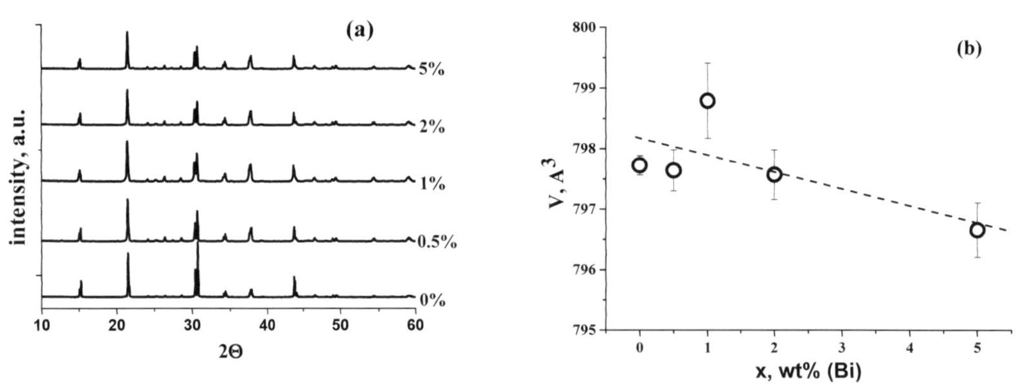

Figure 1. X-ray diffraction patterns of $CsPb_{(1-x)}Bi_xBr_3$ (x = 0.0, 0.5, 0.1, 0.2, 0.5 wt%) (a); the crystal lattice parameters (cell volume) vs Bi-concentration (b).

To demonstrate the incorporation of Bi into the perovskite lattice, the XPS spectra of pure and doped $CsPbBr_3$ samples were also studied. The survey spectra are presented in figure 2.

The peaks related to the binding energies (BE) of ~737.5 eV and ~723.6 eV correspond to Cs 3d3/2 and Cs 3d5/2, respectively. As depicted in figure 4, binding energies of 142.2 eV and 137.6 eV in the spectra of samples correspond to the peaks of Pb4f 5/2 and Pb4f 7/2, respectively. The representative high-resolution XPS spectra for the Bi 4f-doublet is resolved at 158.2 eV (Bi 4f 7/2) and 164.2 eV (Bi 4f5/2) binding energies. The Bi4f 7/2 peak overlaps with the Cs4p 3/2 peak at BE ~ 158.2 eV. The presence of Bi signals and an increase in their intensity with dopant concentration indicated a uniform distribution of Bi and also confirmed the effectiveness of the doping process.

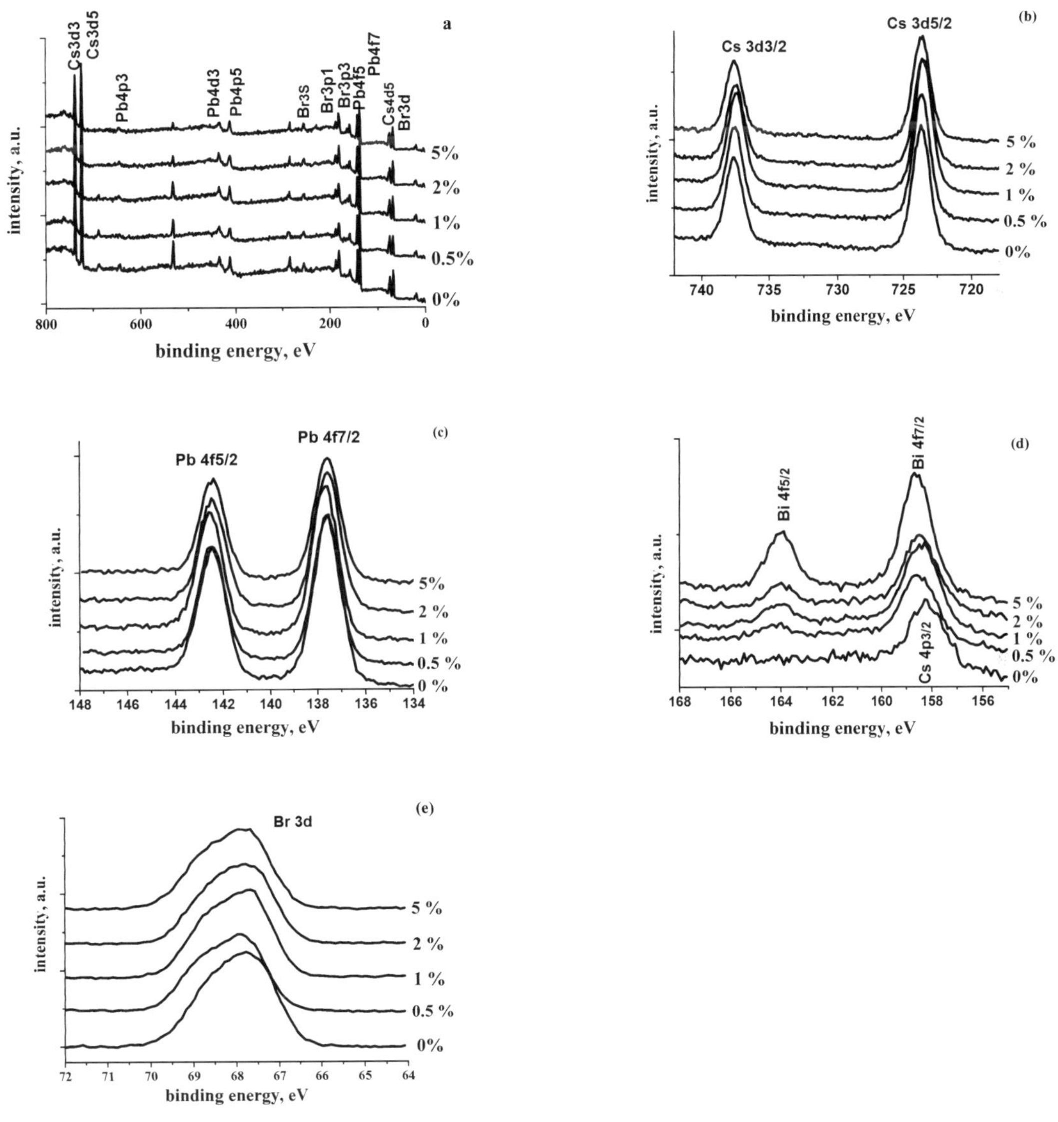

Figure 2. (a) XPS spectra of pure and Bi-doped (0.5, 0.1, 2 and 5 wt%) $CsPbBr_3$ samples (b) XPS spectra of Cs 2d states, (c) Pb 4f states, (d) Bi 4f states, (e) Br 3d states in a perovskite.

Figure 3 (a) shows that incorporation of bismuth leads to a significant red shift of the fundamental absorption edge. The biggest shift takes place at the lowest concentration of bismuth (0.5%) in good agreement with previous studies [13]. A green band around 2.4 eV (515 - 520 nm) has been detected in the emission spectra at room temperature (figure 3(b)). This band assigned to exciton luminescence of $CsPbBr_3$ is well known (for example, see [14]).

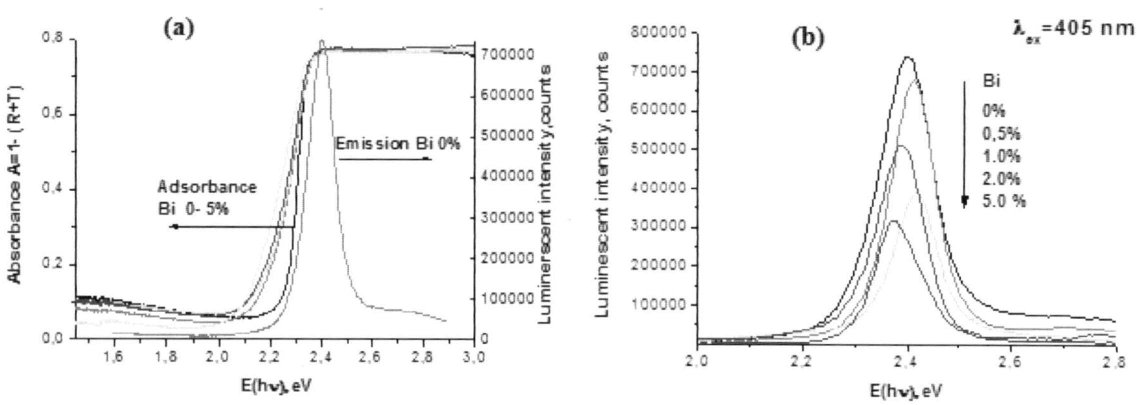

Figure 3. (a) Absorbance spectra of Bi-$CsPbBr_3$ and the emission spectrum of the pure $CsPbBr_3$ sample (right scale). (b) Luminescent emission spectra of the same samples under excitation by monochromatic light (λ_{ex} = 405 nm), recorded at room temperature. Bi concentration: 0, 0.5, 1,0, 2.0 and 5.0 %.

It should be noted that a monotonic reduction of the photoluminescence intensity with increasing dopant concentration is in agreement with the results of [13]. At the same time, the authors of [13] observed a monotonic blue shift of the adsorption fundamental absorption edge with increasing bismuth concentration in the 0.25 – 10% range, as well as a corresponding shift of luminescent band maximums. The authors of [13] attributed these shifts to the Burstein–Moss effect. The present studies do not confirm the results of [13], since we observed a red shift of the fundamental absorption edge (figure 3 (a)) and an irregular shift of luminescent peaks in the 0 -5 % bismuth concentration range for $CsPbBr_3$. The divergence of our results from those presented in [13] can be tentatively related to the size effect, since we have dealt with $CsPbBr_3$ powders (micron size particles) while the authors of [13] studied nanocrystaline thin films. In this regard, it should be noted that a red shift has been observed for millimeter-size $MAPbBr_3$ single crystals [15], as well as for $MAPbCl_3$ millimeter-size single crystals and thin films with 10 μm domains [16]. The red absorption shift for thin films was found to be less than that for $MAPbCl_3$ single crystals.

4. Conclusions

Dispersed $CsPBr_3$ perovskites doped with Bi (0.5 - 5 wt.%) have been synthesized and characterized by means of X-ray diffraction and X-ray photoelectron spectroscopy. It has been shown that Bi3+ cations occupy the positions of lead ions, and the reduction of the lattice cell volume correlates with an increase in dopant concentration. Introduction of Bi into the $CsPbBr_3$ lattice leads to a considerable red shift of the absorption edge of perovskites studied, as well as to a reduction of the exciton luminescent emission. The results concerning the reduction of exciton luminescence with bismuth incorporation are in agreement with those presented in [13]. However, the authors of [13] found a blue shift of the absorption edge in Bi-doped $CsPbBr_3$. A possible explanation for the inconsistency of our results with those presented earlier in [13] may be that nano crystals and microcrystals of $CsPbBr_3$ respond differently to Bi incorporation.

Acknowledgments
This study was supported by the Russian Foundation for Basic Research (grant 17-53-50083).

References

[1] Onoda-Yamamuro N, Matsuo T H and Suga Y 1992 *J Phys Chem Solids* **53** 935
[2] Mitz D, Wang S, Field C and Chess C. Guloy C 1995 *Science* **267** 1473
[3] Sutherland B R and Sargent H 2016 *Nature Photonics* **10** 295
[4] De Wolf S, Holovsky J, Moon S-J, Löper P, Niesen B, Ledinsky M, Haug F-J, Yum J-H and Ballif C 2014 *J Phys Chem Lett* **5** 1035
[5] Li B, Li Y, Zheng C, Gao D Q and Huang W 2016 *RSC Adv* **6** 38079
[6] Benmessaoud I R, Mahul-Mellier A L, Horvath E, Maco B, Spina M, Lashuel H, Hilal A and Forro L 2016 *Toxicol Res* **5** 407
[7] Sun Y-Y, Agiorgousis M, Zhang P and Zhang S 2015 *Nano Lett* **15** 581
[8] McClure E T, Ball M R, Windl W and Woodward PM 2016 *Chem Mat* **28** 1348
[9] Zhang M, Miaoqiang Lyu, Chen P, Hao M, Yun J H and Wang L 2016 *J. Chem. Eng* **11** 392
[10] Filip M, Hillman M, Haghighirad A, Snaith H and Giustino F 2016 *J Phys. Chem Lett* **7** 2579
[11] Volonakis G, Filip M, Haghighirad A, Sakai N, Wenger B, Snaith H and Feliciano Giustino F 2016 *J Phys. Chem Lett* **7** 1254
[12] Slavney A, Hu T, Lindenberg A and Karunadasa H 2016 *J. Am. Chem. Soc.* **138** 2138
[13] Begum R. Parida M, Abdelhady A, Murali B, Ahmed M, Bakr O M and Mohammed O F 2017 *Am Chem Soc* **139** 731
[14] Zhang X, Lin H, Huang H, Reckmeier C, Zhang Y, Choy W C and Rogach A L 2016 *Nano Lett* **16** 1415
[15] Abdelhady A, Saidaminov M I, Murali B, Adinolfi V, Voznyy O, Katsiev K,, Alarousu E, Comin R, Sinatra D, Sargent E Y, Mohammed O F and Bakr O M 2016 *J Phys Chem Lett* **7** 295
[16] Zhang Z, Ren L, Yan H, Guo S, Wang S, Wang M and Jin K 2017 *J Phys Chem* **121** 17436

Measuring the distribution of the charge carrier concentration in delta-doped layers based on diamond

L M Shestakova and V I Zubkov

Department of micro- and nanoelectronics, St. Petersburg State Electrotechnical University "LETI", Prof. Popov str. 5, St. Petersburg 197376, Russia

E-mail: l94shes@yandex.ru

Abstract. Properties of single crystal diamond with four delta-doped layers are studied. The goal of this research is to measure the charge carrier concentration profile. Admittance spectroscopy was used as the principal method to measure the properties of samples at different temperatures and applied biases. The measurements showed that only the first (shallow located) δ-layer was confidently registered, especially at elevated temperatures (350–450 K). In addition, the concentration profiles included a weak response from the second delta layer, significantly shifted from its true location. A general conclusion is made concerning the observed charge carrier distribution. Further study is needed to determine the responses from the deeper delta-doped layers.

1. Introduction

The search for more advanced materials and research of the methods for their application is a relevant task in microwave electronics, power electronics, radioelectronics and photoelectronics, etc. A very promising material for such applications is diamond. It has a relatively high carrier mobility and an exceptionally high thermal conductivity among all semiconductors. The activation energies of impurities in diamond are very high, which leads to a small degree of ionization at room temperature. However, with increasing dopant concentration, the activation energy reduces. On the other hand, an increasing amount of ionized atoms leads to a decreasing carrier mobility which prevents using the bulk-doped diamond as a material for super-high frequency applications. The solution to this problem is doping with a very high concentration of impurities ($\sim 10^{20}$ cm^{-3}) in a few atomic layers, so-called delta-doping. Since the electron wave functions are localized in a wider area than the thickness of a delta layer, the charge carriers move into the undoped region, so the scattering by ionized atoms is reduced. In the future, delta-doped layers may become the main technology for obtaining super high-frequency devices based on semiconductor diamond.

2. Experiment details

Current-voltage and capacitance-voltage characteristics were measured using a computer-controlled setup of admittance spectroscopy. The setup consists of an Agilent E4980A LCR meter, a LakeShore 331S temperature controller and a Janis CCS400/204N probe station with a helium closed cycle cryostat [1]. The equipment allows carrying out measurements in the temperature range from 15 to 450 K and the frequency range 100 Hz ... 2 MHz.

High pressure high temperature (HPHT) (100)-oriented diamond crystals were used as a substrate at which a CVD single-crystal diamond layer with O-terminated surface was grown [2, 3]. The CVD layer contains four delta layers doped by a boron impurity. The distance from the top of the sample and between the delta layers is 60 nm (figure 1). According to the SIMS data, the widths of the δ-layers were about 2.5 nm and the achieved peak intensity of an impurity in the delta layers is more than 10^{20} cm^{-3}. The impurity concentration outside the delta layers was about 10^{16} cm^{-3}.

For the current and capacitance measurements, the platinum Ohmic and Schottky contacts were sputtered on the top of a CVD-layer.

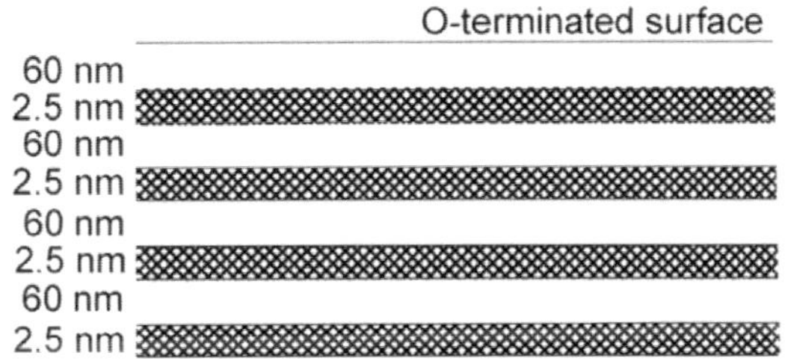

Figure 1. Layer sequence for the investigated sample

3. Results and discussion

We have studied the capacitance and conductance response of the sample in the temperature range 150...450 K. The sample behaves like a dielectric below 230 K, so the capacitance measurements were beyond the capabilities of the equipment used. Further, due to the incomplete impurity ionization in diamond in a wide interval of temperatures, including room temperatures, the total carrier charge under the rectifying barrier is collected from different depths at different temperatures (according to the Gauss theorem). At $T = 300$ K, particularly, the depletion region fully overlapped the delta-doped layers even at zero bias, and the measurements have shown that the concentration fluctuated around $2{\cdot}10^{16}$ cm^{-3}.

As the temperature rises, the depletion region does narrow, so that the distribution of charge carriers in the delta layer region can be registered by capacitance-voltage (*C-V*) measurements. Figure 2 presents the specific capacitance response of the sample at 340 K and the resulting concentration profile manifesting the right-hand edge of the first delta-doped layer.

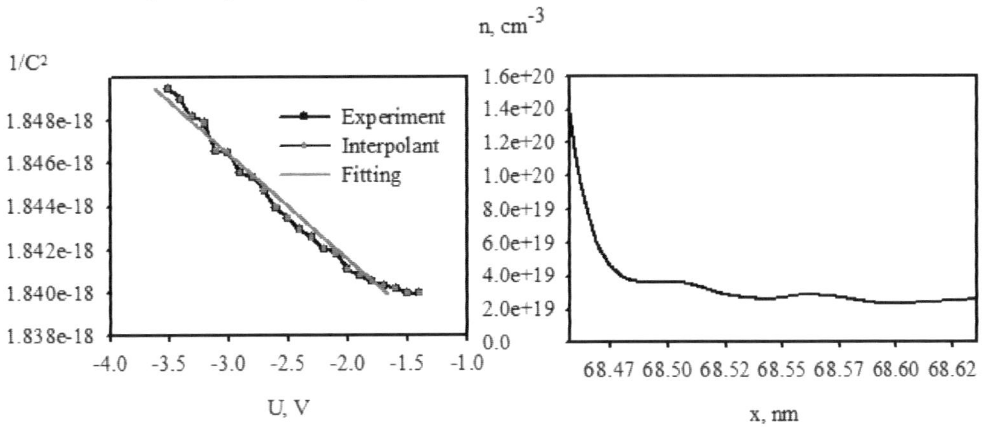

Figure 2. *C-V* characteristics (left) and concentration profiles (right) in a diamond sample with delta-doped layers ($T = 340$ K).

The data about the delta layer can be obtained at temperatures above 430 K, since the cap layer becomes more ionized and the depletion region is narrower (figure 3).

The concentration peak position obtained from the *C-V* measurements (62.7 nm) corresponds fairly well to the technologically specified position, if the hardware error of the *C-V* method is taken into account. The peak amplitude reaches $3.4\cdot10^{20}$ cm^{-3}. Two specific features can be seen in the concentration profiles. Firstly, they demonstrate some apparent "shift" of the delta layer's peak position with temperature. The reason is the different locations of the space charge region (SCR) edge and the true layer position in the structures with heavy doped quantum confined layers. In addition, the profiles also include the weak responses from the deeper delta-doped layers, since in the lateral geometry of the capacitance measurements the smeared edge of SCR can simultaneously test several delta layers located near each other. These phenomena need further investigation.

To support the data obtained in the experiment, we performed preliminary numerical simulations aimed at a self-consistent solution of the Schrödinger and Poisson equations.

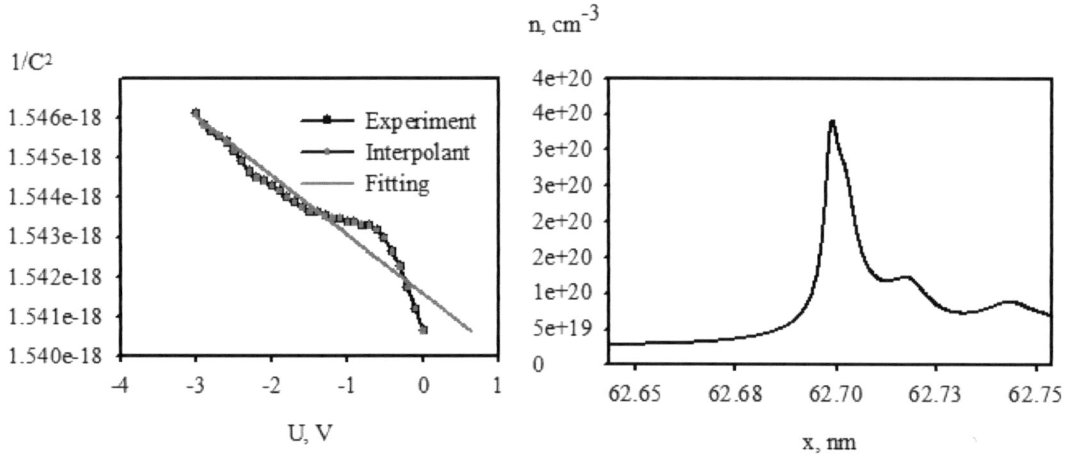

Figure 3. *C-V* characteristics (left) and concentration profiles (right) in a diamond sample with delta-doped layers ($T = 430$ K).

Conclusions

In this study, unique measurements were performed on a multi-delta-doped single-crystal CVD diamond by means of the capacitance-voltage technique in a wide temperature range. As a result, we obtained the distribution of charge carrier concentration in the vicinity of delta-doped layers. The most noticeable feature of this measurement was getting a response from deeper layers. However, there are also some arguable features that need additional investigation. The object of further research will be the detailed simulation of concentration profiles by the self-consistent solution of the Schrödinger and Poisson equations, which represents a separate task due to the complexity of the diamond band structure (namely, a small spin-orbit splitting in the valence band).

References

[1] Zubkov V I, Kucherova O V, Bogdanov S A, Zubkova A V, Butler J E, Ilyin V A, Afanas'ev A V and Vikharev A L 2015 *J. of Appl. Phys.* **118** 145703

[2] Muchnikov A B, Vikharev A L, Gorbachev A M, Radishev D B, Blank V D and Terentiev S A 2010 *Diamond & Relat. Mater.* **19** 432

[3] Vikharev A L, Gorbachev A M, Lobaev M A, Muchnikov A B, Radishev D B, Isaev V A, Chernov V V, Bogdanov S A, Butler J E and Drozdov M N 2016 *Phys. Status Solidi RRL* **10** No. 4 324–7

Determination of the energy structure of recombination centers in heavily doped $Al_xGa_{1-x}N$:Si epitaxial layers with x > 0.5

I V Osinnykh[1,2], T V Malin[1] and K S Zhuravlev[1,2]

[1]Rzhanov Institute of Semiconductor Physics of the Siberian Branch of the Russian Academy of Sciences, 13 Lavrentieva avenue, Novosibirsk 630090, Russia

[2]Novosibirsk State University, 2 Pirogova str., Novosibirsk 630090, Russia

E-mail: igor-osinnykh@isp.nsc.ru

Abstract. The photoluminescence properties of the intensive defect-related emission in heavily doped $Al_xGa_{1-x}N$:Si layers with $x > 0.5$ have been investigated by photoluminescence (PL) spectroscopy. The PL band in AlN was attributed to donor-acceptor (DA) transitions. At the lowest Al content, the impurity band merges with the conduction band and DA transitions are replaced by electron-acceptor transitions involving the same acceptor. The energy structure of recombination centers was obtained using the model of configuration coordinates for $Al_{0.67}Ga_{0.33}N$.

1. Introduction

GaN and its alloys with AlN have emerged as promising semiconductor materials for applications in laser sources emitting in the blue-green spectral range. The creation of effective light-emitting semiconductor devices is impossible without doping the epitaxial layers. The formation of epitaxial *n*-GaN layers using silicon (Si) as a donor does not cause significant difficulties, and the electron concentration can be controllably increased to 10^{20} cm^{-3} [1–4]. However, doping of $Al_xGa_{1-x}N$ becomes less efficient with an increase in Al content (x). The electron concentration becomes significantly lower than the concentration of silicon atoms due to two phenomena: 1) the deepening of the Si donor level from 30 meV in GaN [5] to 250 meV in AlN [6, 7]; 2) the self-compensation of Si caused by the formation of negatively charged cation vacancies (V_{III}) [8, 9]. It is assumed that heavy doping with Si also leads to the appearance of intense broadband luminescence in the visible spectral range that was observed in $Al_xGa_{1-x}N$ epitaxial layers with $x > 0.5$ [10, 11]. This band is attributed to donor-acceptor (DA) and free electron-acceptor (eA) transitions involving the same acceptor. However, the ratio between DA and eA transitions was unclear. The donor was assigned to Si. The acceptor might be V_{III} or its complexes with shallow donors. However, the energy structure of recombination centers has not been completely described. This work is a continuation of previous investigations. In this paper, the new results on the energy structure of recombination centers in heavily doped $Al_xGa_{1-x}N$:Si epitaxial layers with $x > 0.5$ studied by photoluminescence (PL) spectroscopy are presented.

2. Experimental details

We investigated 1.2-μm-thick $Al_xGa_{1-x}N$:Si layers grown by molecular beam epitaxy techniques (using ammonia as a source of active nitrogen) on nitridated (0001)-oriented 400-μm-thick sapphire substrates with an AlN nucleation layer about 300 nm thick. The Al content in the layers was controlled by the Al-to-Ga flux ratio during the growth and varied from 0.5 to 1. The real x value was obtained by secondary-ion mass spectrometry (SIMS) and transmission spectroscopy. The latter method revealed the value of x to be 5-7 % lower than the value produced by SIMS, probably because of band gap narrowing at high impurity concentrations. Therefore, the x value from the SIMS data was assumed to be close to the real values. The mix of 0.7% silane (SiH_4) with nitrogen (N_2) was used as a source of Si. Measurements of photoluminescence (PL) of the studied samples were conducted in a wide temperature range of 5-1200 K using a continuous He-Cd laser with a photon energy E_{ph} = 3.81 eV and a pulsed 4th harmonic of Nd:YLF laser (E_{ph} = 4.71 eV) for excitation. The studied samples were mounted in a helium closed-circuit cryostat for measurements below 300 K and in a custom-made vacuum camera for measurements above 300 K.

3. Results

Figure 1 shows the room-temperature PL spectra of silicon-doped $Al_xGa_{1-x}N$ layers with $x \approx 0.6$ grown at different silane fluxes. The PL spectra of doped $Al_xGa_{1-x}N$ grown at 1 sccm ($N_{Si} \approx 10^{19}$ cm^{-3}) comprise the band-edge ultraviolet (UV) band with maximum energy at E_g = 4.5 eV related to band-to-band transitions, the parasitic defect-related UV band split into two bands with $E_m \approx 3.6$ eV and $E_{em} \approx 3.9$ eV, and the defect-related visible band at $E_{em} \approx 2.2$ eV. An increase in silane flux up to 3 sccm ($N_{Si} \approx 1.5\times10^{20}$ cm^{-3}) leads to a strong decrease in the intensity of the defect-related UV band at 3.6 eV and total reduction of the band-edge band and the defect-related band at 3.9 eV. Thus, transitions through the levels of defects due to Si doping dominate over all other radiative recombination channels.

Figure 2 shows E_{em} of the defect-related PL bands and E_g of the heavily doped $Al_xGa_{1-x}N$ layers with $N_{Si} \approx 1.5\times10^{20}$ cm^{-3}, estimated using the Vegard law:

$$E_g = x_{Al} \cdot E_{gAlN} + (1 - x_{Al}) \cdot E_{gGaN} - b \cdot x_{Al} \cdot (1 - x_{Al}) = 1.0x_{Al}^2 + 1.59x_{Al} + 3.43, \quad (1)$$

where E_{gGaN} = 3.43 eV [12] and E_{gAlN} = 6.015 eV [13] are the GaN and AlN band gaps, respectively, and b = 1.0 eV [14] is the bowing parameter. The parasitic UV band demonstrates the lowest range of blue shift of E_{em} with an increase in x that reveals a stronger deepening of the acceptor energy level than in the case of the visible band.

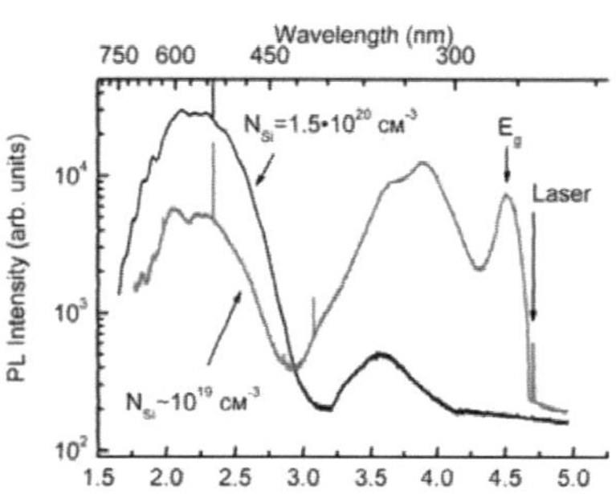

Figure 1. The room-temperature PL spectra of silicon-doped $Al_xGa_{1-x}N$ layers with $x \approx 0.6$ with a silicon concentration (N_{Si}) of about 10^{19} cm^{-3} and 1.5×10^{20} cm^{-3}.

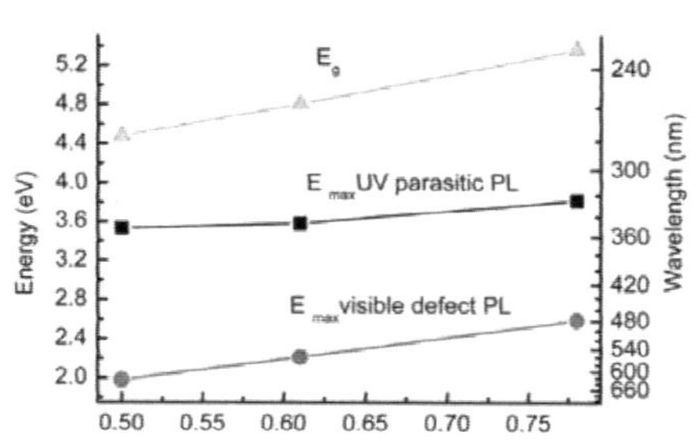

Figure 2. Dependences of the energy position of defect-related P bands (E_{em}) and the band gap (E_g) of heavily doped $Al_xGa_{1-x}N$ layers with $N_{Si} \approx 1.5\times10^{20}$ cm^{-3} on the Al content (x) at T = 300 K.

IOP Conf. Series: Journal of Physics: Conf. Series **993** (2018) 012006 doi:10.1088/1742-6596/993/1/012006

The $Al_{0.67}Ga_{0.33}N$ layer with $N_{Si} \approx 1.5\times10^{20}$ cm^{-3} demonstrates the same behavior of the temperature-dependent visible PL band intensity that was observed before for a lower Si concentration [10], i.e., the PL intensity increases with an increase in temperature from 5 to 270 K, while a further rise in temperature causes PL quenching (see figure 3).

The enhancement and quenching of the PL intensity with temperature are described by the expression:

$$I(T) = A\frac{1+B_0 e^{-\frac{E_0}{kT}}}{1+B_1 e^{-\frac{E_1}{kT}}}, \tag{2}$$

where E_0 and E_1 are the activation energies, and A, B_0, and B_1 are constants. Fitting of the PL quenching curve by equation (2) gives E_0 = 9 meV and E_1 = 160 meV. The enhancement of the PL intensity is explained by the emission of carriers from the states arising due to the band edge perturbations caused by the Al mole fraction fluctuation (Δx). The addition of the second exponential factor with an activation energy E_2 to the denominator [10] would be in excess of the accuracy of the approximation.

The increase in temperature up to 1100 K causes the blue shift of the visible PL band maximum, following a linear law $\Delta E_m \approx 1.2\times kT$ (see figure 4). This linear coefficient of 1.2 was lower than the reported value of 1.7 obtained for $N_{Si} \approx 5\times10^{19}$ cm^{-3} [10].

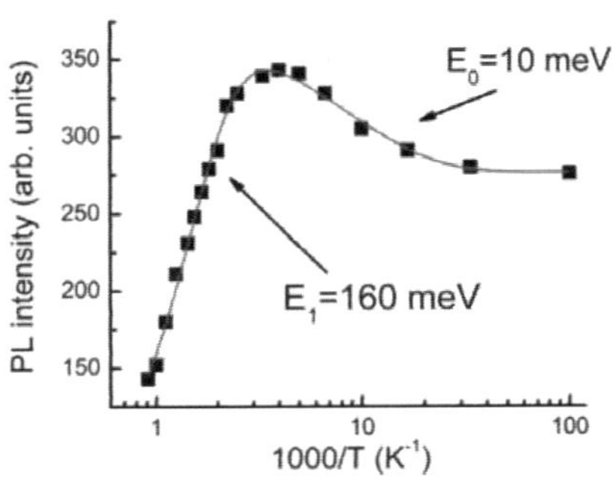

Figure 3. Temperature dependence of the intensity of the visible PL band in the $Al_{0.67}Ga_{0.33}N$ layer with $N_{Si} \approx 1.5\times10^{20}$ cm^{-3}. The solid line depicts the approximation curve (see the text for details).

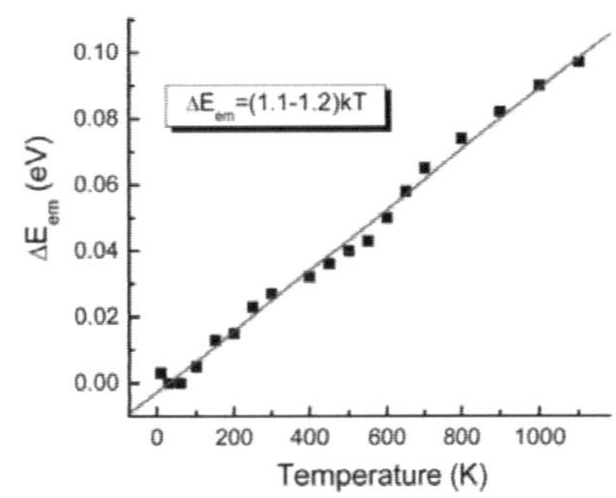

Figure 4. Temperature dependence of the visible PL band maximum in the $Al_{0.67}Ga_{0.33}N$:Si layer. The solid line depicts the approximation curve (see the text for details).

Figure 5 shows the temperature dependence of the PL bandwidth. The full width at half maximum (FWHM or W) of the PL band depends on temperature (T). The temperature broadening of the PL band is described by the expression [15]:

$$W(T) = W(0)\sqrt{\coth\left(\frac{h\nu_e}{2kT}\right)}, \tag{3}$$

where $W(0)$ is the bandwidth at T = 0 and $h\nu_e$ is the local phonon energy. Fitting of the temperature broadening of the PL band by equation (3) gives $W(0)$ = 562 meV and $h\nu_e$ = 62 meV.

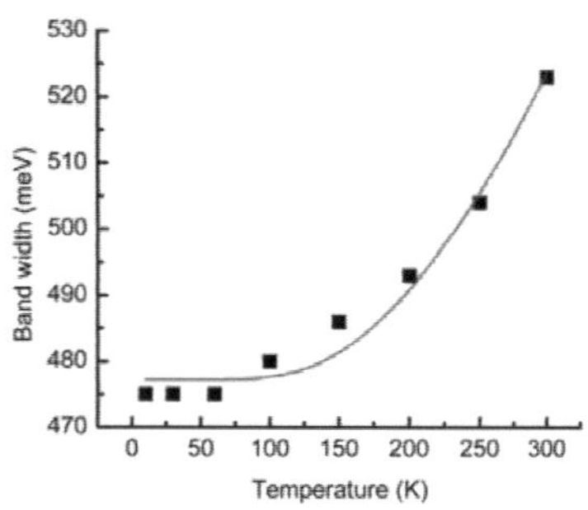

Figure 5. Temperature dependence of the visible PL band bandwidth in the $Al_{0.67}Ga_{0.33}N$:Si layer. The solid line depicts the approximation curve (see the text for details).

4. Discussion

4.1. Type of transitions in heavily doped $Al_xGa_{1-x}N$:Si layers

Broad PL bands are typical for recombination of carriers through deep levels of defects due to strong electron-phonon coupling. The configuration coordinate (CC) model describes such systems well [15, 16]. The measure of the electron-phonon coupling is the Huang–Rhys factor (*S*). In the case of strong electron-phonon coupling, *S* represents the mean number of the emitted phonons for each photon absorption or emission process. Previously, we assumed that *S* is equal for absorption and emission [11] but in general, *S* should be different for both processes. The CC diagram describing optical absorption and emission at defects with strong electron-phonon coupling is shown in figure 6. S_{em} and S_{ab} are the Huang-Rhys factors and $h\nu_g$ and $h\nu_e$ are the photon energies involved in the transitions, while optical energies are E_m and E_{ab}, respectively. We assume that

$$S_{em}h\nu_e = S_{ab}h\nu_g \,. \tag{4}$$

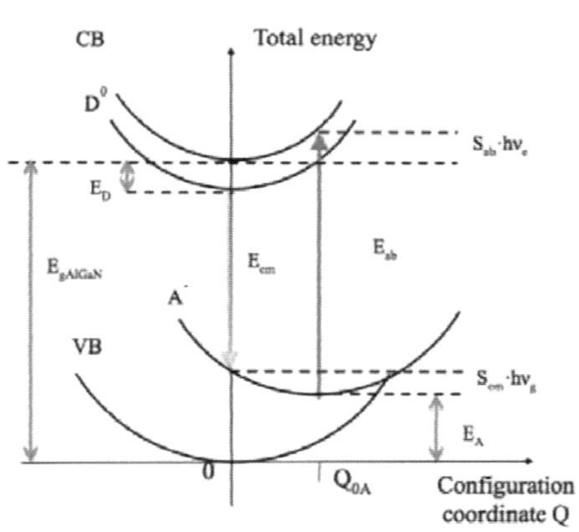

Figure 6. Configuration coordinate diagram describing optical band-to-band absorption and donor-acceptor pair emission at acceptor-like defects with strong electron-phonon coupling. The A^- term corresponds to an occupied state of the defect, and the VB, CB, and D^0 terms correspond to unfilled excited states of the defect where an electron is in the valence band, the conductivity band, and a shallow donor state, respectively. The blue and green arrows indicate optical transitions during excitation and recombination, respectively.

The emission peak energy is given by equations (5) and (6) for DA and eA transitions:

$$E_{em} = E_{gAlGaN} - E_D - E_A - S_{em}h\nu_e \,, \tag{5}$$

$$E_{em} = E_{gAlGaN} - E_D - E_A - S_{em}h\nu_e \,. \tag{6}$$

An important aspect of the applicability of equations (5) and (6) is delocalization of electrons in heavily doped semiconductors. The criterion of heavy doping is expressed by the inequality:

$$N_D a_B^3 > 1\,, \tag{7}$$

where N_D is the donor concentration and a_B is the Bohr radius of the donor state [17]. For GaN, the expression (7) is satisfied at $N_D \approx 7.5\times10^{19}$ cm^{-3} within the framework of the hydrogen atom model [18]. In heavily doped semiconductors, the impurity band merges with the conduction band and electrons are not localized at donor levels. Therefore, only eA transitions should be observed at $N_{Si} \approx 1.5\times10^{20}$ cm^{-3} in cases where Si remains a shallow donor in $Al_xGa_{1-x}N$. Since Si is considered a deep donor with E_D = 250 meV in AlN, it still forms an impurity band below the conduction band and DA transitions take place. A Si donor level deepens at $x \sim 0.85$ [5], therefore equation (4) is applicable in the case of AlN and equation (5) should be used for $x = 0.5$–0.7.

4.2. *Energetic parameters of acceptors in $Al_{0.67}Ga_{0.33}N$:Si*

According to the one-dimensional CC model (see figure 6), the Stock shift of PL should be given by equations (8) and (9) in case of DA and eA transitions:

$$E_{ab} - E_{em} = E_D + 2S_{em}h\nu_e\,, \tag{8}$$

$$E_{ab} - E_{em} = 2S_{em}h\nu_e\,. \tag{9}$$

Equation (9) is acceptable for x = 0.67. The Stock shift E_{ab}-E_{em} ≈ 2.3 eV [10] estimates $S_{em}h\nu_e \approx 1.15$ eV. The temperature broadening of the PL band is given by equation (3). The $W(0)$ is given by

$$W(0) = \sqrt{8\ln 2}\cdot S_{em}h\nu_g / \sqrt{S_{ab}}\,. \tag{10}$$

Equations (3), (4), (9) and (10) allow estimating CC parameters for an acceptor in $Al_{0.67}Ga_{0.33}N$:Si. We obtained $S_{ab} \approx 19.9$, $S_{em} \approx 18.5$, $h\nu_e \approx 62$ meV, $h\nu_g \approx 58$ meV. The differences between S_{em} and S_{ab} and $h\nu_e$ and $h\nu_e$ are at the level of error. After that, we obtain $E_A \approx 1.4$ eV from equation (6). These results follow the previously observed trend: a decrease in Al content leads to a decrease in E_A and an increase in $S_{em}h\nu_e$.

The activation energy E_1 = 160 meV obtained from equation (2) is an order of magnitude less than E_A and close to the previously reported value of 170 meV [10]. Most likely, this activation energy corresponds to internal quenching of luminescence. Due to the fact that the potential curve of the excited state is shifted relative to the potential curve of the ground state, it is possible to intersect these two curves so that, while in an excited state, the system can assume the same configuration as it has in the ground state at a sufficiently large value of the vibrational energy. In this case, nonradiative transitions from the excited state to the ground state occur.

Another feature is the linear blue shift of the maximum of the visible PL band with an increase in temperature, depending on the particulars of the adiabatic potentials of the recombination centers. The CC model predicts the following expression for a temperature-related shift [19]:

$$\Delta E_{CC} = \left[\frac{\nu_g^2 - \nu_e^2}{\nu_e^2} + \frac{8\nu_e^4}{\nu_g^2(\nu_g^2+\nu_e^2)}\right]\left(\frac{E_{ab}(0) - E_{em}(0)}{E_{em}(0)}\right)kT\,, \tag{11}$$

where $E_{ab}(0)$ and $E_{em}(0)$ are the absorption and emission energies at T = 0. Assuming that E_{ab}-E_{em} is about 2.3 eV at T = 0, we obtain $\Delta E_{CC} \approx 2.7kT$. Due to the band-gap reduction, the observed ΔE_m follows the same law but with a less linear coefficient. However, the CC model does not explain the decrease in the linear coefficient from 1.7 to 1.2 with an increase in N_{Si} from 5×10^{19} cm^{-3} to

1.5×10^{20} cm^{-3}. This fact suggests that the theory of heavily doped semiconductors should be taken into account.

5. Conclusions

The new results on the energy structure of recombination centers responsible for the intensive visible emission band in heavily doped $Al_xGa_{1-x}N$:Si epitaxial layers with $x > 0.5$ have been obtained by photoluminescence (PL) spectroscopy. The PL band in AlN was attributed to donor-acceptor (DA) transitions. At the lowest Al content, the impurity band merges with the conduction band and DA transitions are replaced by electron-acceptor transitions involving the same acceptor. Using the model of configuration coordinates, the parameters of recombination centers for $Al_{0.67}Ga_{0.33}N$ were estimated to be $S_{ab} \approx 19.9$, $S_{em} \approx 18.5$, $h\nu_e \approx 62$ meV, $h\nu_g \approx 58$ meV. The acceptor energy level was $E_A \approx 1.4$ eV. The activation energy of internal quenching of luminescence of about 160 meV was obtained. The blue shift of the maximum of the PL band with an increase in temperature exceeded the temperature narrowing of the band gap but decreased with an increase in the Si concentration. This indicates the inadequacy of the configuration coordinate model for heavy $Al_xGa_{1-x}N$ doping.

References

[1] Son K S, Kim D, Cho H K, Lee K, Kim S, Park K and Kim J 2003 *Phys. Stat. Sol. (c)* **0** 2095
[2] Xie J, Mita S, Rice A, Tweedie J, Hussey L, Collazo R and Sitar Z 2011 *Appl. Phys. Lett.* **98** 202101
[3] Sanchez-Garcia M A, Calleja E, Sanchez F J, Calle F, Monroy E, Basak D, Muñoz E, Villar C, Sanz-Hervas A, Aguilar M, Serrano J J and Blanco J M 1998 *Journal of Electronic Materials* **27** 276
[4] Zhuravlev K S , Osinnykh I V, Protasov D Yu, Malin T V, Davydov V Yu, Smirnov A N, Kyutt R N, Spirina A V and Solomonov V I 2013 Phys. Status Solidi C **10** 315
[5] Borisov B, Kuryatkov V, Kudryavtsev Yu, Asomoza R, Nikishin S, Song D Y, Holtz M and Temkin H 2005 *Appl. Phys. Lett.* **87** 132106
[6] Taniyasu Y, Kasu M and Makimoto T 2004 *Appl. Phys. Lett.* **85** 4672
[7] Silvestri L, Dunn K, Prawer S and Ladouceur F 2011 *Appl. Phys. Lett.* **99** 122109
[8] Stampfl C and Van der Walle C G 1998 *Appl. Phys. Lett.* **72** 459
[9] McCluskey M D, Jonson N M, Van de Walle C G, Bour D P, Kneissl M and Walukiewicz W 1998 *Phys. Rev. Lett.* **80** 4008
[10] Osinnykh I V, Malin T V, Plyusnin V F, Suranov A S, Gilinsky A M and Zhuravlev K S 2016 *Jpn. J. Appl. Phys.* **55** 05FG09
[11] Osinnykh I V, Malin T V, Plyusnin V F, Zhuravlev, Ber B Ya and Kazantsev D Yu 2017 *J. Phys.: Conf. Ser.* **816** 012002
[12] Muth J F, Brown J D, Johnson M A L, Yu Zhonghai, Kolbas R M, Cook Jr J W and Schetzin J F 1999 *MRS Internet J. Nitride Semicond. Res.* **4S1** G5.2
[13] Feneberg M, Leute R A R, Neusch B, Thonke K and Bickermann M 2010 *Phys. Rev. B* **82** 075208
[14] Koide Y, Itoh H, Khan M R H, Hiramatsu K, Sawaki N and Akasaki I 1987 *J. Appl. Phys.* **61** 4540
[15] Huang K and Rhys A 1950 *Proc. R. Soc. A* **204** 406
[16] Henry C H and Lang D V 1977 *Phys. Rev. B* **15** 989
[17] Levanyuk A P and Osipov V V 1981 *Sov. Phys. Usp.* **24** 187
[18] Osinnykh I V, Zhuravlev K S, Malin T V, Ber B Ya and Kazantsev D Yu 2014 *Semiconductors* **48** 1134
[19] Shinoya S, Koda T, Era K and Fujiwara H 1964 *J. Phys. Soc. Jpn.* **19** 1157

Photoemission properties of flat and rough GaAs surfaces with cesium and oxygen layers

A G Zhuravlev[1,2], D M Kazantsev[1,2], V S Khoroshilov[1,2], A S Kozhukhov[1] and V L Alperovich[1,2]

[1]Rzhanov Institute of Semiconductor Physics, 630090 Novosibirsk, Russia
[2]Novosibirsk State University, 630090 Novosibirsk, Russia

E-mail: zhuravl@isp.nsc.ru

Abstract. The evolution of surface band bending and the probabilities of electron emission in a vacuum are studied under cesium and oxygen deposition on atomically flat and rough As-rich and Ga-rich GaAs (001) surfaces by means of photoreflectance spectroscopy and photoemission quantum-yield spectroscopy. On the rough surface, as compared to the flat one, the suppression of non-monotonic Cs-induced band bending variations is observed, along with the overall band bending increase. Multiple repeated activations of the GaAs (001) surface by cesium and oxygen followed by vacuum anneals led to an increase in the root-mean-square roughness from $R_q \sim 0.1$-0.2 nm up to ~ 3 nm. As a result, the probability of electron escape into vacuum decreased by half.

1. Introduction

The *p*-GaAs surfaces with adsorbed cesium and oxygen layers are widely used to create photocathodes with a negative effective electron affinity (NEA) [1]. The surfaces of semiconductors with a relatively low (0.2 - 0.4 eV) positive affinity, such as *p*-GaAs with adsorbed cesium, have recently attracted attention due to the possibility of increasing the solar energy conversion efficiency using photon-enhanced thermionic emission [2, 3, 4]. The quantum yield and other photoemission parameters are determined by a set of electronic properties of Cs/GaAs and GaAs (Cs,O) interfaces, including band structure, kinetic and transport parameters, such as true and effective electron affinity, the magnitude and width of surface band bending, surface recombination rate, and the probability of electron escape into vacuum. It is reasonable to assume that these electronic properties should depend on the morphology of the initial GaAs surface and of the interfaces with the adsorbates [5], but the available experimental data on these dependences are scarce. One can start photoemission studies from a smooth GaAs surface with atomically flat terraces separated by monatomic steps. However, the chemical treatment and vacuum anneals aimed at preparing an atomically clean surface, cesium and oxygen deposition, and additional anneals in vacuum for the adsorbate layers removal may substantially change the surface morphology in the course of the two-stage activation procedure [1]. In particular, the chemical removal of oxides in the solution of HCl in isopropyl alcohol retains the overall step-terraced morphology of the GaAs (001) surface, but it leads to the appearance of small-scale roughness with a monolayer amplitude on terraces [6, 7]. Vacuum anneals also lead to the GaAs(001) surface roughening [8]. The literature does not offer sufficient data concerning the influence of cesium and oxygen deposition, as well as of subsequent vacuum anneals on the surface morphology. In this work, we studied how the clean surface preparation, repeated cesium and oxygen depositions, and vacuum anneals, which are necessary for the fabrication of *p*-GaAs(Cs,O) photoemitters with the state of NEA, affected the surface morphology, as well as electronic properties such as surface band bending, electron affinity, and emission probability in vacuum for Cs/GaAs and GaAs(Cs,O) interfaces.

2. Experimental

The experiments were made on *p*-GaAs(001) epitaxial layers and "epi-ready" substrates. The morphology of the surfaces was studied *ex-situ* by atomic force microscopy (AFM). The thermal smoothing of the GaAs(001) surface was performed in conditions close to equilibrium between the surface and Ga and As vapors. The equilibrium conditions during the thermal smoothing were provided by a saturated Ga-GaAs melt placed in a graphite container together with the GaAs sample intended for smoothing [7].

Atomically clean and structurally ordered As-rich and Ga-rich GaAs (001) surfaces were prepared by chemically removing oxides from the solution of HCl in isopropyl alcohol (HCl-iPA) in an inert nitrogen atmosphere, transferring to the ultra-high vacuum setup without air contact, and subsequent annealing in vacuum [9]. The As-rich GaAs surface was obtained by annealing in vacuum at $T = 400$ °C. To obtain the Ga-rich GaAs surface, we also used the effect of a Cs-induced decrease in temperature of the transition from an As-rich to Ga-rich surface at $T = 500$ °C [10]. Cesium and oxygen were deposited from the degassed dispensers.

The band-bending evolution was measured by means of photoreflectance spectroscopy, using the Fourier transform of Franz-Keldysh oscillations. To provide a uniform electric field near the surface, epitaxial UP^+ structures with thin (~ 100 nm) undoped surface layers were used [11]. A modification of the photoreflectance spectroscopy technique to a "real-time" version allowed us to measure the band-bending evolution during the Cs deposition with high precision [12]. The photoemission was measured in the "reflection" geometry from the illuminated surface. The effective electron affinity and the probability of escape of "thermalized" and "hot" electrons into vacuum were determined from the photoemission quantum yield spectra according to the procedure described in [4].

3. Results and discussion

To study the relation between the GaAs(001) surface morphology and the electronic states at the Cs/GaAs(100) interface, we measured the band-bending evolution under the Cs deposition on the smooth step-terraced as-grown UP^+ structures with the root-mean-square $R_q \sim 0.2$ nm (figure 1(a)), on the same epitaxial surfaces subjected to annealing in equilibrium conditions (figure 1(b)), and on the rough surface of the epitaxial structure obtained by annealing in non-equilibrium conditions (figure 1(c)). It is seen in figure 1(a) that the as-grown GaAs surface has a step-terraced morphology with flat terraces separated by monatomic steps, although the steps have a strongly jagged shape, presumably due to the kinetic instabilities during their growth. As seen in figure 1(b) the thermal smoothing in

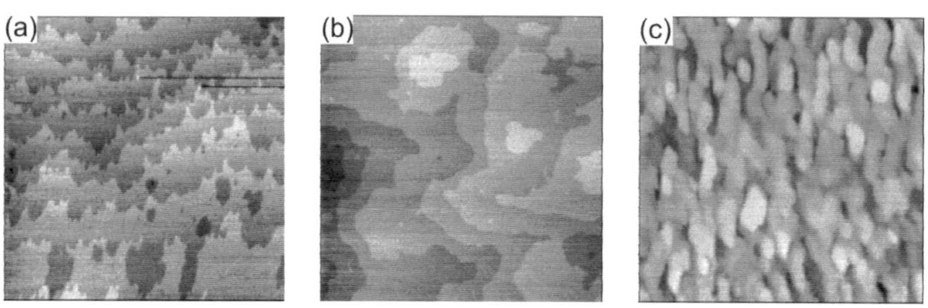

Figure 1. 3×3 μm^2 AFM images of the as-grown GaAs(001) surface (a), the surface subjected to thermal smoothing in the quasi-equilibrium conditions at $T = 650$ °C (b), and the rough surface obtained by annealing in non-equilibrium conditions (c).

equilibrium conditions led to a notable improvement of the step-terraced morphology, in particular, to less jagged steps, although the root-mean-square roughness remained practically the same at $R_q \sim 0.2$ nm. Annealing in non-equilibrium conditions led to the destruction of the step-terraced morphology, the formation of a rough relief with dips and hills, and a roughness increase up to $R_q \sim 1$ nm (figure 1(c)).

The evolutions of band-bending φ_S are shown in figure 2 for the Cs deposition on As-rich (a) and Ga-rich (b) GaAs(001) surfaces. It is seen that the shape of $\varphi_S(\theta)$ is different for the As-rich and Ga-rich surfaces [10, 12]. On the As-rich surface, φ_S increased steeply at small Cs coverages ($\theta \sim 0.1$ monolayer (ML)) and went through a maximum at larger coverages, while on the Ga-rich surface, several distinct maxima and minima ("fine structure") were reproducibly observed [12]. It is seen that the shape of $\varphi_S(\theta)$ was similar for the as-grown sample and for the sample annealed in equilibrium, which is in agreement with the fact that both samples were atomically smooth with their qualitatively similar step-terraced morphology. Thus, the difference in the size and shape of monatomic islands and steps on the initial step-terraced GaAs surfaces did not substantially affect the dose dependences of the band bending. On the contrary, on the rough surface, the shape of $\varphi_S(\theta)$ was distinctly different, with a substantially higher value of φ_S on the clean surface and a reduced amplitude of φ_S variations with the Cs coverage, including a less pronounced "fine structure". A possible reason for the distinction is a much higher defect-induced density of the surface states in the middle of the band gap on the rough surface. These states cause the initial band bending and restrict its variations under the Cs deposition.

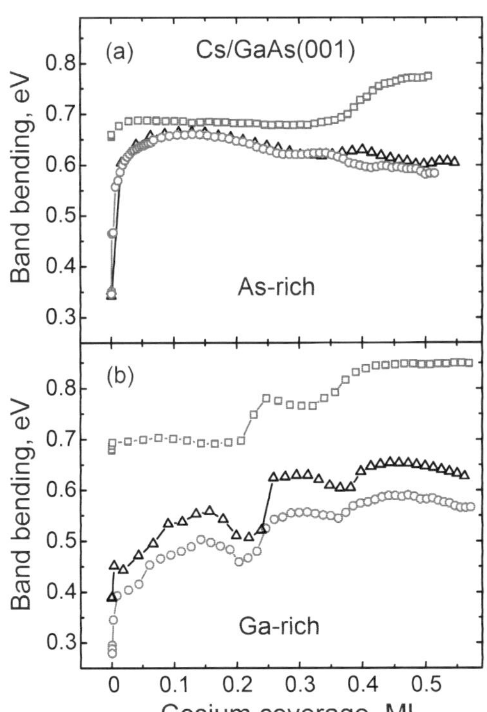

Figure 2. Evolution of the band bending φ_S under the Cs deposition on the as-grown GaAs(001) UP^+ structures (circles), on the structures annealed in equilibrium conditions (triangles), and on the rough (squares) As-rich (a) and Ga-rich (b) GaAs(001) surfaces.

It should be noted that cesium and oxygen deposition followed by annealing in vacuum may cause substantial changes in the GaAs surface relief. These changes should be taken into account when analyzing the relations between the surface morphology and electronic properties of Cs/GaAs and GaAs(Cs,O) interfaces. Figure 3(a) shows the AFM image of the initial as-grown step-terraced surface of an epitaxial *p*-GaAs layer. After removing oxides in HCl-iPA followed by annealing in vacuum at 400 °C, which yields a clean As-rich GaAs(001) surface, 0.5 ML of Cs was deposited; the second

IOP Conf. Series: Journal of Physics: Conf. Series **993** (2018) 012007 doi:10.1088/1742-6596/993/1/012007

anneal was made at 400 °C, after which the sample was retrieved from the vacuum setup and the AFM image shown in figure 3(b) was taken in the air. The image shown in figure 3(c) was obtained on a similar sample by means of the same procedure, except for that both anneals were made at 500 °C, which corresponds to the Ga-rich surface formation [10]. As seen in figure 3(b), the cesium adsorption on the As-rich surface and annealing in vacuum at $T = 400$ °C led to a deterioration in the step-terraced morphology, although the monatomic steps are still visible. A similar procedure with the anneals at a higher temperature $T = 500$ °C led to the appearance of "islands" with diameters of ~ 50 -70 nm, heights of ~ 1 - 3 nm and a density of ~ 10 μm^{-2}. On the surface areas between the islands, the step-terraced morphology disappeared, and the root-mean-square roughness increased from $R_q \sim 0.12$ nm on the initial surface up to $R_q \sim 0.22$ nm.

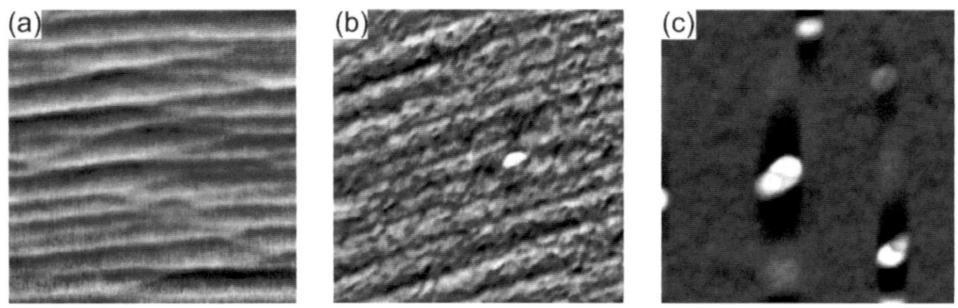

Figure 3. 1×1 μm^2 AFM images of the as-grown GaAs (001) surface (a), after the Cs deposition and annealing in vacuum at $T = 400$ °C (b) and $T = 500$ °C (c).

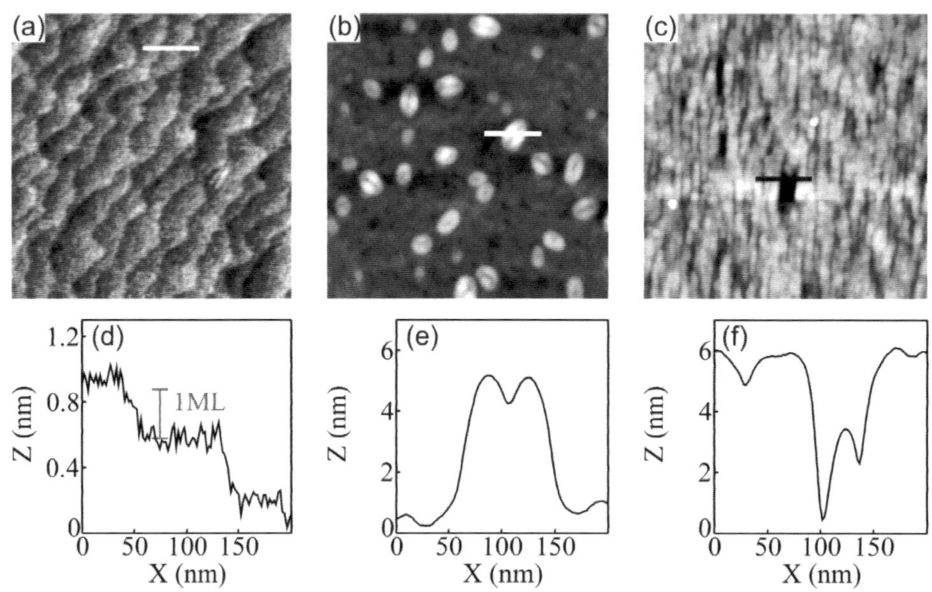

Figure 4. 1×1 μm^2 AFM images of the GaAs(001) substrate (top row) and the *z*-*x* cross sections along the segments denoted in the images (bottom row) after thermal smoothing in quasi-equilibrium conditions at $T = 650$ °C (a, d), activation by cesium and oxygen to the state of NEA and annealing in vacuum at $T = 500$ °C (b, e), and subsequent etching in HCl-iPA (c, f).

Figure 4 shows the AFM images of an "epi-ready" *p*-GaAs substrate, which was first brought to the step-terraced morphology by thermal smoothing in equilibrium conditions (figure 4(a)), and then activated to the state of NEA by cesium and oxygen deposition followed by annealing in vacuum at

$T = 500$ °C (figure 4(b)). The comparison of figures 3(c) and 4(b) shows that similar islands appeared on the surface, although with a larger concentration in the latter case. One can presume that these islands are gallium droplets [13], although we did not determine the chemical composition of the islands directly in this work. Figure 4(c) shows the image of the same sample after treatment in HCl-iPA. Instead of islands, pits of about the same sizes and concentration are seen in the image. This change in the morphology can be interpreted as a result of etching off the Ga droplets and, thus, it supports the assumption about the nature of the droplets.

The results of photoemission studies on the GaAs(Cs,O) surfaces with various morphologies are summarized in figure 5. The measured probabilities of electron escape in vacuum for "thermalized" P_t (a) and "hot" P_h (b) photoelectrons *versus* the effective affinity χ^* are shown here. These dependences were measured under an additional "excessive" oxygen exposure of the GaAs(Cs,O) surfaces "optimally" activated to the state of NEA. The minimum in P_t near $\chi^* \approx 0$ was explained in [14] by the changeover in the emission mechanism under the transition from the positive to the negative effective affinity. It is seen in figure 5 that, for the rough surface with $R_q \sim 3$ nm formed due to multiple cycles of activation and annealing, the escape probabilities are systematically lower than those measured on a smoother surface during the first activation cycle (by a factor of 1.3-2). According to figure 5, the photoemission quantum efficiency of the GaAs(Cs,O) surfaces activated to the state of NEA affinity decreases from 25% at a relatively flat surface to 12% at a rough surface. This decrease is partly due to a smaller absolute value of χ^* (by 15 meV), and partly due to a smaller electron escape probability. In its turn, the decrease in the escape probabilities can presumably be associated with the electron scattering on the rough surface.

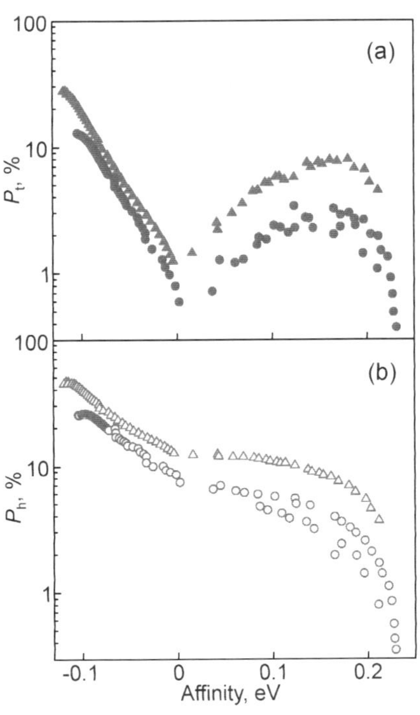

Figure 5. Escape probabilities for thermalized (a) and hot (b) electrons from the GaAs(Cs,O) surface *versus* the effective electron affinity under the oxygen deposition at the surface with R_q~0.5 nm (triangles) and R_q~3 nm (circles).

4. Summary

In conclusion, this work presents experimental evidence of the complex spectra of adatom- and defect-induced surface states on the Cs/GaAs (001) surface. The comparison of the band bending behavior under the Cs deposition on atomically flat and rough surfaces proves that surface roughness produces the defect-induced surface states which restrict the adatom-induced band bending variations. The initial as-grown GaAs surface morphology evolution upon annealing in vacuum and cesium adsorption were studied. It is shown that the surface became rough after annealing in vacuum at $T = 500$ °C, and islands of up to 3 nm in height appeared. The surface roughening under repeated activation by cesium and oxygen and annealing in vacuum led to an increase in the root-mean-square roughness by an order of magnitude up to $R_q \sim 3$ nm. As a result, the electron escape probability and photoemission quantum yield decrease by factors of 1.3 - 2.

Acknowledgments
The quantum yield photoemission spectroscopy studies were supported by the Russian Foundation for Basic Research (Project No. 17-02-00815).

References
[1] Bakin V V *et al* 2007 *e-J. Surf. Sci. Nanotech.* **5** 80.
[2] Schwede J W *et al* 2010 *Nat. Mater.* **9** 762;
[3] Schwede J W *et al* 2013 *Nat. Commun.* **4** 1576.
[4] Zhuravlev A G, Romanov A S and Alperovich V L 2014 *Appl. Phys. Lett.* **105** 251602.
[5] Karkare S and Bazarov I 2011 *Appl. Phys. Lett.* **98** 094104.
[6] Akhundov I O *et al* 2017 *Appl. Surf. Sci.* **406** 307.
[7] Alperovich V L *et al* 2009 *Appl. Phys. Lett.* **94** 101908.
[8] Fan Y, Karpov I, Bratina G, Sorba L, Gladfelter W and Franciosi A 1996 *J. Vac. Sci. Technol. B* **14** 623.
[9] Alperovich V L, Tereshchenko O E, Rudaya N S, Sheglov D V, Latyshev A V and Terekhov A S 2004 *Appl. Surf. Sci.* **235** 249.
[10] Tereshchenko O E, Alperovich V L, Zhuravlev A G, Terekhov A S and Paget D 2005 *Phys. Rev. B* **71** 155315.
[11] Hoof C *et al* 1989 *Appl. Phys. Lett.* **54** 608.
[12] Zhuravlev A G, Scheibler H E, Jaroshevich A S and Alperovich V L 2010 *J. Phys.: Condens. Matter* **22** 185801.
[13] Tersoff J, Jesson D E and Tang W X 2009 *Science* **324** 236.
[14] Zhuravlev A G, Khoroshilov V S and Alperovich V L 2017 *JETP Lett.* **105** 686.

The effects of GaN nanocolumn arrays and thin Si_xN_y buffer layers on the morphology of GaN layers grown by plasma-assisted molecular beam epitaxy on Si(111) substrates

K Yu Shubina, E V Pirogov, A M Mizerov, E V Nikitina and A D Bouravleuv

Nanoelectronics Lab, St. Petersburg Academic University, Khlopina 8/3, 194021 St. Petersburg, Russia

Abstract. The effects of GaN nanocolumn arrays and a thin Si_xN_y layer, used as buffer layers, on the morphology of GaN epitaxial layers are investigated. Two types of samples with different buffer layers were synthesized by PA-MBE. The morphology of the samples was characterized by SEM. The crystalline quality of the samples was assessed by XRD. The possibility of synthesis of continuous crystalline GaN layers on Si(111) substrates without the addition of other materials such as aluminum nitride was demonstrated.

1. Introduction

A^3N wide bandgap semiconductor materials and their ternary alloys have a set of unique electrical, optical and mechanical properties, which are important in modern optoelectronics, high power electronics, and microelectromechanical systems (MEMS). Since natural substrates for A^3N materials are very expensive, silicon carbide (SiC), sapphire (c-Al_2O_3) and silicon (Si(111)) substrates are used for epitaxial growth of III-nitrides.

One of the key technological problems in GaN synthesis on silicon substrates is the dramatic lattice mismatch (16.9%) between GaN and Si(111). Moreover, there is a large difference in their thermal expansion coefficients. These two factors lead to relaxation of elastic strain through the formation of misfit dislocations at the GaN/Si interface. Another important problem is interdiffusion of Si and Ga atoms under growth conditions that are typical for molecular-beam epitaxy (MBE). Therefore, the local etching of a Si(111) substrate is possible with the formation of macroscopic voids during the growth process [1]. Different buffer and transitional layers are used to solve the problems mentioned above. In particular, these are GaN nanocolumn arrays and thin Si_xN_y layers formed on the surface of Si(111) substrates immediately before planar GaN epitaxial growth [2, 3].

2. Experimental details

In this work, the use of GaN nanocolumn arrays and thin Si_xN_y layers as transition or buffer layers for the synthesis of planar epitaxial GaN layers on Si(111) substrates was investigated. The samples of GaN/3D-GaN/Si(111) and GaN/Si_xN_y/Si(111) epitaxial structures were grown by PA-MBE in a Veeco Gen 200 MBE system equipped with a Veeco UNI-Bulb™ RF plasma source. The thickness of planar GaN epitaxial layers was (0.6 - 1.0) μm. The synthesis of epitaxial layers was carried out on semi-insulating ($R > 10$ kΩcm) Si(111) substrates, which were cleaned according to the Shiraki method [4].

The samples of the first type are epitaxial structures with GaN nanocolumn arrays (3D-GaN), which were used as a buffer layer to reduce the dislocation density in the planar GaN layer. To test the procedure of catalyst-free synthesis of GaN nanocolumn arrays (the same procedure was used in [5]), a sample consisting of GaN nanocolumn arrays formed on a Si(111) substrate was grown. SEM images of the cross-section and the surface of this sample are shown in figure 1. The height of individual nanocolumns in the array was about 300 nm. After selection of the growth conditions that allow obtaining catalyst-free GaN nanocolumns on Si(111) substrates, the samples with GaN nanocolumn arrays as a buffer layer were grown.

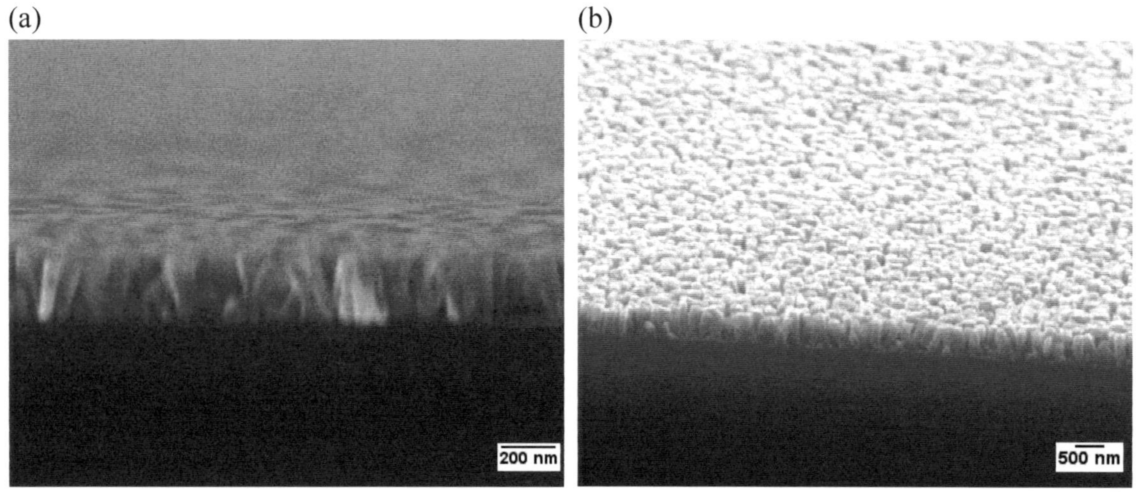

Figure 1. SEM images of the 3D-GaN/Si(111) sample with GaN nanocolumn arrays: cross-section (a) and surface morphology (b).

The samples of the second type are epitaxial structures with a thin Si_xN_y layer, which was used as a buffer layer. The thin Si_xN_y layer was formed during nitridation of a Si(111) substrate in the flux of atomic nitrogen. It is assumed that the use of thin homogeneous continuous Si_xN_y layers allows to avoid the mutual diffusion of Ga and Si atoms, as well as the local etching of the silicon substrate with the formation of macroscopic voids on the GaN/Si(111) interface.

The characteristics of the obtained samples were studied by scanning electron microscopy (SEM) and X-ray diffractometry (XRD). The surface morphology was checked using a Zeiss Supra25 scanning electron microscope. A DRON-8 multifunctional diffractometer was used for X-ray diffractometry experiments.

3. Results and discussion

As a result of the work, two types of samples were obtained (GaN/3D-GaN/Si(111) and GaN/Si_xN_y/Si(111) epitaxial structures).

The SEM micrograph of the sample with GaN nanocolumn array, used as a buffer layer, is shown in figure 2. The height of individual nanocolumns reaches 0.2 μm. On the cross-section a gradual coalescence of GaN nanocolumns into a continuous GaN layer is clearly visible. The surface morphology of the obtained epitaxial layer has a mosaic structure, which is characteristic for GaN epitaxial layers grown on silicon substrates. Thus, as a result of using 3D-GaN buffer layer, continuous crystalline GaN epitaxial layer was obtained on Si(111) substrate with no additional materials used.

In figure 3, the cross-sectional SEM images of GaN/SixNy/Si(111) epitaxial structures are shown. The sample shown in figure 3(a) was obtained with 60 min nitridation, while the duration of the nitridation process for the sample shown in figure 3(b) was 30 min.

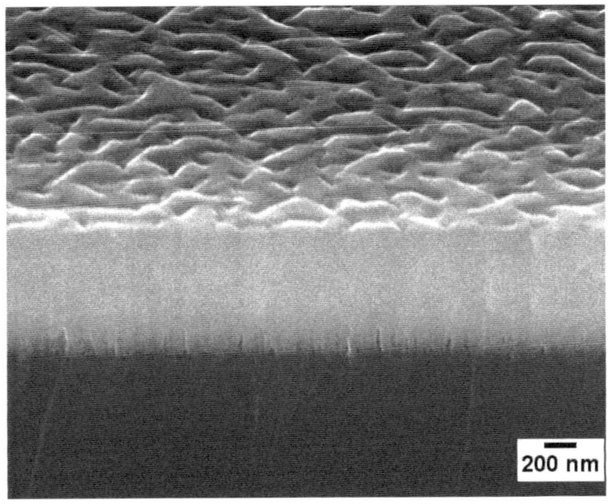

Figure 2. SEM cross-sectional image of the GaN/3D-GaN/Si(111) sample containing GaN nanocolumn array as a buffer layer.

Figure 3(a) shows a continuous GaN epitaxial layer (60 min nitridation of a Si(111) substrate), while the structure of the GaN layer in figure 3(b) is coalesced co-directional GaN blocks (30 min nitridation). Thus, according to [6], the duration of the nitridation process affects the homogeneity and continuity of Si_xN_y and hence the morphology of the GaN epitaxial layer.

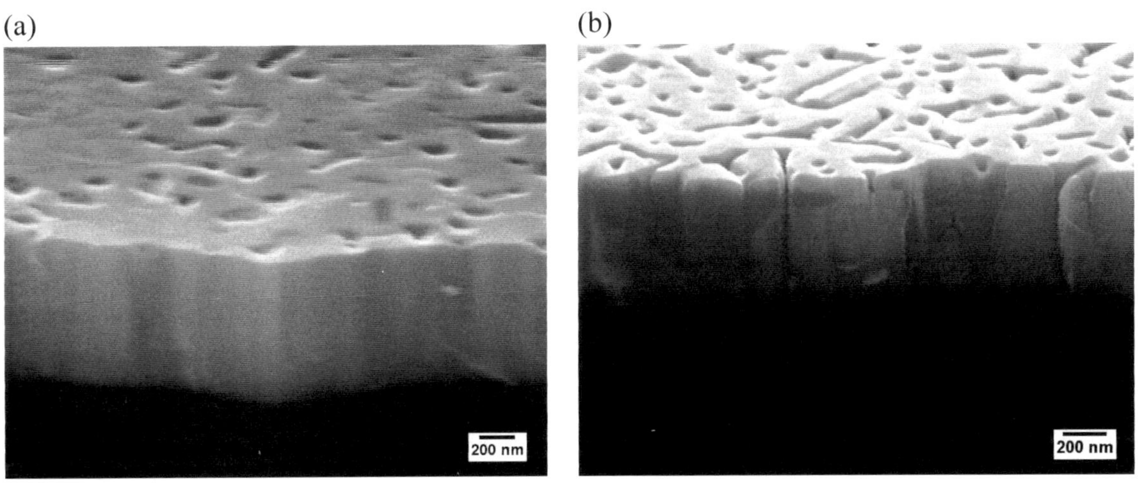

Figure 3. Cross-sectional SEM images of the $GaN/Si_xN_y/Si(111)$ samples grown with nitridation of a Si(111) substrate for 60 min (a) and 30 min (b).

To assess the presence of stresses in GaN layers grown on Si(111) substrates with a thin Si_xN_y layer and arrays of GaN nanocolumns as buffer layers, the samples were examined by X-ray diffractometry. Figure 4 shows the XRD rocking curves for the samples of both types. It can be seen from figure 4 that the position of the peaks corresponding to GaN ($2\theta_{GaN} \approx 34.75$ °) and the broadening of the rocking curves do not vary significantly depending on the buffer layer used. Thus, the use of GaN nanocolumn arrays and a thin Si_xN_y layer as a buffer layer allows growing GaN epitaxial layers on Si(111) substrates that are approximately equal in the crystalline quality. The presence of the defects (dislocations) in the epitaxial GaN layer, as well as its mosaic structure, contribute to the broadening of the rocking curve. The difference in the intensity of the peaks corresponding to GaN directly corresponds to the difference in the thickness of GaN layers.

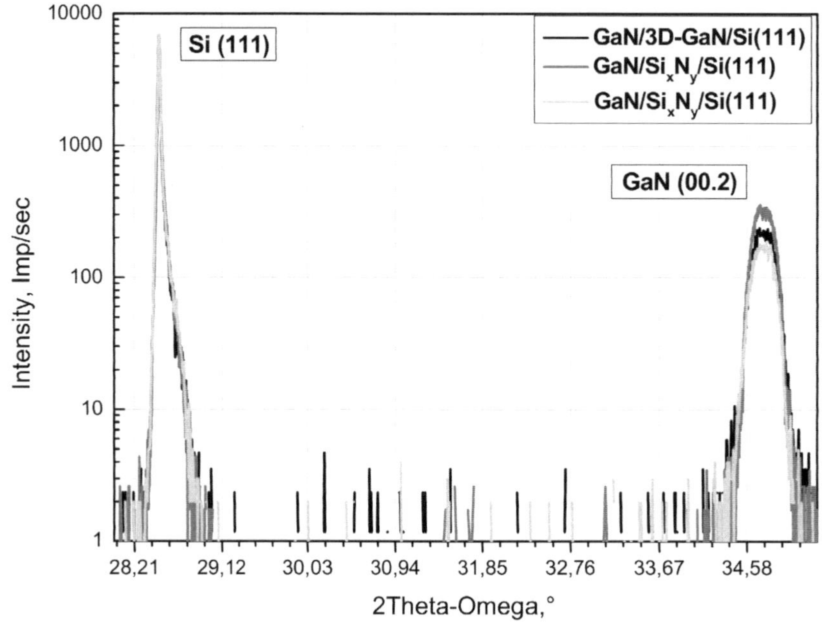

Figure 4. XRD rocking curves for epitaxial structures of GaN/3D-GaN/Si(111) (black), GaN/Si_xN_y/Si(111) with 60 min nitridation (red), GaN/Si_xN_y/Si(111) with 30 min nitridation (green).

It is worth noting that for the sample with 30 min nitridation, the local etching of a Si(111) substrate was observed. For the samples obtained with 60 min nitridation and GaN nanocolumn arrays as a buffer layer, no cases of local etching were detected. Moreover, according to the results of measurements based on the Hall effect, the use of 3D-GaN and thin Si_xN_y buffer layers does not prevent the mutual diffusion of Si and Ga atoms, thus providing unintentional doping.

4. Conclusion

In this work, the effects of GaN nanocolumn arrays and a thin Si_xN_y layer, used as buffer layers, on the morphology of GaN epitaxial layers are investigated. The samples of GaN/3D-GaN/Si(111) and GaN/Si_xN_y/Si(111) epitaxial structures were obtained. The possibility of synthesis of continuous crystalline GaN layers on Si(111) substrates without the use of additional materials such as aluminum nitride was demonstrated. However, neither type of buffer layer makes it possible to avoid the mutual diffusion of Si and Ga atoms with unintentional doping. Moreover, there is a probability of local etching of a Si(111) substrate when using a thin Si_xN_y buffer layer.

Acknowledgments

The results were obtained as a part of Project №4 (the general agreement on research activities №3663-mra of SkolTech and Academic University RAS).

References

[1] Sobanska M, Klosek K, Zytkiewicz Z R, Borysiuk J, Witkowski B S, Lusakowska E, Reszka A and Jakiela R 2012 *Cryst. Res. Technol.* **47** 307-12
[2] Kusakabe K, Kikuchi A and Kishino K 2002 *J. Cryst. Growth.* **237-239**, 988-92
[3] Kumar M, Rajpalke M K, Roul B, Bhat T N, Sinha N, Kalghatgi A T and Krupanidhi S B 2011 *Appl. Surf. Sci.* **257** 2107-10

[4] Ishizaka A and Shiraki Y 1986 *J. Eltrochem. Soc.* **133** 666-71
[5] Mizerov A M, Kladko P N, Nikitina E V and Egorov A Yu 2015 *Semiconductors* **49** 274-77
[6] Yamabe N, Yamamoto Y and Ohachi T 2011 *Phys. Stat. Sol. C* **8** 1552

Zinc oxide hierarchical nanostructures for photocatalysis

O Yukhnovets, A A Semenova, E A Levkevich, A I Maximov, V A Moshnikov

Department of Micro- and Nanoelectronics, St Petersburg Electrotechnical University "LETI", Saint-Petersburg 197376, Russia

E-mail: iuhnovec@gmail.com

Abstract. In this work, we perform the study of zinc oxide hierarchical structures synthesized by the low-temperature hydrothermal method. The paper considers morphological properties of obtained structures. Photocatalytic activity of samples was analysed by methyl orange degradation under UV irradiation. The sufficient decrease in methyl orange has been demonstrated.

1. Introduction

The topic of water and air environment purification from organic impurities is currently relevant. A great alternative to classical refining methods is the photocatalysis [1]. This process can be described as the change of chemical reactions rate under the light excitation in the presence of the photocatalyst which is involved in chemical transformations of the reaction participants [2].

Among the wide variety of materials for photocatalytic performance zinc oxide (ZnO) can be emphasized on due to the ecological safety of formation and utilization, technological ease of preparation [3]. In comparison with the widespread titanium dioxide (a wide bandgap semiconductor with E_g = 3.0-3.3 eV [4]), zinc oxide (E_g = 3.36 eV) is characterized by a higher quantum yield of phototransformation, which is associated with a low degree of charge carriers recombination and high adsorption capacity [1].

In the transition to nanoscale conditions the activity of photocatalytic materials increases. This effect is achieved by increasing the specific surface area and reducing the recombination of charge carriers [3].

With the use of wet chemistry techniques it is possible to obtain different morphologies of ZnO nanostructures: nanoneedles, nanobelts, nanodendrites et al [5]. In this paper, porous zinc oxide nanospheres consisting of 2D-plates were synthesized. These structures are characterized by developed surface and enhanced photocatalytic properties.

2. Experimental section

2.1. Synthesis of ZnO hierarchical structures

Zinc oxide was synthesised under mild hydrothermal conditions. Adjusting pH of a solution, temperature and precursor concentrations it is possible to change the morphology of a structure in a wide range. Among other methods, soft surfactant templating proved to be an effective approach for tailoring complex *3D* hierarchical structures via oriented self-assembly of nanobuilding sub-units [6, 7].

To synthesize zinc oxide hierarchical structures (ZnO HSs) 0.5 ml of 0.5 M zinc nitrate solution ($Zn(NO_3)_2*6H_2O$ was mixed with 30 ml distilled water containing 0.5 g of cetyl-trimethyl-ammonium-bromide (CTAB), followed by adding 5 ml of NaOH solution. The concentration of NaOH solution was estimated on the assumption of $[Zn^{2+}]:[OH^-]=1:10$ ratio. The solution was mixed in an ultrasonic bath (100 W), vigorously stirred for 5 minutes and transferred to the thermostat (90^{0}C) for 2 hours. Products were collected by centrifugation and washed for 3 times in distilled water. ZnO powder then was dried at 80 ^{0}C and annealed at 350 ^{0}C for 20 minutes.

2.2. Photodegradation experiment

To perform photodegradation reaction 10 ppm solution of methyl orange was mixed with photocatalyst and stirred in the dark for 1 hour to reach adsorption-desorption equilibrium between dye and ZnO HSs. The suspension was placed under UV irradiation. To analyse changes of methyl orange concentrations adsorption spectra of a solution were taken at regular time intervals.

3. Results and discussion

3.1. Morphology of ZnO hierarchical structures

Under hydrothermal conditions, ZnO tends to form one-dimensional structures due to the greater growth rate of (0001) plane [6]. It is possible to achieve the formation of complex 3*D*-nanostructures by adjusting synthesis conditions (temperature, precursor concentrations, growth time, pH), using surfactants as a soft template or applying ultrasonic irradiation.

Figure 1 (*a*, *b*) demonstrates the results of synthesis in alkaline media with polyvinylpyrrolidone (PVP). Figure 2 shows the structures obtained under ultrasonic irradiation.

a b

Figure 1. SEM images of ZnO HSs structures grown with PVP.

a *b*

Figure 2. SEM images of ZnO HSs grown under ultrasonic irradiation.

Within the frame of this work the formation and photocatalytic properties of ZnO HSs synthesized with CTAB were considered in details.

Figure 3 (*a*, *b*) shows scanning electron microscopy of hierarchical ZnO nanostructures. It is seen that structures appeared to be porous spheres composed of 2*D* petals merged by the oriented self-assembly process. The exact formation mechanism of hierarchical structures is still undefined. However, it is assumed that the driving force for nanocrystals assembly is the anisotropic hydrophobic attraction and electrostatic interactions derived from surface charges.

At the first stage, the formation of ZnO clusters (30-40 atoms) occurs. Clusters are stabilized by CTAB micelles and aggregate into plates merging to form spherical structures in order to minimize surface energy. Hence, CTAB acts as a coordinating agent for colloidal nanoclusters and enables to grow oriented nanostructures by simple wet chemistry methods [7].

a *b*

Figure 3. SEM images of ZnO HSs grown with CTAB.

3.2. Photocatalytic activity

It is expected that zinc oxide hierarchical structures are to demonstrate enhanced photocatalytic properties due to unique morphology that provides high surface-to-volume ratio, inhibits charge carriers recombination and powder aggregation.

In order to activate the photocatalytic process, the amount of energy should be equal or more than the band gap energy of photocatalysts, so that the electrons of valence band can be excited and moved to the conduction band. This energy is the minimum threshold intensity of radiation for activating and starting the photocatalytic process.

The idea of photocatalytic oxidation is to convert hazardous organic dyes to harmless or less harmful intermediates. The last stage of oxidation is a conversion of organic dye to carbon dioxide and water.

In this paper, the photocatalytic activity of ZnO HSs was analysed by degradation of methyl orange dye under UV irradiation. In order to perform photocatalytic reaction 80 mg of photocatalysis were mixed with 10 ppm methyl orange solution. To analyse changes of methyl orange concentrations adsorption spectra of a solution were taken at regular time intervals. Figure 4 shows the degradation process of dye with the presence of ZnO HSs under UV irradiation. Spectra reveal a feature at 360 nm which corresponds to ZnO nanoparticles adsorption band. It can be seen that methyl orange concentration has sufficiently decreased during the photocatalytic reaction. Besides, the colour of a solution changed from orange to almost clear. This result indicates a great perspective of zinc oxide hierarchical structures for photocatalytic applications.

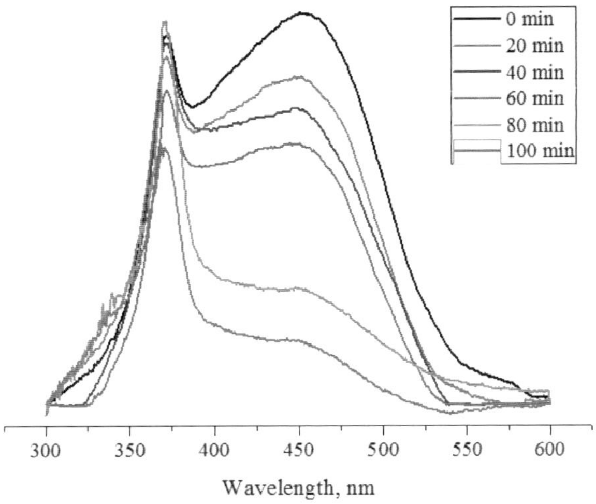

Figure 4. Methyl orange degradation in the presence of ZnO HSs under UV irradiation.

4. Conclusions

Applying such approaches as pH changes, surfactant addition, ultrasonic irradiation enables to obtain complex structures with high surface-to-volume ratio. In this paper, we discuss the formation of 3*D* hierarchical structures by simple low-temperature hydrothermal synthesis with CTAB surfactant added. It is assumed that ZnO colloidal clusters are stabilized by CTAB micelles and aggregate into spheres by oriented self-assembly. The developed morphology of synthesized structures is expected to enhance photocatalytic properties of ZnO. In our experiments, the sufficient decrease in methyl orange dye concentration during photocatalytic reaction has been demonstrated.

Acknowledgments

The investigation was accomplished on account of Russian Science Foundation (project №17-79-20239)

References

[1] Kumar S, Koteswara R R 2015 *The Royal Society of Chemistry* **5** 3306
[2] Савинов Е Н 2000 *Соросовский образовательный журнал* **6**(11) 52
[3] Marschall R Wang L Non-metal doping of transition metal oxides for visible-light photocatalysis 2013 *Catal. Today*
[4] Landmann M, Rauls E, Schmidt W G 2012 *Journal of Physics: Condensed Matter.* **24**(19) 1
[5] Xu S, Wang Z L 2011 *Nano Res.* **4**(11) 1013
[6] Liu B, Zeng H C 2007 *Chem. Mater.* **19**, 5824
[7] Shi Y et al. 2013 *Chem. Mater.* **25** 1000

IOP Conf. Series: Journal of Physics: Conf. Series **993** (2018) 012010 doi:10.1088/1742-6596/993/1/012010

Optimization of conditions for thermal smoothing GaAs surfaces

I O Akhundov[1,2], D M Kazantsev[1,2], A S Kozhuhov[1] and V L Alperovich[1,2]

[1] Rzhanov Institute of Semiconductor Physics, 630090 Novosibirsk, Russia
[2] Novosibirsk State University, 630090 Novosibirsk, Russia

E-mail: akhundov@isp.nsc.ru

Abstract. GaAs thermal smoothing by annealing in conditions which are close to equilibrium between the surface and vapors of As and Ga was earlier proved to be effective for the step-terraced surface formation on epi-ready substrates with a small root-mean-square roughness ($R_q \leq 0.15$ nm). In the present study, this technique is further developed in order to reduce the annealing duration and to smooth GaAs samples with a larger initial roughness. To this end, we proposed a two-stage anneal with the first high-temperature stage aimed at smoothing "coarse" relief features and the second stage focused on "fine" smoothing at a lower temperature. The optimal temperatures and durations of two-stage annealing are found by Monte Carlo simulations and adjusted after experimentation. It is proved that the temperature and duration of the first high-temperature stage are restricted by the surface roughening, which occurs due to deviations from equilibrium conditions.

1. Introduction

Atomically flat semiconductor surfaces are needed for fundamental surface science, fabrication of nanoscale structures and device applications. Smooth surfaces with a small root-mean-square (rms) roughness (comparable to or even smaller than inter-atomic distances) can be obtained by chemo-mechanical polishing (CMP) [1]. However, a surface prepared by CMP is disordered on a microscopic scale due to the mechanical impact during polishing. Annealing at elevated temperatures can effectively reduce this disorder and yield regular step-terraced surfaces with atomically flat terraces separated by steps of monatomic height. Almost perfect step-terraced silicon surfaces are obtained by annealing in vacuum [2,3,4]. For III-V semiconductors, the application of vacuum annealing is hindered by the evaporation of a more volatile V component. To overcome this difficulty and obtain step-terraced GaAs surfaces, annealing in MBE or MOCVD setups under As-contained vapors was used in papers [5,6]. A more efficient and cost-effective technique for GaAs smoothing was proposed in [7]. This technique consists of annealing GaAs substrates in a quartz tube of a liquid phase epitaxy (LPE) setup. The conditions close to equilibrium between the surface and vapors of As and Ga were provided by the presence of a saturated Ga-As melt. This technique proved to be effective for the GaAs step-terraced surface formation on epi-ready substrates with a relatively small root-mean-square roughness ($R_q \leq 0.15$ nm). For such substrates, the step-terraced morphology was formed during about one-hour anneals in the temperature range of 600°C-650°C. The full length of monatomic steps was used for the quantitative characterization of the step-terraced morphology formation. The experimental kinetics of the step length was compared to Monte Carlo simulations within the Kossel crystal model

[8,9]. This comparison yielded the microscopic parameters of the model, namely, the activation energy for surface diffusion, the lateral bond energy, and the adatom desorption energy.

In paper [7] the opportunity to obtain step-terraced GaAs surfaces was proved only for substrates with a sufficiently small initial roughness ($R_q \leq 0.15$ nm). However, there is a need to smooth the surfaces of substrates and epitaxial films with a larger initial roughness. Obviously, smoothing rough surfaces requires higher temperatures and longer annealing times. However, the annealing temperature and duration are restricted by the surface roughening which occurs, supposedly, due to deviations from equilibrium conditions [10,11]. Here we propose to use the two-stage annealing to overcome this difficulty. The first stage should be performed at a high temperature in order to provide an effective mass transfer. The temperature and duration of this stage should be sufficiently high for flattening "coarse" relief features but, on the other hand, not too high in order to prevent the development of thermal roughening. The second low-temperature stage is aimed at "fine" smoothing of the small-scale roughness, including the roughness that is formed at the first stage. The goal of this study is to find the optimal parameters of two-stage annealing both experimentally and using Monte Carlo simulations.

2. Experiment

The anneals were performed in a quartz tube of an LPE setup under a flow of pure molecular hydrogen. The GaAs sample was put in a quasi-closed graphite cassette with a saturated Ga-As melt to provide As and Ga vapors. Before annealing, the surface oxides were removed in HCl-isopropanol solution [12]. The morphology of the initial and annealed GaAs surfaces was studied *ex-situ* by atomic force microscopy (AFM). The details of the anneals and AFM measurements were described earlier [7]. The Monte Carlo simulations of the surface smoothing and step-terrace morphology formation were performed on the Kossel crystal in the standard solid-on-solid (SOS) model [8]. In this model, the complicated processes of III and V components mass transfer and the formation of GaAs unit cells are replaced by a random walk and attachment/detachment of one kind of "atoms" without taking into account the actual atomic structure of the crystal unit cell and surface reconstructions.

3. Results and discussion

As the first step, we tried to smooth a relatively "rough" epi-ready GaAs(001) substrate with $R_q \approx 0.2$ nm in the conditions similar to those used in [7]. Figure 1 shows the AFM images obtained after anneals at T = 625 °C for various durations. Unlike the previous annealing experiments on smoother substrates [7], here, after a one-hour anneal, the surface is far from a regular step-terraced one. It is seen in the 5×5 μm^2 image in figure 1(a) that the surface relief is nonuniform and contains scratch-like trenches of up to ~ 50 – 100 nm in width and ~ 1 – 2 nm (~ 3 – 6 monolayers, ML) in depth. In the 1×1 μm^2 image (see figure 1(b)) one can see the nucleation of jagged atomic steps with a large concentration of islands and dips on terraces. With increasing anneal duration, the step-terraced morphology matures and is clearly seen after an 8-hour anneal, although the steps remain jagged, and some traces of scratches are still visible.

An empirical search for optimal conditions for two-stage anneals is labor-intensive and time-consuming because of a large number of unknown parameters (two annealing temperatures and two durations). Therefore, we performed Monte Carlo simulations to determine the optimal parameters of the two-stage anneals. The dimensionless step length $l_S = L_S/L_0$ was used for quantitative characterization of the step-terraced morphology formation [7,8]. Here L_S is the full length of monatomic steps on a certain area, and L_0 is the length of straight monatomic steps on the ideal vicinal surface corresponding to this area. The step length proved to be a more adequate parameter than the rms roughness for the description of the surface morphology evolution of initially smooth substrates. The simulated kinetics of the step length for various annealing temperatures is shown in figure 2(a). It is seen that, after the initial steep decrease, the step length kinetics slows down and saturates at the equilibrium value l_S. The values of the equilibrium step length were determined from the inflection points of the kinetics (marked by arrows in figure 2(a)) and are shown by crosses versus annealing duration in figure 2(b). In the temperature range of 525 °C – 725 °C, the equilibrium step length increases by 20% with increasing temperature due to thermal fluctuations. The temperature increase of 100 °C speeds up the smoothing kinetics by one or two orders of magnitude. It is also seen that, according to the simulations, the use of two-stage anneals can substantially reduce the total duration required to reach a certain step length. In particular, $l_S = 1.2$ can be obtained by a 4000-min one-stage anneal at 525 °C, as compared to the 20-min two-stage anneal at 725 °C followed by the second stage at 525 °C.

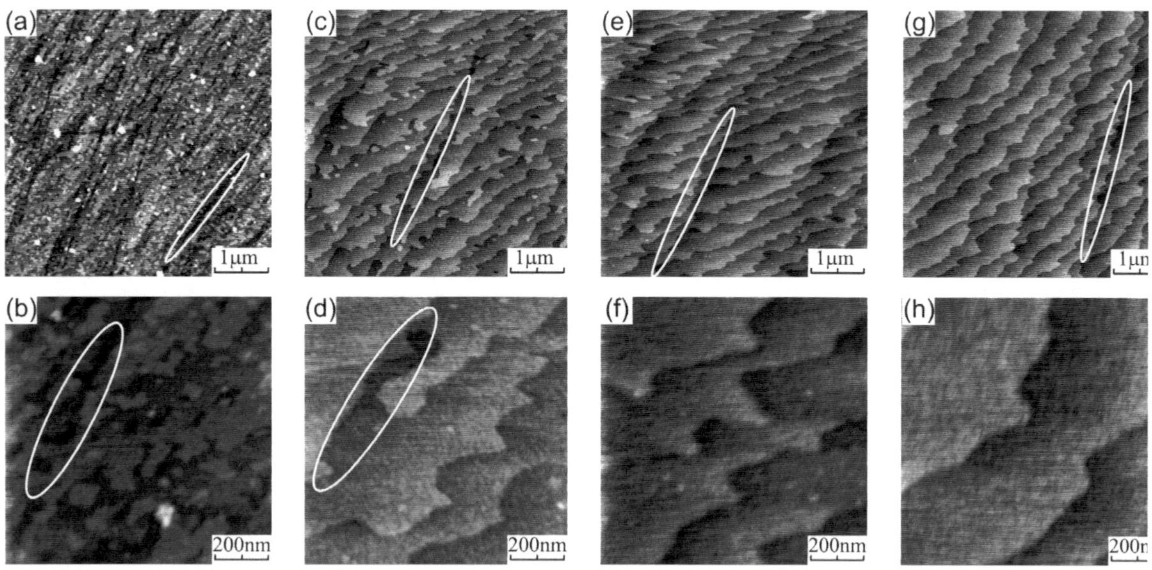

Figure 1. Smoothing kinetics of the "rough" substrate. The AFM images of 5×5 μm² and 1×1 μm² (top and bottom rows, respectively) obtained after the anneals at 625 °C, and successively increasing durations: 1 hour (a, b); 2 hours (c, d); 4 hours (e, f); 8 hours (g, h). White oval lines denote the "scratches" on the surface.

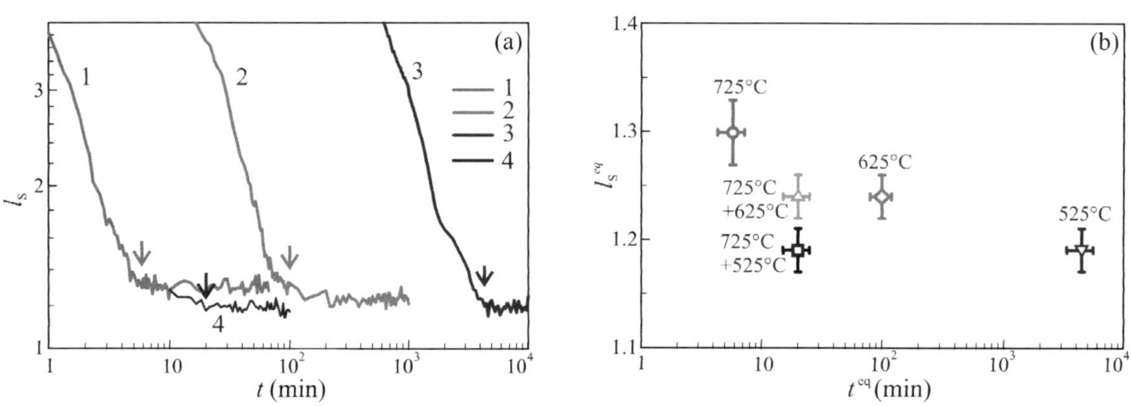

Figure 2. Monte Carlo simulation of surface smoothing by one-stage and two-stage anneals.
(a) Step length kinetics under one-stage anneals at 725 °C (red curve #1), 625 °C (green, #2) and 525 °C (dark blue, #3). The second stage of the two-stage anneal at 725 °C and 525 °C is shown by the black curve #4. The inflection points of the kinetics t^{eq}, which corresponds reaching the equilibrium step length l_S^{eq}, are indicated by the arrows. (b) The equilibrium step length l_S^{eq} as a function of the moments of time t^{eq} for the one-stage and two-stage anneals at various temperatures. Two temperatures are given near the points corresponding to the two-stage anneals.

Thus, according to the simulations, the first stage of thermal smoothing should be carried out at the highest possible temperature. The experimental realization of this idea is complicated by the changeover from surface smoothing to surface roughening, which occurs at high temperatures [7,10,11]. This roughening consists in the destruction of the step-terraced morphology and the formation of islands (in the case of deviation towards sublimations) or pits (deviation towards growth). These islands (or pits) are, presumably, the surface spots at which the sublimation or growth, respectively, are suppressed, so their height (or depth) increases with respect to the overall surface level [10,11]. One should not allow the development of roughness, which can be smoothed only by CMP. Therefore, the upper limits for the temperature and duration of the first high-temperature stage are restricted by the roughening process. To find these limits, we performed one-hour anneals of a step-terraced GaAs(001) surface at successively increasing temperatures. The results are shown in figure 3. It is seen from figure 3(a, d) that the anneal at 700 °C led to the twists in the step shape, which occasionally caused the local step to merge into double steps. The temperature increase up to 750 °C led to the formation of elongated "finger-like" steps and the formation of islands with a height of 1 – 3 ML (figure 3(b, e)). At 775 °C, further step bunching and the formation of higher islands were observed (figure 3(c, f)). The formation of multilayer islands is supposedly due to the step flowing around some spots at which the sublimation is inhibited [10,11].

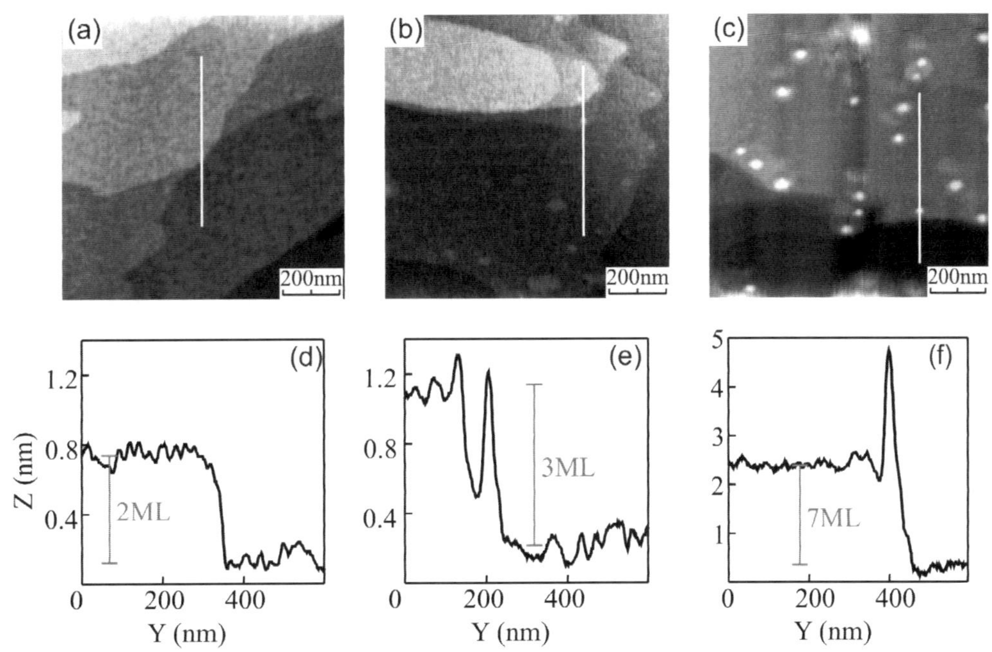

Figure 3. AFM images taken after one-hour anneals at (a) 700 °C (a); (b) 750 °C; (c) 775 °C. The respective *z-y* cross-sections along the white lines in panels (a-c) are shown in figures (d-f).

The analysis of the simulation (figure 2) and experimental (figure 3) results allowed us to choose the temperature $T = 725$ °C for the first stage of the two-stage anneal. On the one hand, this temperature is sufficiently high for the effective smoothing of the roughness with $R_q \sim 0.2$ nm and, on the other hand, not high enough to cause thermal roughening. To determine an optimal duration of the first stage, we performed anneals at 725°C for various durations (not shown). It was found that small durations of about 1 min, followed by a quick cooling down to 500 °C within 1–3 minutes, were insufficient to form a well-defined step-terraced morphology (a large concentration of monatomic islands that did not incorporate into vicinal steps was observed on terraces). However, at durations of

~ 30 min, pronounced step bunching was observed. Therefore, durations of ≤ 15 min should be chosen for the first stage of the two-stage smoothing experiments.

The results of the two-stage and one-stage smoothing experiments are presented in figure 4. It is seen from figure 4 (c, d) that the one-stage anneal at 625 °C did not form a well-defined step-terraced morphology. In distinction, the two-stage anneal (15 min at 725 °C plus one hour at 625 °C) yielded a much more perfect step-terraced surface (figure 4(e, f)). Interestingly, the two-stage anneal with reduced durations (1 min at 725 °C followed by a slow cooling down to 500 °C within ~ 15 min) yielded essentially the same well-defined step-terraced surface (figures 4(g, h)). The only difference is the presence of a small concentration of monatomic islands on the terraces after the latter two-stage anneal (figure 4(g)).

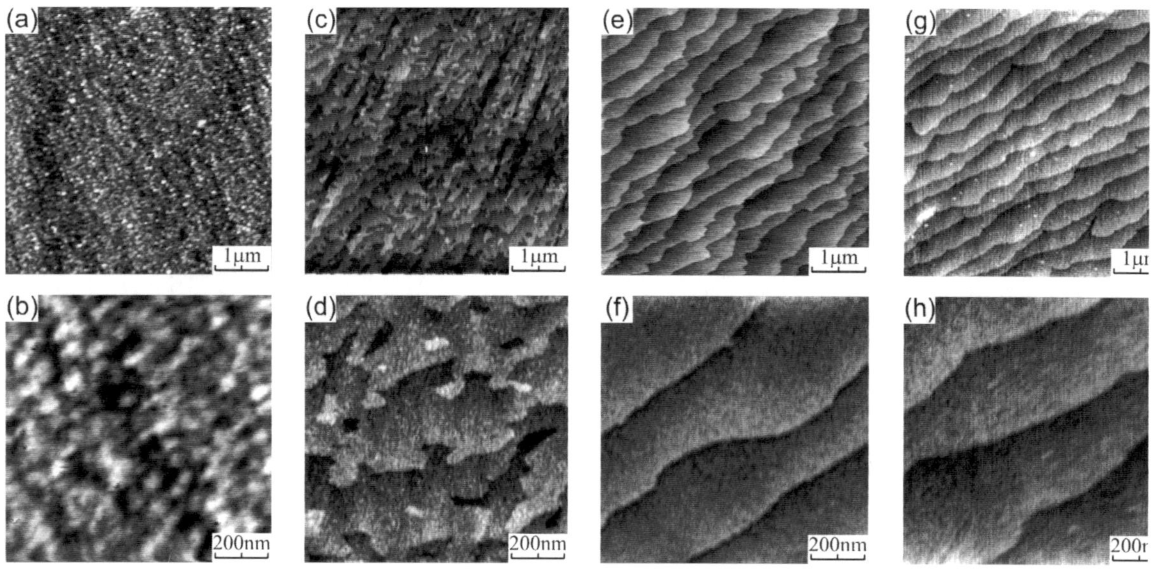

Figure 4. The AFM images of 5×5 μm^2 (top row) and 1×1 μm^2 (bottom row) measured on the initial GaAs(001) substrate surface (images (a, b), after one-hour anneal at 625 °C (c, d), after two-stage anneal (15 min at 725 °C plus one-hour at 625 °C) and after another two-stage anneal (one min at 725 °C with a subsequent slow cooling down to 500 °C within ~ 15 min (g, h).

4. Summary

Thus, it was proved both by Monte Carlo simulations and experimentally that the efficiency of thermal smoothing GaAs substrates can be improved using the two-stage anneals. This means that the substrates with a larger initial surface roughness can be brought to the step-terraced morphology and the smoothing durations can be considerably reduced. The surface roughening, which occurs at high temperatures due to deviations from equilibrium conditions, should be taken into account and it restricts the temperature and duration of the first high-temperature stage of thermal smoothing.

Acknowledgments

The thermal smoothing experiments were supported by RFBR (grant No. 16-32-00220). The AFM surface relief measurements and Monte Carlo simulations were supported by RSF (grant No. 14-22-00143).

References

[1] Allwood D A, Cox S, Mason N J, Palmer R, Young R and Walker P J 2002 *Thin Solid Films* **412** 76
[2] Jeong H-C and Williams E D 1999 *Surf. Sci. Rep.* **34** 171

[3] Latyshev A V, Aseev A L, Krasilnikov A B and Stenin S I 1989 *Surf. Sci.* **213** 157
[4] Rodyakina E E, Kosolobov S S, Sheglov D V, Nasimov D A, Song S A and Latyshev A V 2004 *Phys. Low-Dim. Struct.* **1-2** 9
[5] Ding Z, Bullock D W, Thibado P M, LaBella V P and Mullen K 2003 *Phys. Rev. Lett.* **90** 216109
[6] Epler J E, Jung T A and Schweizer H P 1993 *Appl. Phys. Lett.* **62** 143
[7] Alperovich V L, Akhundov I O, Rudaya N S, Sheglov D V, Rodyakina E E, Latyshev A V and Terekhov A S 2009 *Appl. Phys. Lett.* **94** 101908
[8] Kazantsev D M, Akhundov I O, Karpov A N, Shwartz N L, Alperovich V L, Terekhov A S and Latyshev A V 2015 *Appl. Surf. Sci.* **333** 141
[9] Kazantsev D M, Akhundov I O, Shwartz N L, Alperovich V L and Latyshev A V 2015 *Appl. Surf. Sci.* **359** 372
[10] Akhundov I O, Kazantsev D M, Kozhuhov A S and Alperovich V L 2016 *J. Phys. Conf. Ser.* **741** 012042
[11] Kazantsev D M, Akhundov I O, Kozhukhov A S and Alperovich V L 2017 *J. Phys. Conf. Ser.* **816** 012008
[12] Alperovich V L, Tereshchenko O E, Rudaya N S, Sheglov D V, Latyshev A V and Terekhov A S, 2004 *Appl. Surf. Sci.* **235** 249

IOP Conf. Series: Journal of Physics: Conf. Series **993** (2018) 012011 doi:10.1088/1742-6596/993/1/012011

Influence of GaAs substrate properties on the congruent evaporation temperature

A A Spirina[1,2], **A G Nastovjak**[1] **and N L Shwartz**[1,2]

[1]Rzhanov Institute of Semiconductor Physics, 630090, Novosibirsk, Russia
[2]Novosibirsk State Technical University, 630073, Novosibirsk, Russia

E-mail: spirina.anna.alex@gmail.com

Abstract. High-temperature annealing of GaAs(111)A and GaAs(111)B substrates under Langmuir evaporation conditions was studied using Monte Carlo simulation. The maximal value of the congruent evaporation temperature was estimated. The congruent evaporation temperature was demonstrated to be dependent on the surface orientation and concentration of surface defects.

1. Introduction

High-temperature substrate annealing in vacuum is called free or Langmuir evaporation. Depending on the temperature range, evaporation can be congruent, with equal numbers of A^{III} and B^{V} desorbed atoms, and non-congruent, when particles of group V evaporate more intensively, and the excess atoms of group III are assembled into clusters and form droplets on the surface. The transition from one evaporation mechanism to the other occurs at the temperature referred to as congruent. The renewal of interest in Langmuir evaporation of $A^{III}B^{V}$ materials is associated with the formation of group III metal drops on the surface. Liquid metal droplets are used in the implementation of a number of technologies for nanostructure growth. It is known that metal droplets can act as a growth catalyst for semiconductor nanowires [1]. Metal droplets are the basis of droplet epitaxy. When crystallizing such droplets in an arsenic flux, nanostructures of various morphologies can be obtained [2]. Drops of liquid metal found their application in plasmonics as an active medium [3]. Langmuir evaporation of GaAs substrates has been studied for a long time [4–7]. It was experimentally shown that the congruent evaporation temperature depends on the surface orientation [5, 7]. However, the literature data on the congruent evaporation temperature are ambiguous. For example, the congruent evaporation temperature of Ga(001) was found to be 663° C in [5] and only 630° C in [7].. A detailed analysis of the processes occurring during Langmuir evaporation of $A^{III}B^{V}$ materials is possible using Monte Carlo simulation. In this paper, the Monte Carlo model for Langmuir evaporation of GaAs substrates is described. Using this model, the reasons for the ambiguous determination of the congruent temperature are analyzed.

2. Monte Carlo Model

The simulations were carried out using the SilSim3D program package [8]. A four-component system was considered in the modeling of GaAs high-temperature annealing: Ga(*s*), Ga(*l*), As, As_2 (gallium in solid and liquid states, and atomic and molecular arsenic, respectively). The possible elementary events included in the model and schematically presented in figure 1 are: diffusion of components over the surface, Ga(*s*), Ga(*l*), and As_2 desorption from the surface, As_2 dissociation and formation, phase transformations (melting, dissolution, crystallization), and As dissolution and diffusion in a liquid metal. The probability of any of these events depends on the temperature in accordance with the expression $P = A\cdot\exp(-E_{act}/k_B T)$, where A is the pre-exponential factor, E_{act} is the activation energy of the process, k_B is the Boltzmann constant, T is the temperature.

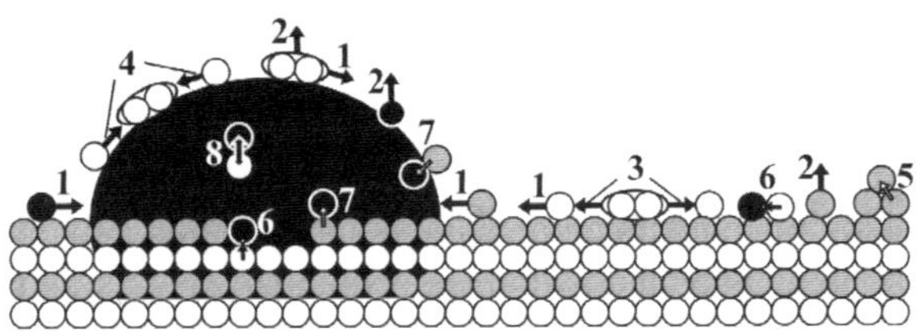

Figure 1. Schematic representation of elementary events of the model: 1 – particle diffusion along the surface; 2 – Ga and As_2 desorption; 3 – As_2 dissociation; 4 – As_2 formation; 5 – Ga melting; 6 – Ga crystallization; 7 – Ga dissolution; 8 – As dissolution in a liquid metal and diffusion through a liquid drop. As is marked in white, Ga(*s*) in gray, and Ga(*l*) in black; As_2 is shown in an ellipse.

The energy parameters of the model were chosen on the basis of available literature data and special computational experiments. When choosing the energies of covalent interaction of particles, the correlation between the binding energy of atoms and the melting point of the material was taken into account [9]. The single-bond energy values were estimated by an empirical calculation, taking into account the electronegativity of the metal and As atoms [10]. The diffusion activation energy in liquid Ga of 0.7 eV was taken according to the experiment [11]. The activation energies of Ga dissolution in a liquid drop and Ga crystallization at a drop-crystal interface in the form of GaAs were chosen by fitting the simulated and experimental solubility of GaAs in liquid Ga [12]. By fitting the model and experimental [13] temperature dependences of the equilibrium gallium pressure over the Ga(*l*)-As solution surface, it was possible to estimate the liquid gallium desorption characteristics: the energy barrier for sublimation E_{sub} and the covalent interaction energies responsible for the gallium droplet surface tension. The model takes into account the decrease in the surface tension of a metal drop when atomic arsenic is dissolved in it. The equilibrium pressure of molecular arsenic depends on its concentration on the surface and on the As_2 desorption rate. The same characteristics determine the Langmuir evaporation rate. The molecular arsenic concentration is determined by the rates of As_2 decomposition and formation, that is, by the activation energies of these reactions. The activation energies of the reactions were estimated from the agreement of the model and experimental results of such processes as epitaxial growth, evaporation under equilibrium conditions and Langmuir evaporation. The parameters characterizing the As_2 desorption rate were chosen from the agreement of these calculation results with the experiment. A more detailed justification for choosing the values of energy parameters can be found elsewhere [14].

The simulation of high-temperature annealing of GaAs substrates was performed with the following values of energy parameters: the binding energy between atoms $E_{Ga\text{-}As} = 0.9$ eV, $E_{As\text{-}As} = 0.7$ eV, $E_{Ga\text{-}Ga} = 0.6$ eV; the activation energies of gallium dissolution and crystallization $E_{dis} = 1.25$ eV, $E_{cr} = 1.2$ eV; the activation energy of molecular arsenic formation and decay $E_{form} = 3.0$ eV; $E_{dec} = 1.2$ eV.

3. Simulation results

The simulation of Langmuir evaporation of GaAs substrates was carried out on vicinal surfaces with two surface orientations: (111)A and (111)B. In the annealing temperature range below T_c, layer-by-layer or congruent evaporation was observed; that is, arsenic and gallium evaporated at the same rate. In this temperature range, the evaporation rate is limited by the gallium evaporation rate. At temperatures above T_c, molecular arsenic evaporated more intensively than gallium atoms. The calculated temperature dependences of Ga and As_2 desorption rates at high-temperature annealing of

GaAs substrates are shown in figure 2. The excess of metal atoms on the surface at $T > T_c$ leads to the formation of metal clusters, which causes the droplet nucleation. On the GaAs (111)A surface, gallium droplets were formed near the steps because of the maximum concentration of Ga atoms in this region. On the GaAs (111)B surface, the droplet nucleation starts on the point vacancy defects formed due to the high volatility of arsenic. The Ga desorption rate is the same during annealing of GaAs (111)A and GaAs (111)B substrates, in contrast to the As_2 desorption rate.

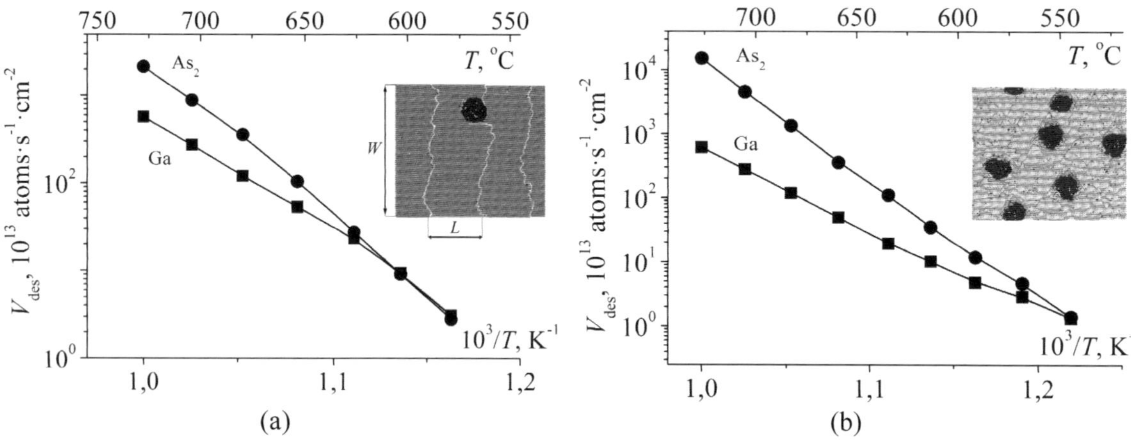

Figure 2. Temperature dependences of Ga and As_2 desorption rates in the Langmuir evaporation of GaAs(111)A (a) and GaAs(111)B (b) substrates. The inserts show the top views of model surfaces at $T = 1000$ K, $L = 20$ nm, $W = 40$ nm.

Figure 3 shows the temperature dependence of the ratio of evaporating arsenic to gallium fluxes. It can be seen that, at temperatures below T_c (850 K for GaAs (111)B and 920 K for GaAs (111)A), Ga and As atoms are removed from the surfaces in equal amounts. With an increase in the annealing temperatures ($T > T_c$), the arsenic evaporation increases more intensively on the (111)B surface than on (111)A. This is explained by the crystallographic structure of substrates (the upper layer of GaAs(111)B consists of arsenic atoms) and the volatile properties of As_2.

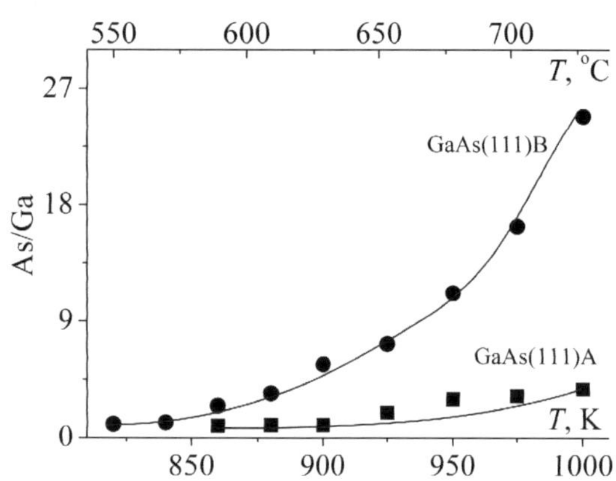

Figure 3. Temperature dependence of the evaporating As/Ga flux ratio during annealing of GaAs(111)A and GaAs(111)B substrates; $L = 20$ nm, $W = 40$ nm.

The obtained congruent evaporation temperatures of model GaAs(111)A and GaAs(111)B substrates turned out to be smaller than the experimental values ($T_c = 950$ K for (111)A, $T_c = 900$ K for (111)B [5] and $T_c = 930$ K for (111)A, $T_c = 890$ K for (111)B [7]). The difference between the model and experimental T_c may have multiple explanations. The influence of the model surface sizes on the congruent temperature was analyzed. To realize this, the simulation was carried out on vicinal substrates with different widths W (step lengths) and length L of terraces (distance between steps). With an increase in the terrace length, the congruent temperature increased (figure 4(a)). With increasing L, the step density on the substrate surface decreased. A decrease in the step density leads to a decrease of Ga adatom concentration and, as a result, the droplets are formed at a higher temperature. When the distance between the steps reached the double gallium diffusion length before the evaporation, the steps stopped interacting with each other and T_c ceased to change. Under these conditions, surface vacancy islands were formed on the terraces during annealing. A decrease in the length of steps W leads to an increase in the congruent temperature (figure 4(b, c)). Such a dependence of T_c on the step length was observed at $W < 40$ nm. To eliminate the influence of the geometric factor on the simulation results, the calculations were carried out at $W > 40$ nm. As can be seen in figure 4(a), a noticeable decrease in the congruent evaporation temperature is observed when the surface deviates from the (111) direction at an appreciable angle (more than 0.1°), and the annealing of surfaces close to the singular ones points to an overestimated T_c value ($T_c = 1000$ K for GaAs(111)A and $T_c = 900$ K for GaAs(111)B), in comparison with the experiment.

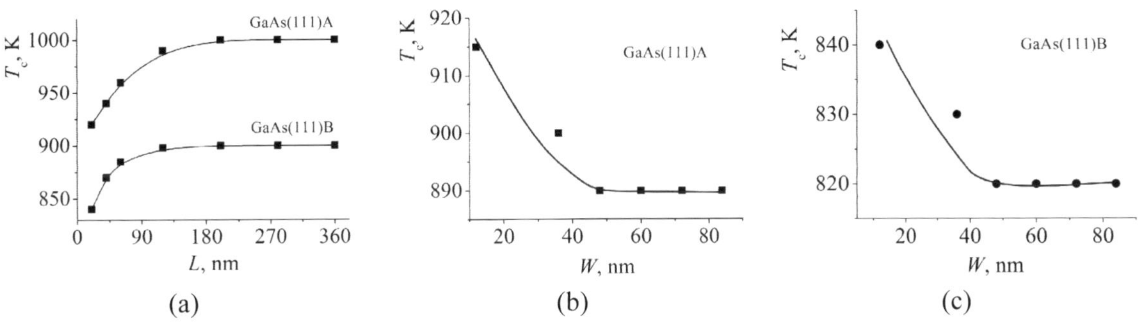

Figure 4. The congruent evaporation temperature *versus* terrace length (a) and the terrace width (b, c) of vicinal GaAs(111)A and GaAs(111)B surfaces.

The discrepancy between the experimental and calculated T_c may be due to the presence of defects on the real surface. Surface defects should lead to an increase in the concentration of gallium adatoms and, consequently, to an increase in the probability of drop formation. Therefore, when considering the ideal model surface, the congruent evaporation temperature is likely to be slightly overestimated. Figure 5(a) shows the congruent temperature as a function of the concentration of point surface defects. It is seen that T_c initially decreases with the increasing defect concentration, but when n_{def}* reaches a certain concentration, T_c ceases to depend on n_{def}. Point defects provoke the formation and the subsequent enlargement of vacancy islands that increase the concentration of gallium adatoms on the surface. However, for high defect concentration, when the average distance between them becomes less than the gallium and arsenic diffusion lengths, some of the islands are built up by embedding of Ga and As atoms in the vacancy island steps, and the original defect disappears from the surface. The model surface image with the initial density of point defects $n_{def} > n_{def}$* after annealing is shown in figure 5(b). One can see that out of eight original point defects, there are only three that resulted in the formation of triangular vacancy islands, and the number of drops on the surface is even smaller (two).

It should be noted that the desorption rate can vary with time. Figure 6(a) demonstrates the dependence of the total number of atoms (gallium and arsenic) evaporating from the vicinal GaAs(111)B surface on the annealing time. It can be seen that this dependence is nonlinear, indicating a change in the desorption rate with time. This fact was noted in the experimental work on the InAs Langmuir evaporation [15], where the material evaporation rate ceased to depend on time only after the entire surface had been covered with a liquid metal film. Figures 6(b-d) show the surface images at different moments of the annealing process. These moments are indicated in figure 6(a) by the arrows.

It can be seen that the desorption rate becomes practically constant when almost all the surface is covered with liquid Ga.

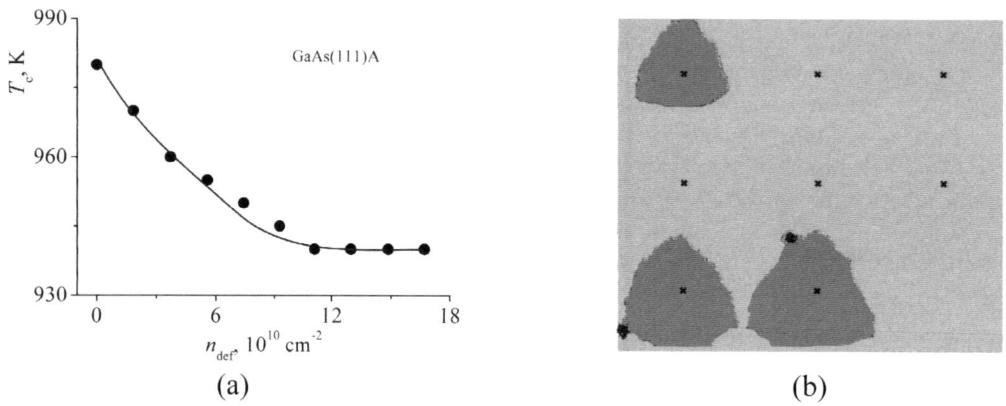

Figure 5. (a) The congruent evaporation temperature of the singular GaAs (111)A substrate as a function of the concentration of point surface defects; (b) top view of the GaAs(111)A surface after annealing at $T = 970$ K; the crosses mark the positions of original defects.

The analysis of simulation results showed that the dispersion in congruent temperature values in different experiments can be related not only to inaccuracy in measuring the substrate temperature, but also to the substrate itself: the degree of deviation from the singular surface and the concentration of surface defects. In addition, measuring the desorption rate at the initial annealing stage, before the surface is completely covered by a liquid metal, can also lead to inaccuracy in determining T_c.

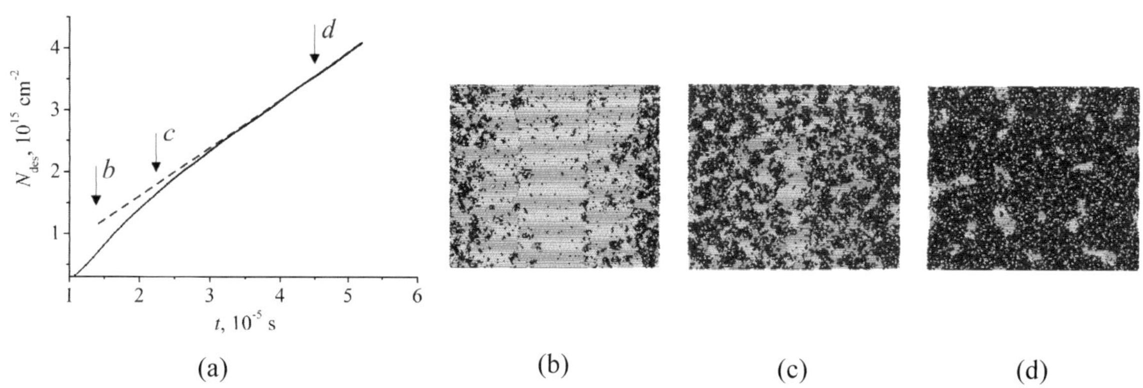

Figure 6. (a) Dependence of the total number of evaporating gallium and arsenic atoms from the GaAs(111)B surface on the annealing time at $T = 1200$ K; (b-d) the model surface top views after different annealing times.

4. Conclusions

The processes of high-temperature annealing of GaAs(111)A and GaAs(111)B substrates under Langmuir evaporation conditions were studied using Monte Carlo simulations. The congruent and noncongruent evaporation temperature ranges were determined. The maximum congruent evaporation temperature was estimated. The formation of metal droplets on the (111)A surface began near the steps of the vicinal substrates and, on the (111)B surface, the droplets were formed and randomly distributed over the entire substrate surface. The desorption rate dependence on the annealing time and the kinetics of the surface morphological evolution during noncongruent evaporation were analyzed. It was found that the congruent temperature value depends on the substrate orientation, step density on the vicinal surface and point defect concentration on the singular surface.

Acknowledgments
This work was supported by the RFBR (grant 16-31-00120) and by the Russian Academy of Sciences Programs.

References

[1] Spirkoska D, Abstreiter G and Fontcuberta i Morral A 2009 *Semicond. Sci. Technol.* **24** 113119
[2] Thelander C, Martensson T, Ohlsson B, Larsson M, Wallenberg L and Samuelson L 2003 *Appl. Phys. Lett.* **83** 10 2052
[3] Blaber M G, Engel C J, Vivekchand S R C, Lubin S M, Odom T W and Schatz G C 2012 *Nano Lett* **12** 5275
[4] Lou C Y and Somorjai G A 1971 *J. Chem. Phys.* **55** 4554
[5] Goldshtein B, Stozak D J and Ban V S 1976 *Surf. Science.* **57** 733
[6] Jian-yun Shen J and Chatillon C 1990 *J. of Cryst. Growth.* **106** 543
[7] Kanjanachuchai S and Euaruksakul C 2014 *ACS Appl. Mater. Interf.* **5** 7709
[8] Zverev A V, Zinchenko C Yu, Shwartz N L and Yanovitskaja Z Sh 2009 *Nanotech., in Russian* **4** 215
[9] Kulish U M 1987 *Izv. Akad. Nauk SSSR, Neorg. Mater., in Russian* **23(5)** 850
[10] Pauling L and Pauling P 1975 *Chemistry* (San Francisco: W H Freeman) p 792
[11] Gorokhov V A, Dedegkaev T T, Ilyin Y L, Moshnikov V A, Petrov A S, Sosov Y M and Yaskov D A 1984 *Cryst. Res. and Tech.* **19** 1465
[12] Hall R N 1963 *J. of the Electrochem. Society.* **5** 385
[13] Pupp C, Murray J and Pottie R 1974 *J. of Chem. Thermodynam.* **6** 123
[14] Vasilenko M A, Neizvestny I G and Shwartz N L 2015 *Comput. Mat. Sci.* **102** 286
[15] Khukhryanskii Yu P, Verem'yanina L N, Kombarova I V, Nikishina I V and Sisoev O I 1997 *Russian J. of Phys. Chem.* **71** 771

Experimental study of surface plasmon-phonon polaritons in GaAs-based microstructures

A I Galimov[1], V A Shalygin[1], M D Moldavskaya[1], V Yu Panevin[1], G A Melentyev[1], A A Artemyev[1], D A Firsov[1], L E Vorobjev[1], G V Klimko[2], A A Usikova[2], T A Komissarova[2], I V Sedova[2] and S V Ivanov[2]

[1] Department of Physics of Semiconductors & Nanoelectronics, Peter the Great St. Petersburg Polytechnic University, 29 Polytechnicheskaya str., St. Petersburg 195251, Russia
[2] Ioffe Institute, 26 Polytechnicheskaya str., St. Petersburg 194021, Russia

E-mail: shalygin@rphf.spbstu.ru

Abstract. Optical properties of a heavily-doped GaAs epitaxial layer with a regular grating at its surface have been experimentally investigated in the terahertz spectral range. Reflectivity spectra for the layer with a profiled surface drastically differ from those for the as-grown epilayer with a planar surface. For *s*-polarized radiation, this difference is totally caused by the electromagnetic wave diffraction at the grating. For *p*-polarized radiation, additional resonant dips arise due to excitation of surface plasmon-phonon polaritons. Terahertz radiation emission under significant electron heating in an applied pulsed electric field has also been studied. Polarization measurements revealed pronounced peaks related to surface plasmon-phonon polariton resonances of the first and second order in the emission spectra.

1. Introduction

One of the most promising phenomena in plasmonics is the surface electromagnetic waves at a semiconductor/dielectric interface. There are three types of such waves: surface plasmon polaritons (SPP), surface phonon polaritons (SPhP) and surface plasmon-phonon polaritons (SPPhP) [1]. In the particular case of the semiconductor/vacuum interface, the surface waves can occur only in the spectral ranges where the dielectric permittivity of semiconductor ε satisfies the condition $\mathrm{Re}(\varepsilon) < -1$ [2]. A possible type of the surface wave in a polar semiconductor is determined by the relation between the plasma frequency $\omega_\mathrm{P} = \sqrt{\dfrac{4\pi e^2 N_e}{\varepsilon_\infty m_e}}$ (here N_e and m_e are the concentration and the effective mass of free electrons, respectively, e is the elementary charge, ε_∞ is the high-frequency permittivity) and the frequency of transverse optical phonons ω_TO. In semiconductors with a low doping level (when $\omega_P < \omega_\mathrm{TO}$), one can observe "pure" SPP and SPhP, the former being in the spectral range $0 < \omega < \omega_\mathrm{P}$ and the latter in the range $\omega_\mathrm{TO} < \omega < \omega_\mathrm{LO}$, where ω_LO denotes the frequency of longitudinal optical phonons. In heavily-doped semiconductors (when $\omega_\mathrm{P} > \omega_\mathrm{TO}$), surface waves represent mixed plasmon-phonon modes on the *semiconductor surface*, which are referred to here as SPPhP. These types of surface waves can be observed in a wide spectral range from

0 to ω_+, excluding a narrow interval around the point ω_-, where ω_+ and ω_- denote the frequencies of the mixed plasmon-phonon modes in a *bulk semiconductor* [3].

The direct excitation of the SPPhP by external electromagnetic radiation at the planar semiconductor/vacuum interface is impossible. The excitation becomes possible when spatial inhomogeneity is introduced at the interface. For instance, the surface-relief grating can allow coupling between the incident radiation and the SPPhP excitations. In our previous papers [4, 5], we investigated optical properties of heavily-doped *n*-GaN epitaxial layers with a surface-relief grating. In the terahertz reflectivity spectra, pronounced dips due to the SPPhP excitation have been observed. Under a pulsed lateral electric field, the selective terahertz radiation emissions at the frequencies of SPPhP resonances have been observed from the profiled *n*-GaN epitaxial layer.

In the present paper, we investigate SPPhP at the *n*-GaAs/vacuum interface. The electron mobility in GaAs is by one order of magnitude higher than in GaN, which provides an opportunity to observe higher-quality SPPhP resonances. We study optical properties of a heavily-doped *n*-GaAs epitaxial layer with a surface-relief grating and reveal specific features related to SPPhP excitations in the reflectivity and radiation emission spectra in the terahertz spectral range. For comparison, we perform similar optical experiments on the reference samples with a planar *n*-GaAs surface.

2. Experimental technique

In accordance with our preliminary simulations [6], we fabricated an *n*-GaAs-based microstructure suitable for the observation of SPPhP in the terahertz spectral range. The microstructure was made of a 7.6-μm-thick *n*-GaAs:Si epitaxial layer with an electron concentration $n \sim 2\cdot10^{18}\,\text{cm}^{-3}$, grown by molecular beam epitaxy on a semi-insulating GaAs substrate. A set of samples containing a pair of Au/Ge/Ni/Au electrical contacts was fabricated. A regular grating was patterned by conventional photolithography and wet chemical etching in the area between contacts. The grating period a was 54 μm, while the depth of the grating grooves was about 3 μm. A few reference samples with a planar surface between contacts were also prepared.

Properties of the epilayer were characterized by means of the Hall effect and conductivity measurements (at temperatures T = 77 and 300 K) and analysis of current-voltage characteristics (at T = 4.2, 77 and 300 K) and reflectivity spectra (at T = 300 K). It was found that free electrons are not frozen even at liquid helium temperature, which indicates impurity band formation. The electron concentration is practically constant and equal to $1.9\cdot10^{18}\,\text{cm}^{-3}$ across the entire temperature range examined. The electron mobility μ at room temperature is equal to 2300 $\text{cm}^2/\text{V}\cdot\text{s}$ and slightly increases (by about 9%) under microstructure cooling to 77 K.

We investigated the terahertz reflectivity spectra of the samples under equilibrium conditions at room temperature. Measurements were carried out by means of a Fourier spectrometer operating in a rapid-scan mode. Terahertz radiation emission under significant electron heating by a pulsed electric voltage applied to the contacts was also studied at cryogenic temperatures. For these studies, we used a Fourier spectrometer operating in a step-scan mode. The experimental methods are described in more detail in [4].

3. Results and discussion

The experimental reflectivity spectra under slightly oblique incidence of radiation in the frequency range of 2–19 THz (with photon energies of 8–80 meV) are presented in figure 1. The spectra for the sample with a surface-relief grating (the incidence plane was oriented perpendicular to the grating ridges) demonstrate a lower reflectivity for both *p*- and *s*-polarized radiation in comparison with the reflectivity of the reference sample (measured for unpolarized radiation under any orientation of the incidence plane). This decrease is due to radiation diffraction at the grating. The diffracted beams of non-zero orders are deflected by rather large angles and propagate beyond the angular aperture of the photodetector ($\Delta\theta = 4°$) which collects the zero-order diffracted beam only (for more details, see [5]). Nevertheless, all three spectra show two common features, namely a sharp dip at the photon energy $\hbar\omega_-$ and the reflection edge near the phonon energy $\hbar\omega_+$. Both of these features are characteristic of the *bulk* plasmon-phonon polaritons (see [3]) in the *n*-GaAs epilayer. The fingerprints of the *surface* plasmon-phonon polariton excitation that we are interested in are clearly seen in the reflectivity spectrum of the grating sample for *p*-polarized radiation (solid line in figure 1). The spectrum demonstrates a few additional dips in comparison with the spectrum for *s*-polarized radiation (dashed line in figure 1). It is known (see, e.g., [4]) that only the radiation having a component of the electric field perpendicular to the grating ridges (i.e. *p*-polarized radiation in the above-mentioned geometry of the experiment) can excite the surface waves. This causes dissipation of the radiation energy and is seen as a decrease in reflectivity.

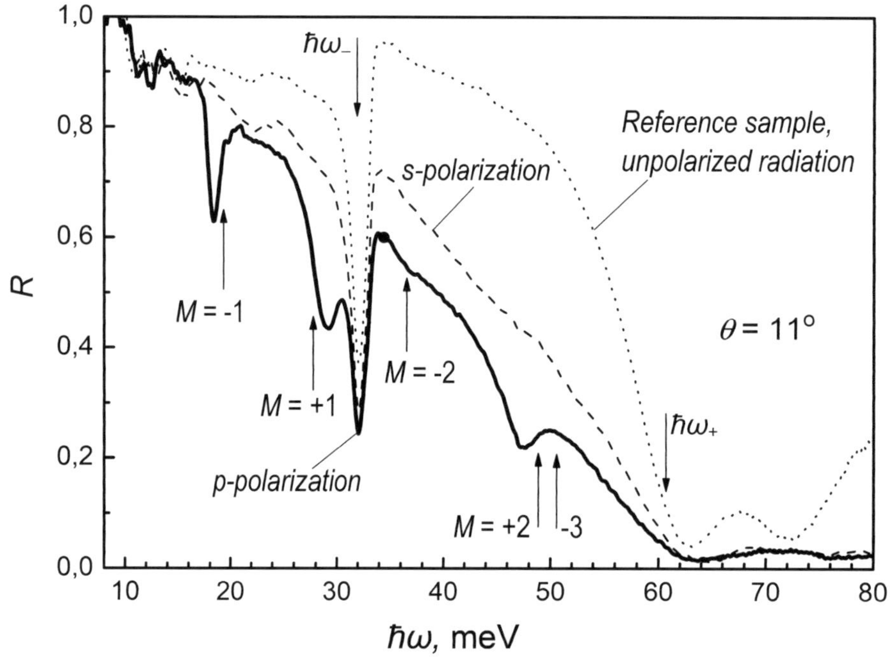

Figure 1. Reflectivity spectra for the *n*-GaAs epitaxial layer with a surface-relief grating (solid and dashed curves) and for the reference sample with a planar surface (dotted line). Conditions: temperature $T = 300$ K, angle of incidence $\theta = 11°$, angular aperture of the radiation beam $\Delta\theta = 4°$, spectral resolution 0.25 meV. The arrows show the simulated spectral positions of the bulk plasmon-phonon modes ($\hbar\omega_-$ and $\hbar\omega_+$) and SPPhP resonances of different orders ($M = \pm1, \pm2, -3$).

We used the SPPhP model considered in [6] to interpret the experimental spectra. Following this model, the dielectric permittivity can be simulated as

$$\varepsilon(\omega) = \varepsilon_\infty \left[1 + \frac{\omega_{LO}^2 - \omega_{TO}^2}{\omega_{TO}^2 - \omega^2 - i\omega\gamma} - \frac{\omega_P^2}{\omega^2 + i\omega\tau^{-1}}\right] \tag{1}$$

where γ is the damping constant for TO phonons, and τ is the electron relaxation time determined by the mobility μ: $\tau = \mu m_e / e$. For numerical calculations, we used the above-mentioned results of the epilayer characterization and $\gamma = 0.006\omega_{TO}$. The dispersion law for SPPhP at the semiconductor/vacuum interface is described as [1, 6]

$$k_{SPPhP}^2 = \left(\frac{\omega}{c}\right)^2 \frac{\varepsilon(\omega)}{\varepsilon(\omega)+1}, \tag{2}$$

where we assume that the angular frequency ω is the complex quantity, while the wave vector k_{SPPhP} is the real quantity. In this case, the real and imaginary parts of the frequency correspond to the eigenfrequency and decrement of SPPhP oscillations, respectively. SPPhP with a certain frequency ω can be converted into terahertz photons with the same frequency (and vice versa) by means of a regular surface grating, which provides an opportunity to meet the phase-matching condition for the SPPhP and photon [1, 6]:

$$k_{\mathrm{SPPhP}} = k_{\mathrm{THz}} \sin\theta + M \cdot k_{\mathrm{G}} , \tag{3}$$

where $k_{\mathrm{THz}} = \frac{\omega}{c}$ is the wave vector of a terahertz photon in vacuum, θ is the incidence angle, $k_{\mathrm{G}} = \frac{2\pi}{a}$ is the reciprocal vector of the grating, a is the grating period, and $M = \pm1, \pm2, \pm3 \ldots$ is the order of the SPPhP resonance under their decay (or excitation).

Spectral positions of possible SPPhP resonances simulated in accordance with equations (1) – (3) for the particular conditions of our reflectivity measurements on the *n*-GaAs/vacuum interface are shown in figure 1 by up arrows. One can see that in the experimental reflectivity spectrum for *p*-polarized radiation, one dip is located close to the theoretically expected position of the SPPhP resonance of the +1 order, while another dip corresponds to the –1 order, and the third dip can be considered as a common resonance for M = +2 and –3. At the same time, in the experimental spectrum, there is no pronounced dip in the vicinity of the theoretically simulated SPPhP resonance of the –2 order. It should be emphasized that very similar features were observed earlier for the reflectivity of the *n*-GaN/vacuum interface with a surface-relief grating [4, 5]. Namely, the dips corresponding to photoexcitation of SPPhP modes of the orders +1, –1, +2 and –3 were also observed, while the mode with M = –2 was not excited. Such behavior was theoretically justified in [4] using a differential method with an explicit integration scheme for the analysis of electromagnetic fields in the near-field zone of the grating.

Our findings in the study of terahertz radiation emission from the regularly profiled *n*-GaAs/vacuum interface when applying a lateral electric field seem to be more significant than the previous results obtained in a similar study on the *n*-GaN/vacuum interface [4]. Firstly, in *n*-GaAs we are able to provide substantial heating of free electrons under a rather small increase of the lattice temperature, while in *n*-GaN we heated electrons and the lattice simultaneously to the same temperature. As a result, using short electric field pulses (~ 3 μs), we provided much stronger electron heating in the *n*-GaAs epilayer compared to the *n*-GaN one. Secondly, in the present work we additionally study the polarization dependence of terahertz radiation emission that provides a deeper insight into the physics of SPPhP (see figure 2).

Terahertz radiation was detected by a silicon bolometer in the direction perpendicular to the *n*-GaAs/vacuum interface. The photoresponse at a repetition frequency of the electric field pulses of 1.6 Hz was measured by a lock-in amplifier. The *photoresponse spectra* for the samples with a surface-relief grating and reference samples are qualitatively different. The emission spectrum of the reference sample with a planar surface is totally determined by "hot" bulk plasmon-phonon-polariton modes. Analysis of the polarization dependence of the photoresponse spectra for the grating sample reveals resonant contributions to terahertz emission due to radiative decay of non-equilibrium SPPhP excitations. As a result of SPPhP decay, additional terahertz photons with the electric field **E** parallel to the reciprocal vector of the grating $\mathbf{k_G}$ (we call this radiation polarization "active") are emitted. However, no additional photons with $\mathbf{E} \perp \mathbf{k_G}$ (we call this polarization "passive") are emitted. The ratio of the photoresponse signals for active and passive polarizations demonstrates the selective amplification of the terahertz emission in the vicinity of the SPPhP resonance of the first order ($M = \pm1$) by about 3 times. For the SPPhP resonance of the second order ($M = \pm2$), the amplification is about 2 times. Both polarizations have the same radiation intensity in the vicinity of the bulk plasmon-phonon mode $\hbar\omega_-$. The presented data on terahertz emission are preliminary. To determine the actual *spectral density* of the terahertz radiation emission from the investigated samples,

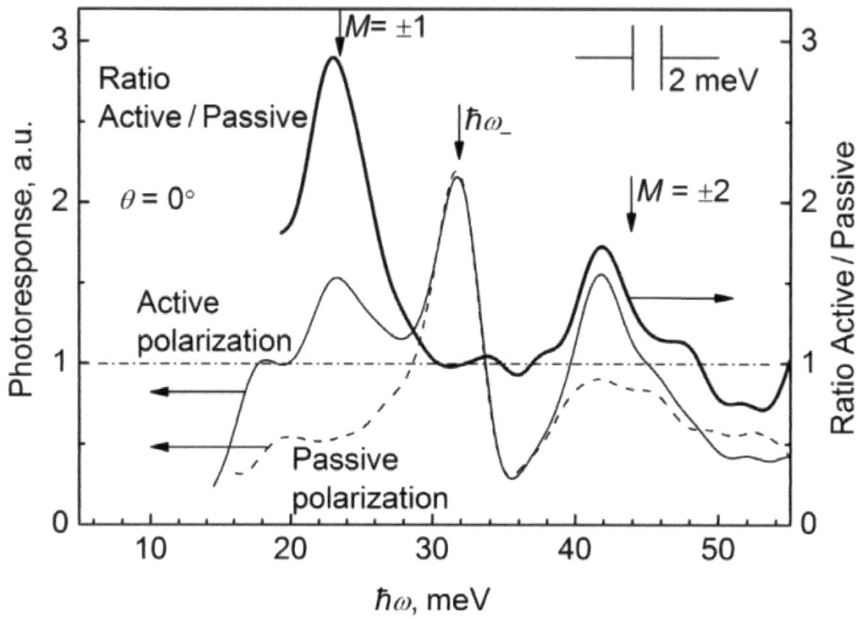

Figure 2. Polarization dependence of terahertz radiation emission from the n-GaAs epilayer with a surface-relief grating under electron heating by electric field (experiment). Conditions: temperature $T = 5$ K, electric field $E = 618$ V/cm, $\theta = 0°$, $\Delta\theta = 16°$. The arrows denote the simulated spectral positions of the SPPhP resonances ($M = \pm 1$ and ± 2) and low-frequency plasmon-phonon mode ($\hbar\omega_-$).

the spectral measurements of the transmission of the optical path of the installation (including a beamsplitter) should be carried out. The detectivity spectrum of the photodetector should be measured as well. This will be done at the next stage of the research.

4. Conclusions

We have investigated the interaction of surface plasmon-phonon polaritons at a regularly profiled n-GaAs/vacuum interface with terahertz photons. The spectra of the terahertz reflectivity and emissivity in a wide frequency range have been measured and their polarization dependencies have been examined. Resonant dips related to photoexcitation of SPPhP modes of the orders +1, –1, +2 and –3 have been revealed in the reflectivity spectra for p-polarized radiation. Under electron heating and non-equilibrium SPPhP excitation by a pulsed lateral electric field in the profiled n-GaAs epitaxial layer, terahertz emission has been detected in the spectral range of 3.5–13 THz (15–55 meV). In addition to unpolarized radiation emission related to "hot" *bulk* plasmon-phonon polariton modes, pronounced peaks related to *surface* plasmon-phonon polariton resonances of the first and second orders have been found in the emitted radiation with the electric field polarized perpendicularly to the ridges of the surface-relief grating. It has been shown that the selective amplification of terahertz emission in the vicinity of the SPPhP resonance of the first order is rather high (about 3 times).

The results of our studies can be applied for the development of portable sources of terahertz radiation operating under electrical pumping.

Acknowledgements
The authors are grateful to I.S. Makhov for the help in the experiments. The work at St. Petersburg Polytechnic University was supported in part by the Russian Foundation for Basic Research (grant

IOP Conf. Series: Journal of Physics: Conf. Series **993** (2018) 012012 doi:10.1088/1742-6596/993/1/012012

18-02-00848) and the Ministry of Education and Science of the Russian Federation (state assignment 3.933.2017/4.6).

References

[1] Karakachian H and Kazan M 2017 *J. Appl. Phys.* **121** 093103

[2] Raether H 1988 Surface Plasmons on Smooth and Rough Surfaces and on Gratings *Springer Tracts in Modern Physics* vol. 111 (Berlin: Springer-Verlag)

[3] Orlov E Yu, Melentev G A, Shalygin V A and Suihkonen S 2017 *J. Phys.: Conf. Series* **816** 012004

[4] Melentev G A *et al* 2016 *J. Appl. Phys.* **119** 093104

[5] Shalygin V A, Melentev G A, Vorobjev L E, Panevin V Yu, Firsov D A, Korotyeyev V V, Lyaschuk Yu M, Kochelap V A and Suihkonen S 2017 *J. Phys.: Conf. Series* **864** 012073

[6] Melentev G A, Shalygin V A, Firsov D A, Vinnichenko M Ya and Vorobjev L E 2017 *J. Phys.: Conf. Series* **917** 062038

IOP Conf. Series: Journal of Physics: Conf. Series **993** (2018) 012013 doi:10.1088/1742-6596/993/1/012013

Intraband light absorption by holes in InGaAsP/InP quantum wells

N V Pavlov and G G Zegrya

Ioffe Institute, 194021, Russian Federation, St. Petersburg, Politekhnicheskaya st. 26

E-mail: pavlovnv@mail.ru

Abstract. A microscopic analysis of the mechanism of intraband radiation absorption by holes with their transition to a spin-split band for quantum wells based on InGaAsP/InP solid solutions is performed within the framework of the four-band Kane model. The calculation is made for two polarizations of the incident radiation: along the crystal growth axis and in the plane of the quantum well. It is shown that this process can be the main mechanism of internal radiation losses for quantum well lasers. It is also shown that the dependence of the absorption coefficient on the width of the quantum well has a maximum at a well width from 40 to 60 A.

1. Introduction

At present, semiconductor lasers with quantum wells (QWs) based on InGaAsP/InP compounds with a generation wavelength of 1.3 and 1.55 μm have been widely used in optical communication lines [1-3]. However, these structures are characterized by strong intraband absorption of radiation, which leads to a disruption of generation [4,5]. The intraband absorption increases with decreasing thickness of the active region, and the absorption coefficient can reach several tens of reciprocal centimeters.

The mechanisms of intraband absorption of radiation in QW lasers have been studied both theoretically and experimentally for many years [6]. Experimental results [7,8] show that the coefficient of intraband absorption of laser radiation is substantially higher than that predicted by the theory [9]. One of the candidates for explaining these results is the process of intraband radiation absorption by holes with the transition to the spin-split (so) zone. In this paper, we use a modification of the four-band Kane model [10] proposed by Polkovnikov and Zegrya [11,12], which is based on the use of the 8×8 ***kP*** Hamiltonian. It allows us to obtain explicit analytical expressions for energy spectra and wave functions of charge carriers, as well as matrix elements of transitions. The authors proposed a modification of this method, allowing to take into account the elastic stresses arising in mismatched heterostructures.

The aim of this work is to calculate the coefficient of intraband radiation absorption by holes with their transition to the so-zone for QWs based on A_3B_5 semiconductors, and also to study the dependence of the results obtained on the electromagnetic wave polarization, temperature, hole concentration, and QW width. The calculations are made for the InGaAsP/InP heterostructure, which is widely used in fabricating semiconductor lasers with a radiation wavelength of 1.55 μm. The structure parameters are taken from [13].

2. Basic Equations

We used the 8×8 ***kP*** Hamiltonian [11] that takes into account the interaction with higher bands and the terms arising from elastic stresses up to quadratic terms by the wave vector but neglects the relativistic linear terms and the term with the heavy-electron mass.

For our calculations, we choose the following representation of basis wave functions:

$$|s\uparrow\rangle, |s\downarrow\rangle, |x\uparrow\rangle, |x\downarrow\rangle, |y\uparrow\rangle, |y\downarrow\rangle, |z\uparrow\rangle, |z\downarrow\rangle,$$

where the spinors s and x, y, z are s-type and p-type Bloch functions with an angular momentum of 0 and 1, respectively. The s-functions describe the conduction band states and the p-functions describe the valence band states at the Γ-point. The arrows indicate the spin direction. The carrier wave function ψ can be represented in the form:

$$\psi = \mathbf{\Psi}|\boldsymbol{u}\rangle = \Psi_s|s\rangle + \mathbf{\Psi}_p|\boldsymbol{p}\rangle,$$

where $\mathbf{\Psi}$, Ψ_s, $\mathbf{\Psi}_p$ are spinors.

The Kane equations in the spherical approximation near the Γ-point are given by the following expressions:

$$(E_C - E)\Psi_s - i\hbar\gamma\nabla\mathbf{\Psi}_p = 0,$$

$$(E_V - \delta - E)\mathbf{\Psi}_p - i\hbar\gamma\nabla\Psi_s + \frac{\hbar^2}{2m_0}(\gamma_1 + 4\gamma_2)\nabla(\nabla\mathbf{\Psi}_p) -$$

$$\frac{\hbar^2}{2m_0}(\gamma_1 - 2\gamma_2)[\nabla\times[\nabla\times\mathbf{\Psi}_p]] + i\delta[\boldsymbol{\sigma}\times\mathbf{\Psi}_p] = 0. \quad (1)$$

Here E_C and E_V are the energies of the conduction and valence band edges, γ is the Kane matrix element with a dimension of velocity, $\delta = \Delta_{so}/3$ is the spin-orbit splitting constant, m_0 is the free electron mass, γ_1 and $\gamma_2 = \gamma_3$ are the generalized Luttinger parameters, $\boldsymbol{\sigma}$ is the Pauli matrices, $\boldsymbol{k}$ is the wave vector, and E is the energy.

The absorption coefficient per one QW in the framework of the Kane model can be calculated from the formula:

$$\alpha_p^d = \frac{2\gamma^2 e^2}{\hbar^2 c}\frac{1}{n\omega a}\sum_{k_h}\sum_{k(0)} m_r^d \sigma_p^d f(E_h) \quad (2)$$

for the transitions into the discrete spectrum of so-holes and

$$\alpha_p^c = \frac{\gamma^2 e^2}{2\pi\hbar^2 c}\frac{m_h}{n\omega a}\sum_{k_h}\int_{E_h^{\min}}^{E_h^{\max}} \frac{\sigma_p^c f(E_h)}{k\left|\frac{\partial E(k^2)}{\partial k^2}\right|} dE_h \quad (3)$$

for the transitions into the continuous spectrum of so-holes.

Here, the index p denotes the polarization of light and takes on a value *per* for a wave polarized along the crystal growth axis (i. e. when the amplitude of the vector potential A_0 is parallel to the x axis) and the value *par* for a wave polarized in the QW plane, n is the refractive index, ω is the optical transition frequency, a is the QW width, k and k_h are the wave vector x components of so- and heavy holes, respectively, E and E_h are the energies of so- and heavy holes, respectively, $f(E)$ is the Fermi-Dirac distribution function, $m_h = \frac{m_0}{\gamma_1 - 2\gamma_2}$ is the heavy hole effective mass, and $m_r = \frac{m_h m_{so}}{|m_h - m_{so}|}$, m_{so} is the so-hole effective mass. The value of σ is determined by the overlap integrals of the wave functions:

$$\sigma_{per} = 2\pi\left|I_x^{j,a}\right|^2,$$

$$\sigma_{par} = \pi\left(\left|I_y^{j,s}\right|^2 + \left|I_z^{j,s}\right|^2\right), \quad (4)$$

where $I_i = \int_{-a/2}^{a/2} \Psi_{h,i}^* \Psi_s dx$.

3. Results

Figures 1(a) and 1(b) show the frequency dependences of α^{d}_{per} and α^{d}_{par} at different temperatures in the InGaAsP/InP heterostructure at a = 80 A and a hole concentration p = 10^{12} cm^{-2}. It can be seen that more sharp peaks with a larger amplitude are observed at T = 150 K than at T = 300 K. It is also clear that the transition energy corresponding to the absorption peaks increases with temperature. It is also noticeable that the radiation polarized in the plane of QW is absorbed more efficiently than the radiation of the transverse polarization.

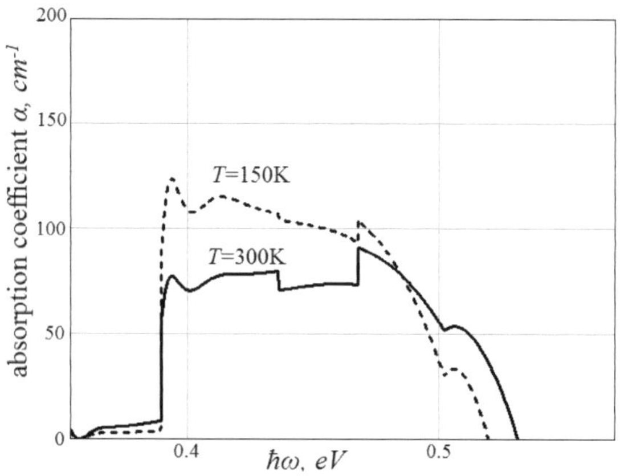

Figure 1(a). Dependence of the absorption coefficient α^{d}_{par} for the discrete levels of so-holes on the incident radiation frequency for different temperatures in the InGaAsP/InP heterostructure with a QW width a = 80A and a hole concentration p = 10^{12} cm^{-2}. The solid curve shows the calculation result for T = 300 K, and the dotted curve shows the result for T = 150 K.

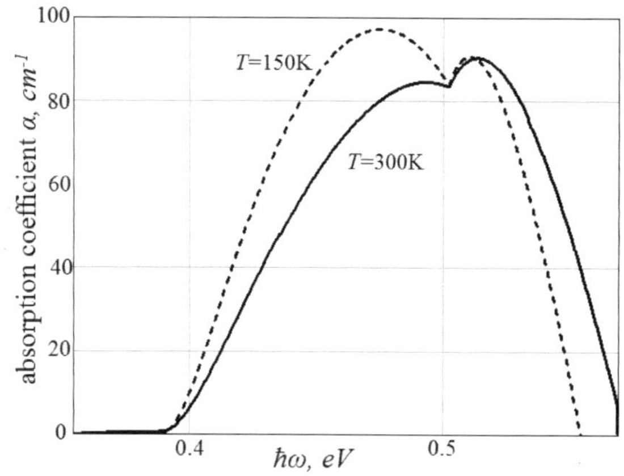

Figure 1(b). Dependence of the absorption coefficient α^{d}_{per} for the discrete levels of so-holes on the incident radiation frequency for different temperatures in the InGaAsP/InP heterostructure with a QW width a = 80A and a hole concentration p = 10^{12} cm^{-2}. The solid curve shows the calculation result for T = 300 K, and the dotted curve shows the result for T = 150 K.

Figures 2(a) and 2(b) show the frequency dependences of α^{c}_{par} and α^{c}_{per} at different temperatures in the InGaAsP/InP heterostructure at a = 80 A and a hole concentration p = 10^{12} cm^{-2}. First of all, it should be noted that the absorption maximum with the transition to the continuous spectrum is observed at a transition energy of about 0.7 eV, which is close to the value of the optical transition energy at a wavelength of 1.55 μm. Hence, the light absorption by heavy holes with the transition to the continuous spectrum of so-holes can make a large contribution to the value of internal radiation losses and explain the large value of the internal loss coefficient observed in experiment [7,8]. Another observation is that the predominance of absorption of radiation polarized in the QW plane is much stronger for the transitions to the continuous spectrum than for the transitions to the discrete spectrum.

Figure 3 shows the dependence of the absorption coefficient at the generation wavelength on the QW width at room temperature and a hole concentration p = 10^{12} cm^{-2}. It can be seen that the maximum absorption is observed at values of the QW width from 40 to 60 A, and the absorption is much weaker for values less than 30 A and greater than 80 A.

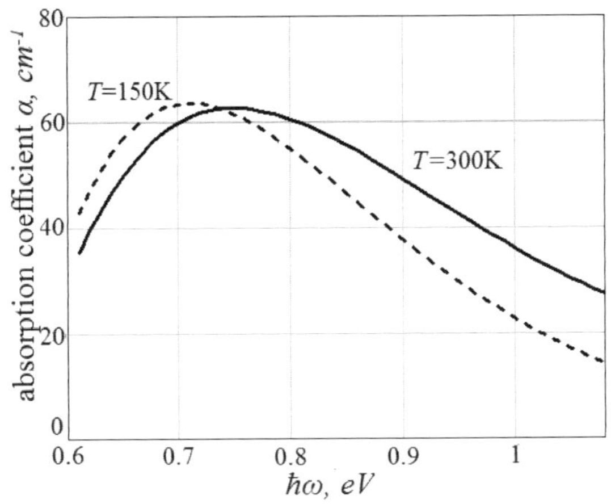

Figure 2(a). Dependence of the absorption coefficient α^{c}_{par} for the continuous spectrum of so-holes on the incident radiation frequency for different temperatures in the InGaAsP/InP heterostructure with a QW width a = 80A and a hole concentration p = 10^{12} cm^{-2}. The solid curve shows the calculation result for T = 300 *K*, and the dotted curve shows the result for T = 150 *K*.

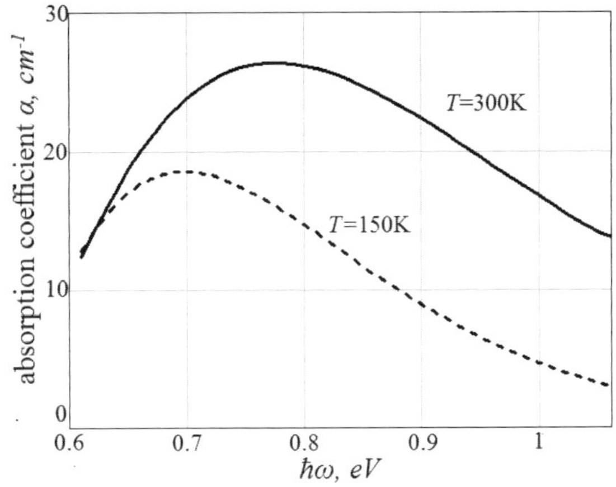

Figure 2(b). Dependence of the absorption coefficient α^{c}_{per} for the continuous spectrum of so-holes on the incident radiation frequency for different temperatures in the InGaAsP/InP heterostructure with a QW width a = 80A and a hole concentration p = 10^{12} cm^{-2}. The solid curve shows the calculation result for T = 300 *K*, and the dotted curve shows the result for T = 150 *K*.

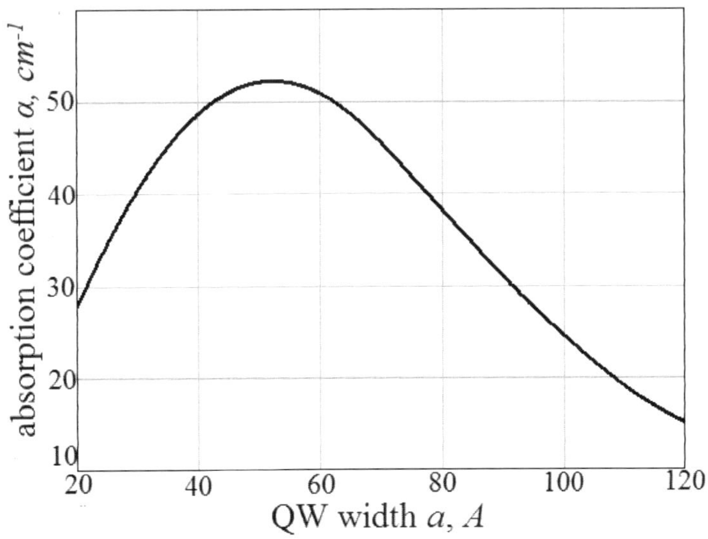

Figure 3. Dependence of the absorption coefficient to the continuous spectrum of spin-split holes on the width of a quantum well in the InGaAsP/InP heterostructure at room temperature and a hole concentration 10^{12} cm^{-2}.

4. Conclusions

The mechanism of intraband radiation absorption by holes with a transition to the so-band for QWs based on A_3B_5 semiconductors is analyzed microscopically. The cases of the transverse and longitudinal polarizations of the incident radiation are considered separately. The analysis is performed for two cases: transitions to a discrete spectrum and to a continuous spectrum of so-holes. It is shown that the intraband absorption can be the main mechanism of internal optical losses for semiconductor lasers on QWs.

In quantum wells, the absorption that occurs with the transition to the discrete spectrum of so-holes is stronger than the absorption that occurs with the transition to the continuous spectrum. The frequency dependence of the absorption coefficient corresponding to the transitions to the discrete spectrum of so-holes has pronounced maxima, the position and magnitude of which depends on the temperature. With increasing temperature, the peak shifts towards higher energies, and the peak value of the absorption coefficient decreases. Also, the absorption of radiation polarized in the plane of the quantum well predominates over the absorption of the radiation of the transverse polarization.

The value of the absorption coefficient with the transition to the discrete spectrum decreases with increasing value of the energy of the forbidden band of the material. For the transitions to the continuous spectrum, its pronounced dependence on the width of the forbidden band is not observed. In this case, the inverse proportion of the absorption coefficient to the transition energy plays a more significant role.

Acknowledgements
This work is supported in part by the RFBR grant № 16-08-01130-A.

References

[1] Thijs P J A, van Dongen T, Tiemeijer L F and Binsma J J M 1994 *J. Lightwave Technol.* **12** 28
[2] Piprek J, Abraham P and Bowers J E 2000 *IEEE J. Quantum Electron.* **36** 366
[3] Silver M and O'Reilly E P 1995 *IEEE J. Quantum Electron.* **31** 1193
[4] Agrawal G P and Dutta N K 1993 *Long-Wavelength Semiconductor Lasers* (Van Nostrand Reinhold, New York)
[5] Zory P S 1993 *Quantum Well Lasers* (Academic Press)
[6] Rosencher E and Levine B 1992 *Intersubband Transitions in Quantum Wells* (Plenum, New York)
[7] Ackerman D A, Shtengel G E, Hybertsen M S, Morton P A, Kazarinov R F, Tanbun-Ek T and Logan R A 1995 *IEEE Select. Topics Quantum Electron.* **1** 250
[8] Phillips A F, Sweeney S J, Adams A R and Thijs P J A 1998 IEEE *Select. Topics Quantum Electron.* **5** 401
[9] Asryan L V, Gun'ko N A, Polkovnikov A S, Zegrya G G, Suris R A, Lau P.-K and Makino T 2000 *Semicond. Sci. Technol.* **15** 1132
[10] Kane E O 1957 *J. Phys. Chem. Sol.* **1** 249
[11] Zegrya G G and Polkovnikov A S 1998 *J. Exp. Theor. Phys.* **86** 815
[12] Polkovnikov A S and Zegrya G G 1998 *Phys. Rev. B.* **58** 4039
[13] Zegrya G G, Pikhtin N A, Skrynnikov G V, Slipchenko S O andTarasov I S 2001 *Semiconductors* **35** 962

IOP Conf. Series: Journal of Physics: Conf. Series **993** (2018) 012014 doi:10.1088/1742-6596/993/1/012014

Structural and optical characteristics of GaAs films grown on Si/Ge substrates

A V Rykov[1], M V Dorokhin[1], P S Vergeles[2], N V Baidus[1], V A Kovalskiy[2], E B Yakimov[2] and O A Soltanovich[2]

[1]Research Institute of Physics and Technology, Lobachevsky State University, Nizhni Novgorod, 603950, Russia
[2]Institute of Microelectronics Technology, Russian Academy of Sciences, Chernogolovka, 142432, Russia

rikov@nifti.unn.ru

Abstract. A GaAs/AlAs heterostructure and a GaAs film grown on Si/Ge substrates have been fabricated and studied. A Ge buffer on a silicon substrate was fabricated using the MBE process. A3B5 films were grown by MOCVD at low pressures. Photoluminescence spectroscopy was used to define the optical quality of A3B5 films. Structural properties were investigated using the electron beam induced current method. It was established that despite a rather high density of dislocations on the epitaxial layers, the detected photoluminescence radiation of layers indicates the acceptable crystalline quality of the top GaAs layer.

1. Introduction

Silicon is the most commonly used material for semiconductor devices due to its natural abundance, lesser weight, and better thermal conductivity. Next in popularity are GaAs and other A3B5 semiconductors, which are highly valued for their optoelectronic applications and high carrier mobility. The creation and investigation of A3B5 epitaxial structures on Si is a rapidly developing trend [1-2]. This trend is related to the solution for one of the major problems in modern functional microelectronics: creating optoelectronic elements of integrated microcircuits based on silicon. As a particular example, hybrid A3B5 heterolasers compatible with the complementary metal-oxide-semiconductor technology (CMOS) on Si provide the transition from copper wires to optical interconnects in high-performance processors [3].

The problem of creating these structures lies in the difference between the crystal lattices and thermal expansion coefficients of A3B5 and Si. Because of these differences, direct epitaxial growth of GaAs on Si is accompanied by the formation of a large number of defects that reduce the quality of a structure. Therefore, the fundamental task of the A3B5/Si heterostructure growth is a search for growth regimes that ensure an improvement in the quality of layers grown, namely, a reduction in the quantity of defects and dislocations that negatively affect the operation of devices built on the basis of such structures. This requires a combination of improvements to the appropriate technology with the use of advanced techniques for the in-depth study of A3B5 layers grown on Si substrates. In this paper, we report on the growth of GaAs-on-Si heterostructures and the study of these structures using the electron beam induced current (EBIC) measurements technique. The EBIC method combines high-resolution visualization of electrically active defects in materials or structures with measurement of the diffusion length of minority charge carriers [4]. The latter is a parameter that characterizes the quality of a semiconductor material and determines the main characteristics of many semiconductor devices.

2. Fabrication of A3B5/Ge/Si structures

The A3B5/Ge heterostructures were grown on a part of a 2-inch epi-ready (100)-oriented Si substrate with 4° off-cut toward (110). During the first stage, Si substrates were prepared as Ge intermediate layers were deposited via molecular beam epitaxy (MBE) [5]. First of all, a 50-nm Ge layer was deposited at 275 °C. A lower temperature was used to exclude elastic-strain relaxation via the formation of three-dimensional islands. In this case, relaxation occurs via the formation of a large number of misfit dislocations, and the layer with a high density of threading dislocations (> 10^{10} cm^{-2}) is formed. Then the growth temperature was increased to 600°C and a 1-μm-thick Ge layer was grown, which improved the crystalline quality. In the end, cyclic annealing was used (850°C/2 min, 550°C/2 min, 5 repetitions) to lower the density of threading dislocations after formation of the structure.

During the second stage, A3B5 films were grown using a MOCVD system (AIX 200RF) at 50 mbar of pressure. The Si/Ge substrate was heated up to 650 °C in H_2 atmosphere for oxide removal before epitaxial growth of A3B5. Then a 20-nm-thick AlAs nucleation layer was grown at 650 °C. This layer also serves as a barrier to diffusion of Ge and As through the heterojunction [6]. After that, a pair of 100-nm GaAs:C and 20-nm AlAs layers were alternately deposited, with 5 repetitions in total. The carrier concentration in the GaAs:C layer amounted to $1 \cdot 10^{18}$ cm^{-3}. Next, a 1-μm GaAs:Si layer was grown with a carrier concentration of $2 \cdot 10^{18}$ cm^{-3}. Trimethylgallium $Ga(CH_3)_3$, trimethylaluminum $Al(CH_3)_3$ and arsine AsH_3 were used as precursors.

To analyze the quality of epitaxial layers, the optical characteristics of samples were investigated by photoluminescence (PL) spectroscopy alongside with structural investigation by EBIC. For the EBIC measurements, the Al Schottky contact was deposited to the top of the sample, and the Ohmic contact was applied to the back of the sample. Details of EBIC technique are described elsewhere [4].

3. Photoluminescence study

The photoluminescence measurements were carried out with a Nanometrics RPM-2000 system that allows scanning of samples with a resolution down to 0.1 mm. The results of measurements are the PL maps that visualize the distribution of basic PL parameters over the area of a scanned sample. The parameters are the wavelength of the PL peak maximum, the spectral width of the PL peak (FWHM or full width at half maximum), the intensity of the spectra maximum, etc. PL was excited by an Nd:YAG laser with a wavelength of 532 nm and an optical power density of 386.4 W/cm^2.

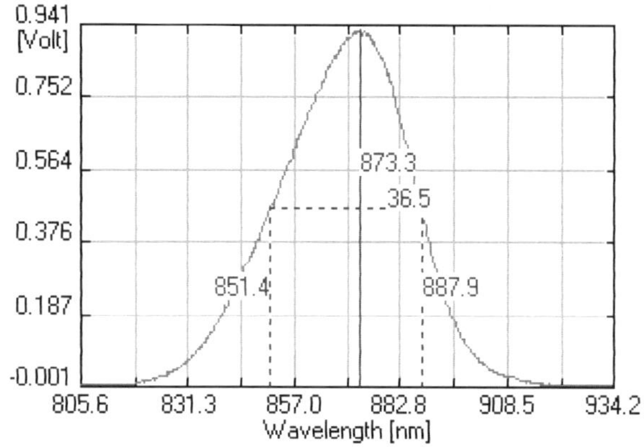

Figure 1. PL spectra of the GaAs/AlAs/Ge/Si sample.

Figure 2. The wavelength peak map of the GaAs/AlAs/Ge/Si sample.

IOP Conf. Series: Journal of Physics: Conf. Series **993** (2018) 012014 doi:10.1088/1742-6596/993/1/012014

The PL spectrum with a maximum at 873.3 nm corresponding to GaAs is shown in figure 1. Figure 2 shows the wavelength peak map. PL maps demonstrate the uniformity of the grown GaAs top layer. For instance, FWHM distribution amounts to 36.5±0.3 nm. This amount is higher than the average FWHM amount for PL peaks of GaAs grown on native substrates using the same MOCVD system (about 23 nm for layers with a comparable level of doping). This is explained by the presence of non-radiative recombination centers and charged defects.

4. Structural characteristics

The primary evaluation of the surface quality was performed by etching the GaAs surface using standard selective dislocation etchants and direct counting of etch pits using an optical microscope. An example of an etched GaAs surface image taken with a LEIKA DM 4000M microscope is shown in figure 3. Average dislocation pit density was calculated for 5 fields of view and amounted to $2 \cdot 10^5$ cm^{-2}.

A more in-depth and comprehensive study was conducted using the EBIC method. The EBIC scan of the sample is presented in figure 4. One can see a complex of visualized bulk defects, which is probably due to the complexity of the structures including a number of interfaces. In particular, the black dots in the EBIC image correspond to the electrically active threading dislocations. The dislocation density estimated from the EBIC image is about 10^8 cm^{-2}.

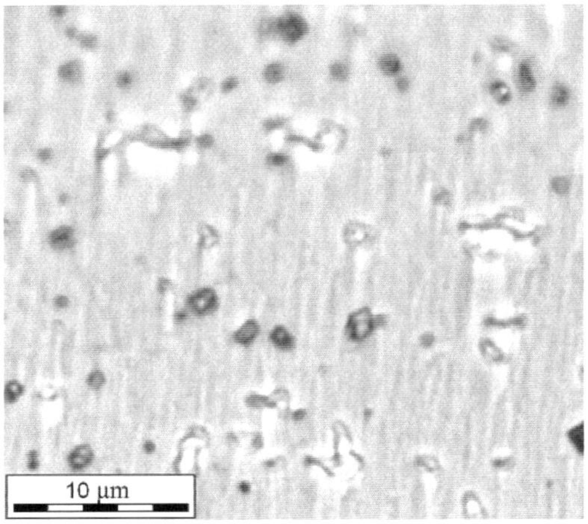

Figure 3. Etched GaAs surface image taken with an optical microscope.

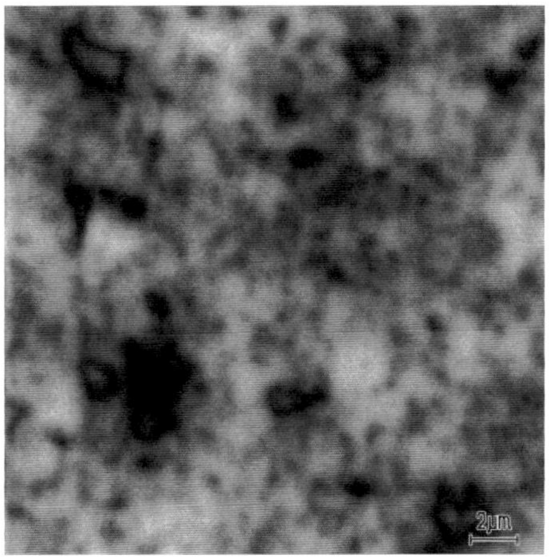

Figure 4. The EBIC image of the investigated sample measured under the electron beam with an accelerating voltage of 35 kV.

The revealed disagreement between the optical and EBIC estimations of the dislocation density is supposedly due to different regions of data collection used for the two methods. The optical data were collected from the top GaAs:Si layer, whereas the EBIC measurements may visualize the dislocations which grow from the Si substrate through the Ge buffer layer and are blocked at numerous GaAs/AlAs interfaces. Preliminary capacitance-voltage data confirm that the EBIC signal is most likely collected from the Ge buffer layer (with a hole concentration of about $3 \cdot 10^{17}$ cm^{-3}) in the vicinity of the Si/Ge interface. To validate this hypothesis, some additional experiments are required. However, a relatively high photoluminescence signal (figure 1) is another indication of the good quality of the top GaAs layer.

In conclusion, we have grown and investigated a GaAs/AlAs heterostructure and a GaAs film on Si/Ge substrates. Although the measurements reveal a relatively high dislocation density from the entire heterostructures, the top layer is probably of better crystalline quality and thus can be used as an active region for GaAs-based semiconductor devices.

Acknowledgments
This study was supported by the Russian Foundation for Basic Research (projects 16-07-01102, 17-37-80008), Ministry of Education and Science of Russian Federation (project 8.175.2017/PP).

References
[1] Bolkhovityanov Yu B and Pchelyakov O P 2009 *The Open Nanoscience Journal* **3** 20
[2] Buzynin Y, Shengurov V, Zvonkov B, Buzynin A, Denisov S, Baidus N, Drozdov M, Pavlov D and Yunin P 2017 *AIP Advances* **7** 015304
[3] Aleshkin V Ya, Baidus N V, Dubinov A A, Krasilnik Z F, Nekorkin S M, Novikov A V, Rykov A V, Yurasov D V and Yablonskiy A N 2017 *Semiconductors* **51**(5) 663
[4] Shmidt N M, Vergeles P S and Yakimov E B 2007 *Semiconductors* **41**(4) 491
[5] Yurasov D V, Bobrov A I, Daniltsev V M, A Novikov A V, Pavlov D A, Skorokhodov E V, Shaleev M V and Yunin P A *2015 Semiconductors* 2015 **49** (11) 1415
[6] Chia C K, Dong J R, Chi D Z, Sridhara A, Wong A S W, Suryana M, Dalapati G K, Chua S J, and Lee S J *2008 Appl, Phys. **Lett.*** **92** 141905

IOP Conf. Series: Journal of Physics: Conf. Series **993** (2018) 012015 doi:10.1088/1742-6596/993/1/012015

Anomalous Nernst-Ettingshausen effect in δ<Mn>GaAs/InGaAs ferromagnetic semiconductor heterostructures

Yu M Kuznetsov, M V Dorokhin, A V Kudrin and O V Vikhrova

Physical Technical Research Institute of Lobachevsky State University of Nizhny Novgorod, Nizhny Novgorod, 603950 Russia

E-mail: yurakz94@list.ru

Abstract. In the present work, the magnetic properties of GaAs with a ferromagnetic delta layer of manganese were studied using the Nernst-Ettingshausen effect. It is shown that at temperatures below the Curie point the dependence of the Nernst-Ettingshausen voltage on the magnitude of the external magnetic field is nonlinear, which is known as the anomalous Nernst-Ettingshausen effect. In the present study, it is established that the nonlinear dependence becomes linear above the Curie temperature.

1. Introduction

Diluted magnetic semiconductors (DMS), such as (A3,Mn)B5, are promising materials for the creation of spintronic devices. However, a major obstacle to implementing DMS is a typically low value of the Curie temperature that determines low operating temperatures for DMS-based devices [1]. To increase the value of the Curie temperature, it is of great importance to analyze the ferromagnetic ordering mechanisms and to define the factors that influence ferromagnetism. In this paper, we investigate the magnetic properties of a ferromagnetic semiconductor heterostructure containing a δ-<Mn> layer in a GaAs matrix using the Nernst-Ettingshausen effect. The latter measurements have proven to be a good tool to analyze the magnetic properties at the ferromagnetic-to-paramagnetic transition point [2]. In particular, the presence of the anomalous Nernst-Ettingshausen effect at temperatures below the Curie point is shown. The results of the measurements are discussed with the assumption of spin-dependent scattering in the magnetized ferromagnetic δ-<Mn> layer.

2. The anomalous Nernst-Ettingshausen effect

The Nernst-Ettingshausen effect (NE) is based on the same principle as the Hall effect: the separation of carriers moving from one end of the sample to the other by an external magnetic field [2]. However, unlike the Hall effect, the carrier motion in NE is triggered via the application of a temperature gradient at the ends of the investigated structure (see figure 1).

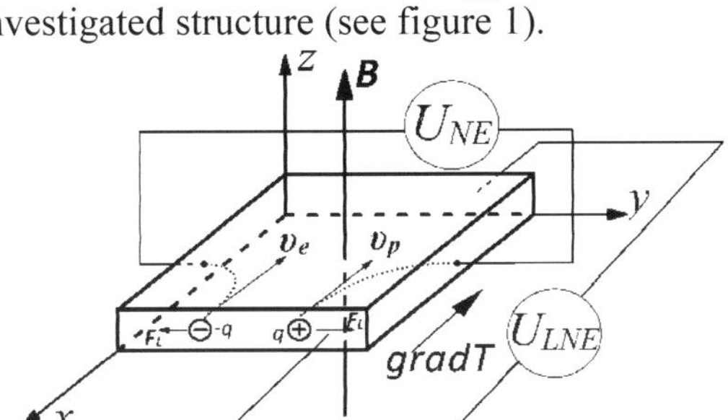

Figure 1. Schematic of the Nernst-Ettingshausen effect measurements. v_e is the electron velocity, v_p is the hole velocity, U_{NE} is the transverse Nernst-Ettingshausen voltage, and U_{NEL} is the longitudinal Nernst-Ettingshausen voltage.

To analyze the anomalous NE effect, one can use similar formulas as for the Hall effect. In ferromagnetic samples, the dependence of the Hall voltage on the magnetic field is nonlinear, which is explained by spin-dependent scattering of carriers [3]. The formalism used in discussing the Hall effect is also valid for the NE effect. In particular, an analogous expression can be used to determine the NE electric field, according to [2]:

$$\boldsymbol{E} = S\nabla T - Q_0[\boldsymbol{B}\nabla T] - Q_s[\boldsymbol{M}\nabla T], \tag{1}$$

where S is the Seebeck coefficient (which characterizes the thermo-voltage emerging due to the asymmetry of contacts relative to the heat flux and the direction of the external magnetic field), ∇T is the temperature gradient in the structure under study, Q_0 is the Nernst-Ettingshausen constant, $\boldsymbol{B}$ is the external magnetic field, Q_s is the anomalous Nernst-Ettingshausen constant, and $\boldsymbol{M}$ is the magnetization of the sample.

3. Experimental method

The investigations were made within an epitaxial ferromagnetic multilayer film grown on a semi-insulating GaAs (100) substrate. First, a sequence of layers consisting of a GaAs epitaxial layer, a 10-nm $In_xGa_{1-x}As$ quantum well ($x \approx 0.16$), and a 3-nm GaAs layer was grown by metal-organic vapor phase epitaxy at 650 °C. At the next stage, the growth temperature was decreased and a δ<Mn> layer followed by a 4-nm GaAs cap layer were grown by the pulsed laser deposition technique. The details of the growth process were described elsewhere [4]. The schematic view of the structures is depicted in figure 2.

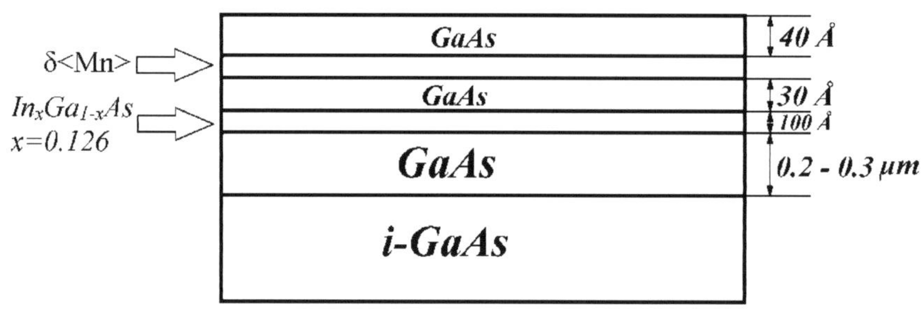

Figure 2. Schematic view of the structure under study.

To measure the Hall Nernst-Ettingshausen effects, four Ohmic contacts were formed on the surface of the sample in such a way that one pair of contacts was located along the heat flux (D and C) and the other was across it (A and B, see figure 3). The sample was attached to a dielectric plate with low thermal conductivity. A resistor was installed on the surface of the sample, which played the role of a heater. The thermal contact between the resistor and the surface of the sample was ensured by a heat conducting paste which also provided the electrical isolation of the resistor and the sample. The opposite side of the structure rested against a massive metal radiator, which was used for removing heat. The space between the face of the sample and the metal radiator was filled with a thermal paste to create a thermal contact and provide dielectric insulation. The entire system was mounted on a closed-cycle cryostat holder such that the metal radiator was in good thermal contact with the holder, i.e. with one end of the sample, whereas the sample substrate was thermally isolated from the holder. The holder temperature was varied within the range of 10 - 300 K. The sample was placed between the poles of an electromagnet, so that a constant magnetic field was applied perpendicular to the sample surface.

To measure thermomagnetic effects, it is necessary to create a temperature gradient in the sample. This can be achieved by passing an electric current through the resistor F-E. To control the temperature at the sample ends, we have used A-B contacts as a detector. For this reason, the temperature dependence of the resistance between A-B contacts was recorded (figure 4). In the operating regime, the temperature of the A-B region of the sample was considered as the average temperature (T_{av}) and its value was estimated from the $R(T)$ dependence shown in figure 3. The temperature of the cold end (T_{cold}), provided there is good thermal contact, can be equated to the temperature of the holder. The temperature of the hot end (T_{hot}) thus can be calculated by the formula:

$$T_{hot} = 2T_{av} - T_{cold} \tag{2}$$

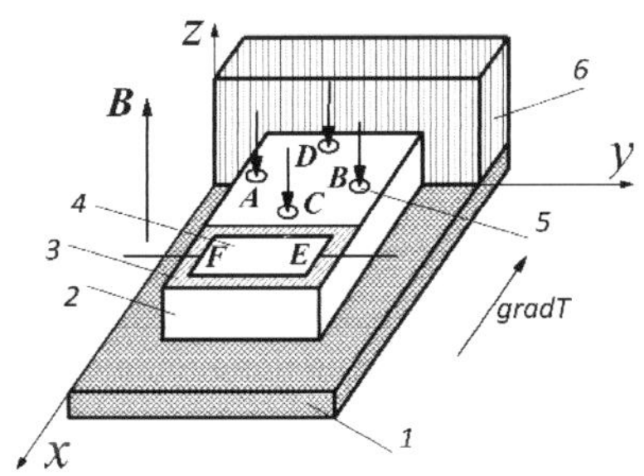

Figure 3. Schematic view of mounting the sample on the holder. The heat flow propagates along the structure. 1 – carrier holder, 2 – sample, 3 – layer of dielectric thermal paste, 4 – resistor heater, 5 - contact areas, 6 – radiator.

The value of the Seebeck voltage was measured by connecting a voltmeter to the contacts C-D. Measurements of the voltage at the contacts A-B allowed us to investigate the transverse Nernst-Ettingshausen effect. We emphasize that the NE voltage was several orders of magnitude lower than the one used to determine the A-B contact resistance. Thus, the NE effect did not affect the measurements of the resistance and the T_{av}. The measurements of the voltages and currents were carried out using a Keithley 2401 universal measuring instrument.

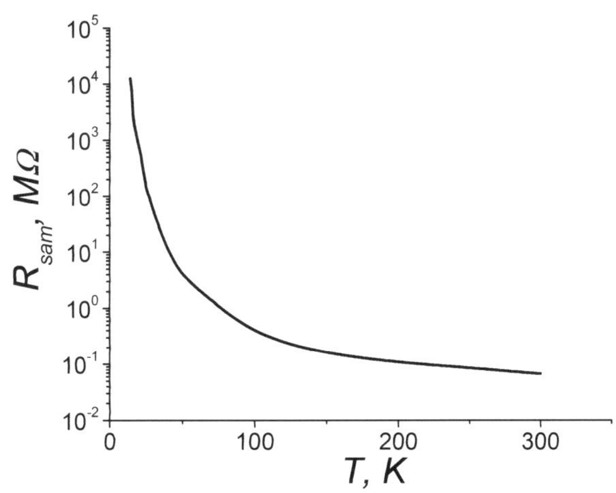

Figure 4. Dependence of the resistance between A-B contacts on the temperature of the cold end.

4. Results of the experiment

Figure 4 illustrates the temperature dependence of the resistance for the studied structure. From the dependence obtained, it can be seen that at temperatures around 15 K the resistance reaches tens of GΩ, which does not allow us to carry out the Hall effect measurements. However, since the dependence is strongly nonlinear, the resistance at temperatures about 50 K is 1 MΩ.

Unlike the Hall effect, the measurements of the Nernst-Ettingshausen effect could be performed and the results are shown in figure 5. In the low temperature region (figure 5 (a)); T_{cold} = 16 K, T_{av} = 64.4 K, T_{hot} = 112.8 K), a nonlinear dependence of the NE voltage on the magnetic field was observed. When the holder temperature (i.e. the cold end temperature) was increased up to 25 K and further to 50 K, the dependence of the NE voltage on the magnetic field became linear, so it could be fitted by a linear function (see figure 5 (b) and figure 5 (c), respectively).

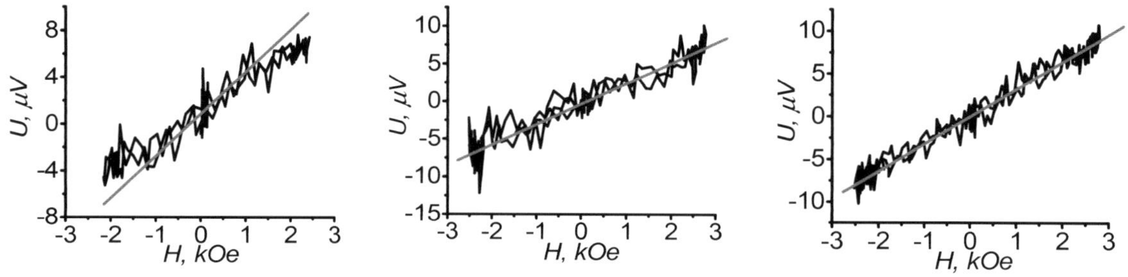

(a) (b) (c)

Figure 5. Experimentally obtained curves of the transverse Nernst-Ettingshausen voltage as a function of the external magnetic field: (a) – the anomalous Nernst-Ettingshausen effect (T = 17.00 K); (b),(c) – the ordinary (linear) Nernst- Ettingshausen effect ((b) – T=25 K; (c) – T = 50 K).

The linear magnetic field dependence of the NE voltage is typical for the Nernst-Ettingshausen effect and can be attributed to different trajectories of carriers moving from the cold end to the hot end of the sample and vice-versa. Thus, the excess charge is accumulated in the left part of the sample and the electric field can be measured between contacts A and B [4]. A nonlinear $U(H)$ dependence, i.e. the anomalous Nernst-Ettingshausen effect, is usually attributed to spin-dependent scattering of carriers in magnetized ferromagnetic samples [2]. An interesting fact is that the average temperature of the sample is much higher than the Curie point, hence no abnormality should arise. Although both the average temperature and the temperature of the hot end are above the Curie point, the temperature of the cold end remains below the Curie point. So, one can assume that the anomalies of the NE effect are due to spin-dependent scattering in the cold region of the sample that remains below the Curie point.

The result obtained gives us a tool to measure ferromagnetic properties of highly resistive samples for which the measurements of the anomalous Hall effect are rather complicated and not always realizable. The application of a temperature gradient creates thermally generated carriers that participate in the conductivity and the formation of thermomagnetic voltages. Moreover, the analysis performed at the ferromagnetic-to-paramagnetic transition point can supply some new information on the ferromagnetic ordering in diluted magnetic semiconductors.

5. Conclusions

The measurements of the anomalous Nernst-Ettingshausen effect were performed on a ferromagnetic δ<Mn>/GaAs/InGaAs sample. It has been shown that the anomalous Nernst-Ettingshausen effect can be observed whenever the cold end of the sample is kept below the Curie point, although the average temperature is way above the Curie point. This gives one a powerful tool for studying ferromagnetism in samples that are highly resistive below the Curie point.

Acknowledgments

The authors would like to acknowledge Dr. B.N. Zvonkov for the fabrication of the investigated samples.
This study was supported by the Russian Science Foundation (project 17-79-2017).

References

[1] Dietl T 2010 *Nature Materials* **9** 965.
[2] Campbell I A 1979 *J. of Magnetism and Magn. Mater.* **12** 31.

[3] Nagaosa N, Sinova J, Onoda S, MacDonald A H and Ong N P. 2010 *Rev. Mod. Phys.* **82** 1539.
[4] Zvonkov B N, Vikhrova O V, Danilov Yu A, Demina P B, Dorokhin M V, Podol'skiĭ V V, Demidov E S, Drozdov Yu N and Sapozhnikov M V. 2008 *J. Opt. Technol.* **75** 389.

IOP Conf. Series: Journal of Physics: Conf. Series **993** (2018) 012016 doi:10.1088/1742-6596/993/1/012016

Non-equilibrium hole capture to excited acceptor states in quantum wells due to optical scattering

G R Balakhtar, A N Sofronov, M Ya Vinnichenko, D A Firsov and L E Vorobjev

Peter the Great St.Petersburg Polytechnic University, 195251 St.Petersburg, Russia

Email: sofronov@rphf.spbstu.ru

Abstract. The probability of optical-phonon-assisted scattering of non-equilibrium holes to excited acceptor states in GaAs/AlGaAs quantum wells is considered. In the model we used, a well-known expression for the probability of intersubband optical scattering of two-dimensional carriers was expanded to the case when the final hole state is the excited state of an acceptor center. The temperature dependences of the probability of hole capture to excited acceptor states with simultaneous optical phonon emission are calculated. The result allows one to estimate the contribution of optical scattering to the experimentally observed terahertz electroluminescence due to intracenter carrier transitions.

1. Introduction

The binding energies of shallow impurities in doped semiconductors and semiconductor nano-structures correspond to the terahertz (THz) spectral range. Currently, the terahertz radiation emission associated with the radiative capture of non-equilibrium carriers on impurity centers in doped quantum well (QW) structures is extensively investigated. Terahertz photoluminescence associated with donor states was experimentally observed in *n*-GaAs/AlGaAs QWs [1] under interband optical excitation in the case of both spontaneous and stimulated near-IR luminescence due to transitions from the ground donor state to the valence band. Low-temperature terahertz electroluminescence of acceptors in *p*-GaAs/AlGaAs QWs under conditions of impurity breakdown in a strong lateral electric field was reported in [2, 3]. In this case, the integral THz emission intensity sharply rises when the applied electric field exceeds the impurity breakdown threshold corresponding to an increase in lateral conductivity by several orders of magnitude, as reported in [3]. The experimentally observed spectra of the acceptor-related terahertz emission demonstrate both optical transitions of hot holes to acceptor states and intracenter optical transitions between excited and ground states of an acceptor, as it was shown in [2].

The binding energy of a ground acceptor state in QWs can exceed the energy of an optical phonon, whereas the binding energy of an excited state is usually smaller. As a result, the intensity of the non-radiative capture of non-equilibrium holes to a ground acceptor state will be dramatically reduced, whereas the efficient occupation of excited acceptor states can be reached by the process of hot holes scattering from the lowest subband to the excited acceptor state with the emission of an optical phonon. This situation will facilitate hole optical transitions from the excited levels to the ground one. Therefore, the purpose of this research is to calculate the probability of non-equilibrium hole capture to the excited acceptor level with simultaneous optical phonon emission.

2. Results

2.1. The wavefunctions of excited acceptor states

In this work, we use the two-dimensional hydrogen-like model to describe the impurity state wavefunctions in QW. One can consider these functions $\xi(\vec{r})$ as a series expansion in the Bloch waves of the lowest valence subband $u_{\vec{k}}(\vec{r})$:

$$\xi(\vec{r})=\sum_{k}C(\vec{k})u_{\vec{k}}(\vec{r})\exp(i\vec{k}\vec{r}), \tag{2.1.1}$$

with the serial coefficients given by the Fourier transform of the envelope wavefunction for impurity states $\psi(\vec{r})$ in $\vec{r}$-space:

$$C(\vec{k})=\frac{1}{\sqrt{S}}\int\psi(\vec{r})\exp(-i\vec{k}\vec{r})d^2\vec{r}, \tag{2.1.2}$$

where $\vec{k}=\{k_x,k_y\}$ is the two-dimensional wave vector corresponding to the in-plane motion of holes, $\vec{r}=\{x,y\}$ is the two-dimensional radius vector, and S is the normalizing area. In the framework of the model we use, these envelope wavefunctions for the first excited states can be expressed as *p*-like states of a two-dimensional hydrogen atom with a principle quantum number $n=1$ and magnetic quantum numbers $m=0,\pm1$, and have the form [4]:

$$\psi_0(\vec{r})=\frac{1}{\sqrt{2\pi}}\frac{4}{a_0 3\sqrt{3}}\left(1-\frac{4r}{3a_0}\right)e^{-2r/3a_0}, \tag{2.1.3}$$

$$\psi_{\pm1}(\vec{r})=\frac{1}{\sqrt{2\pi}}e^{\pm i\varphi}\frac{16}{a_0 9\sqrt{6}}\frac{r}{a_0}e^{-2r/3a_0}, \tag{2.1.4}$$

where a_0 is the Bohr radius of a hydrogen-like impurity, and φ is the polar angle.

Thus, the serial expansion coefficients can be calculated using the Fourier transform of the excited states of the wave functions for a two-dimensional hydrogen-like impurity. After some transformations and analytical integration over *r*, we obtained these coefficients for excited acceptor states in the form:

$$C_0(\vec{k})=\frac{1}{\sqrt{S}}\frac{1}{\sqrt{2\pi}}\frac{4a_0}{3\sqrt{3}}\int_{\varphi}\frac{3A(\varphi)-8}{3A^3(\varphi)}d\varphi, \tag{2.1.5}$$

$$C_{\pm1}(\vec{k})=\frac{1}{\sqrt{S}}\frac{1}{\sqrt{2\pi}}\frac{32a_0}{9\sqrt{6}}\int_{\varphi}\frac{e^{\pm i\varphi}}{A^3(\varphi)}d\varphi, \tag{2.1.6}$$

where $A(\varphi)=2/3+ik_x a_0\cos\varphi+ik_y a_0\sin\varphi$, which cannot be further analytically simplified. The envelope wavefunctions of a two-dimensional hydrogen-like impurity and their Fourier transforms are plotted in figure 1.

2.2. The probability of scattering of non-equilibrium holes to excited acceptor states

The scheme of the intraband scattering processes with emission of optical phonons is shown in figure 2. The process marked with the arrow *a* is a pure intraband transition with both initial k_i and final k_f states in the first valence subband. The two-dimensional matrix element for these intraband transitions M_{2D} is calculated by P.J. Price [5]. The process marked with the arrow *b* is the transition of interest, when the initial state is in the valence band, and the final state is the excited acceptor level.

Since we expanded the final state in the Bloch wave series according to (2.1.1), it is natural to write the squared matrix element for the transition *b* as a product of the squared pure intraband matrix element $|M_{2D}|^2$ and the squared expansion coefficient $|C|^2$:

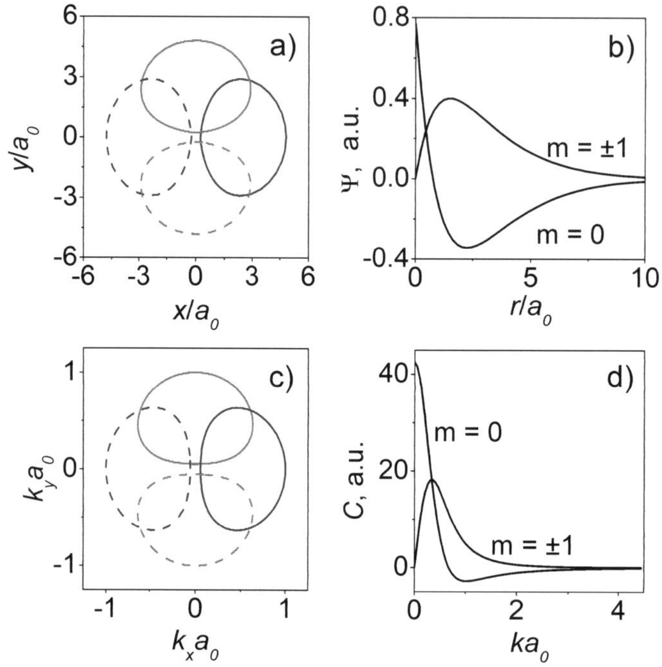

Figure 1. Radial and angular parts of the two-dimensional hydrogen atom wavefunctions (*a, b*) and the Fourier transforms (*c, d*) of the excited acceptor states, respectively. The angular parts are shown as contour plots of common p_x–like combination (the blue lines) and p_y–like combination (the red lines) of the corresponding functions with $m = \pm 1$; in other words, $(\psi_{+1} + \psi_{-1})/\sqrt{2}$ and $(\psi_{+1} - \psi_{-1})/\sqrt{-2}$). The solid and dashed lines correspond to the positive and negative levels for contour plots, respectively.

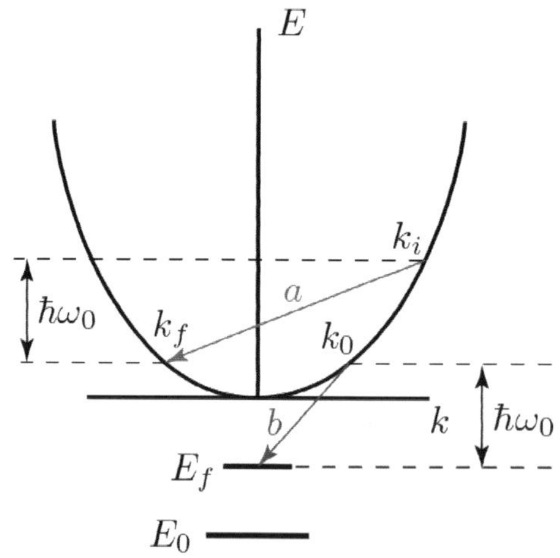

Figure 2. Scheme of the intraband transition (*a*) and scattering to excited acceptor states (*b*). Energy of an optical phonon is $\hbar\omega_0$.

$$|M|^2 = |M_{2D}|^2 \cdot |C_{0,\pm 1}|^2 . \tag{2.2.1}$$

Taking the expression for the matrix element for pure intraband polar optical scattering with the emission of an optical phonon from [5] and assuming the number of equilibrium phonons to be negligible, we can now write the probability of scattering of a hole from the state k_i in the valence subband to the state k_f, that forms the excited acceptor level with the envelope $C(k_f)$ in the form:

$$W_{0,\pm1}\left(\vec{k}_i,\vec{k}_f\right)=\frac{e^2\omega_0}{2\pi}\left(\frac{1}{\varepsilon_\infty}-\frac{1}{\varepsilon_0}\right)\left|C_{0,\pm1}\left(\vec{k}_f\right)\right|^2\delta\left(E_i-E_f-\hbar\omega_0\right)\int_{-\infty}^{\infty}\frac{|I(q)|^2}{Q^2+q^2}dq, \tag{2.2.2}$$

where ε_0 and ε_∞ are the static and high-frequency dielectric constants of the QW material, respectively, ω_0 is the optical phonon frequency, $E_i=\frac{\hbar^2k_i^2}{2m_{hh}}$ is the initial energy , $E_f=-\frac{4}{9}E_0$ is the final energy (independent of k_f), and E_0 is the Rydberg constant for a three-dimensional hydrogen-like impurity in a QW material. The values of q and $Q = k_f - k_i$ stand for the phonon wavevector projection to the growth axis and the length of a two-dimensional phonon wavevector in a QW plane, respectively. The form-factor $|I(q)|^2$ describes the modified momentum conservation law for scattering of two-dimensional electrons on three-dimensional phonons and can be expressed in terms of a z-dependent part of the hole envelope wavefunctions in QW [5].

In order to calculate the total probability of the hole capture to the acceptor excited state, one should perform summation of the probability (2.2.2) over all phonons and average it over the distribution function f of holes in the valence band:

$$\tau_{0,\pm1}^{-1}\sim\int f\left(\vec{k}_1\right)\cdot\sum_Q W_{0,\pm1}\left(\vec{k}_i,\vec{k}_f\right)d^2\vec{k}_i. \tag{2.2.3}$$

After summation over all the phonons, the δ-function in (2.2.2) allows one to integrate (2.2.3) over k_i since the argument of the δ-function is independent of k_f. As a result, the total probability of the hole capture becomes proportional to the amount of holes involved in scattering due to the energy conservation:

$$\tau_{0,\pm1}^{-1}\sim f\left(k_0\right),\quad k_0=\sqrt{2m_{hh}\left(E_f+\hbar\omega_0\right)}/\hbar. \tag{2.2.4}$$

The proportionality coefficient includes the integration over angles and was calculated numerically.

2.3. *Time of the hole capture to excited acceptor states*

Let us now discuss the conditions of the experiment in [2,3]. One can estimate the non-equilibrium hole concentration p in the valence band to be equal to the doping level N_A since there is a full ionization of acceptors under conditions of impurity breakdown in a lateral electric field. This concentration can be expressed in terms of the distribution function $f(E)$ and the 2D density of states $g_{2D}(E)$:

$$p=\int_0^\infty g_{2D}(E)f\left(E\right)dE. \tag{2.3.1}$$

The non-equilibrium distribution function $f\left(E\right)$ can be approximated in the concept of hot carriers for a relatively high carrier density. Under this assumption, we can write:

$$f\left(E\right)=\left[\exp\left(\frac{E-\mu}{k_BT_h}\right)+1\right]^{-1}, \tag{2.3.2}$$

where T_h is the hot hole temperature that describes the average hole energy, k_B is the Boltzmann constant, and μ is the quasi-Fermi-level that can be calculated from (2.3.1).

The dependences of the calculated reciprocal capture time on the hot hole temperature are shown in figure 3. At low temperatures, the contribution of optical phonon scattering is negligible due to an exponentially small amount of holes in the valence band that have enough energy to emit optical phonons. However, at high temperatures, the optical phonon contribution is significant and the probability of hole capture to an acceptor state drastically increases. For comparison, the red dotted line shows the typical high limit value for the cascade capture probability [6-8] dominating only at low temperatures.

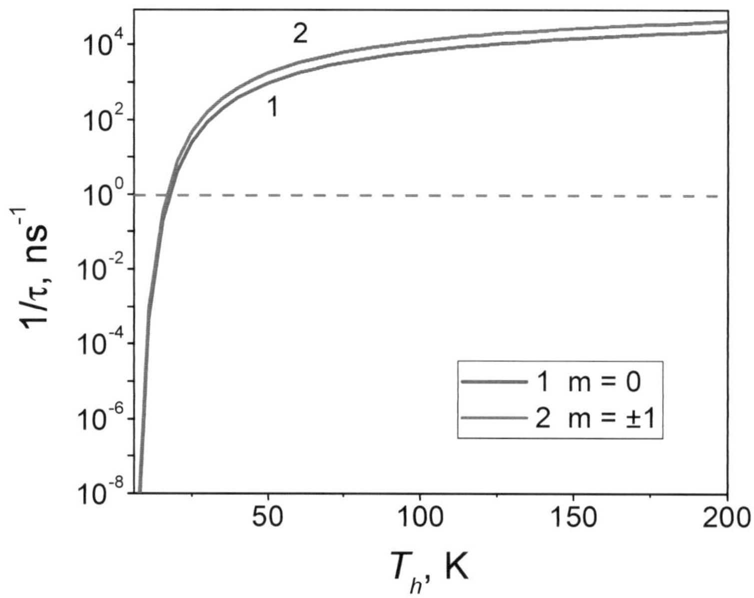

Figure 3. Dependences of the inversed capture time on the hot hole temperature for GaAs:Be QWs with $N_A = 3.2\times10^{11}$cm^{-2}. QW width is equal to 3.8 nm.

Thus, the optical phonon contribution is significant for an increase in the intensity of terahertz emission related to intracenter hole transitions.

Acknowledgements

This work was supported by the Russian Foundation for Basic Research (grant 16-32-60085) and the Ministry of Education and Science of the Russian Federation (state assignments, project 3.933.2017/4.6).

References

[1] Makhov I S *et al* 2017 *Superlattices and Microstructures* **112** 79-85
[2] Firsov D A *et al* 2015 *Semiconductors* **49** 28-32
[3] Sofronov A N *et al* 2017 Proceedings of the 25th Int. Symp. "Nanostructures: Physics and Technology" p 128
[4] Haug H and Koch S W 2009 Quantum Theory of the Optical and Electronic Properties of Semiconductors (World Scientific Pub Co Inc) pp 235-267
[5] Price P J 1981 *Annals Of Physics* **133** 217
[6] Lax M 1960 *Phys. Rev.* **119** 1502
[7] Abakumov V N, Perel V I and Yassievich I N 1991 *Nonradiative Recombination in Semiconductors* (North Holland/Elsevier)
[8] Aleshkin V Y and Gavrilenko L V 2017 *Semiconductors* **51** 1444

Impurity-assisted terahertz photoluminescence in quantum wells under conditions of interband stimulated emission

I S Makhov[1], V Yu Panevin[1], D A Firsov[1], L E Vorobjev[1], A N Sofronov[1], M Ya Vinnichenko[1], N A Maleev[2] and A P Vasil'ev[3]

[1] Peter the Great St. Petersburg Polytechnic University, 195251 St. Petersburg, Russia
[2] Ioffe Institute, 194021 St. Petersburg, Russia
[3] Submicron Heterostructures for Microelectronics Research and Engineering Center of the RAS, 194021 St. Petersburg, Russia

E-mail: MakhovIS@spbstu.ru

Abstract. Terahertz and near-infrared photoluminescence under conditions of interband stimulated emission are studied in *n*-GaAs/AlGaAs quantum well laser structure. The observed terahertz emission is related to the optical transitions of nonequilibrium electrons from the first electron subband and excited donor states to donor ground states in quantum wells. The opportunity to increase the intensity of impurity-assisted terahertz emission due to interband stimulated emission with the participation of impurity centres is demonstrated.

1. Introduction

In recent decades, the terahertz spectral range (0.1-10 THz) has attracted considerable interest because terahertz devices, particularly terahertz radiation sources, may find their application in various fields of science and technology. Thus, terahertz devices can be used in spectroscopy, medicine, monitoring of environment and food quality, wireless communications, etc [1]. The terahertz radiation mechanism based on optical transitions of nonequilibrium charge carriers with the participation of shallow impurity states in semiconductors is one of the promising for the creation of terahertz radiation sources.

Recently, spontaneous terahertz emission associated with the radiative transitions of nonequilibrium electrons from the first electron subband and donor excited states to donor ground states was observed in *n*-GaAs/AlGaAs quantum wells (QWs) under interband optical excitation [2, 3]. The intensity of terahertz radiation in such systems is determined by the depopulation rate of the ground donor state in QW, which occurs due to the spontaneous radiative recombination of electrons from the donor ground state and holes from the first heavy hole subband. Thus, the increase in the depopulation rate for the donor ground state in QW should result in an increase of the terahertz radiation intensity. A possible way to increase the depopulation rate for the donor ground state is to realize stimulated emission between the donor ground state and the valence subband. A similar approach was used earlier in InGaAs/AlGaAs quantum dots [4], where stimulated interband recombination led to an increase of mid-infrared radiation intensity associated with intraband interlevel optical transitions of charge carriers in quantum dots. In this work, we present the results of the investigation of impurity-assisted terahertz photoluminescence (PL) under conditions of interband stimulated emission in doped QW laser structure.

2. Sample and experimental setups

The structure for optical studies was MBE grown on a semi-insulating GaAs substrate and contained 10 quantum wells formed with 7.6 nm thick GaAs layers and separated with 5 nm $Al_{0.3}Ga_{0.7}As$ barrier layers. Quantum wells were doped with silicon (donor) in the 2.6 nm central region of each QW up to a surface impurity concentration of $5\cdot10^{10}$ cm^{-2}. In order to realize interband stimulated emission, QWs were embedded into a symmetric waveguide formed with 0.6 μm thick $Al_xGa_{1-x}As$ graded layers (x varied from 0.4 to 0.6). The sample substrate was polished off to the thickness of about 120 μm. The four "as-cleaved" facets of the sample formed the total internal reflection optical cavity with dimensions of about 400×400 μm^2.

The sample was mounted in a closed cycle cryostat, which allowed to cool the sample down to about 5 K. The powerful interband optical excitation of the QW layers as well as the barrier and waveguide layers of the sample was attained with a frequency doubled Nd:YAG solid state pulsed laser ($\lambda = 532$ nm, 0.25 μs pulse duration, 8 kHz repetition rate). For the terahertz luminescence experiment, the laser beam was additionally chopped at 360 Hz with the aim to use a lock-in technique to measure the detector signal.

The terahertz PL spectra were studied in a step-scan mode of a vacuum FTIR spectrometer equipped with a 6 μm thick Mylar multilayer beamsplitter. The terahertz photoluminescence intensity was measured with a liquid helium cooled silicon bolometer. The bolometer photoresponse was measured by a lock-in amplifier SR510 on a pump laser chopper frequency. Near-infrared photoluminescence spectra were measured from the facet of the sample using a monochromator with a holographic grating with 1200 groves per mm and a liquid nitrogen cooled silicon CCD camera.

In order to reveal the influence of interband stimulated emission on impurity-related terahertz emission in QWs, the integrated terahertz photoluminescence intensity and near-infrared photoluminescence spectra were measured simultaneously. A terahertz sensitive Ge:Ga photodetector was mounted opposite the surface of the sample at a distance of about 12 mm, in this way integrated terahertz emission was measured in "face to face" mode, when the excited sample and the detector were mounted in a cryostat at the same cold finger. The surface of the Ge:Ga detector was covered with two cold filters: 1 mm thick high resistivity compensated Ge (Sb, Au) and a layer of black polyethylene (about 100 μm thick) with the aim to block pumping laser radiation scattered from the surface of the sample, as well as near-infrared radiation of the sample. After being pre-amplified and filtered by a current preamplifier SR-570, the detector photoresponse was measured by a lock-in amplifier SR-510 on a pump laser chopper frequency.

3. Experimental results and discussion

Spectral measurements of the terahertz PL intensity in our sample with doped QWs at a low optical pumping power (about 7 W/cm^2) show that terahertz photoluminescence spectrum contains two broad emission bands with maxima at the photon energies $h\nu$ of about 10 and 20 meV. Note that a terahertz photoluminescence signal from the GaAs semi-insulating substrate (without QW layers) was not detected. Thus, according to the results of theoretical calculations of the energy spectrum of donor states in QWs from the Ref. [5], the binding energy of a donor impurity in the 7.6 nm wide QW is about 15 meV and the photoluminescence band near the photon energy 20 meV is associated with the capture of nonequilibrium electrons from the first electron subband $e1$ to the ground impurity state $1s$. The terahertz emission band near a photon energy of about 10 meV could be associated with electron transitions from the excited donor state $2p_{x,y}$ to the ground donor state $1s$ in QW. These terahertz optical transitions are marked with the arrows $h\upsilon_{\mathrm{THz}}$ in the optical transitions scheme presented in figure 1.

Investigations of near-infrared photoluminescence spectra at different levels of interband optical excitation show broad emission bands associated with radiative recombination of free and bound excitons, radiative recombination of electrons and heavy holes via ground and excited donor states ($1s$-$hh1$ and $2p_{x,y}$-$hh1$ optical transitions shown in figure 1), as well as via impurity states of residual acceptor [6] (the $e1$-A optical transition shown in figure 1) in QWs. At relatively high optical pumping levels three narrow stimulated emission lines associated with radiative recombination of electron-hole pairs via excited and ground donor states and ground acceptor states in QWs arise (these stimulated optical transitions are shown in figure 1 with the arrows $h\upsilon_{\mathrm{NIR}}^{\mathrm{st}}$). The threshold value of the optical pumping power for interband lasing is approximately the same for all stimulated emission lines and is about 370 W/cm^2.

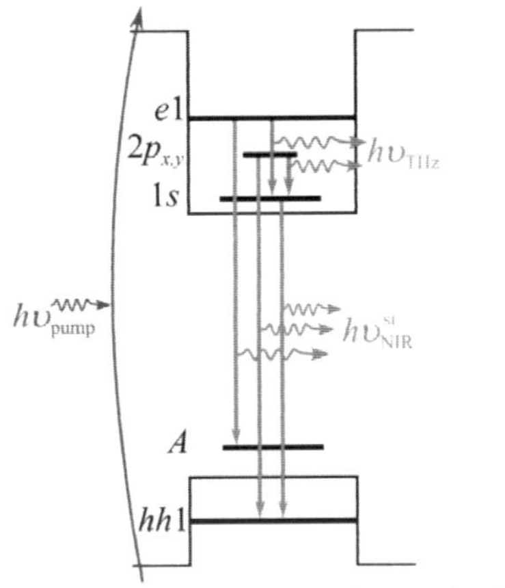

Figure 1. Scheme of optical transitions in QWs in conditions of interband stimulated emission under interband photoexcitation.

The dependence of the integrated terahertz emission intensity on the optical pumping power is presented in figure 2 by dots. At low levels of optical pumping, the intensity of terahertz radiation increases with the optical pumping increase by the square root law. This dependence is in good agreement with results of the experimental studies for terahertz photoluminescence in bulk GaAs doped with donor impurities [7]. We do not expect any difference in such dependence due to 2D-quantization phenomena. At relatively high optical pumping intensities exceeding 2 kW/cm^2, this dependence changes its behavior to linear (the corresponding pumping power is shown with the arrow labeled "terahertz intensity increase" in figure 2). We connect this change in the dependence of the integrated terahertz intensity on the optical pumping power with the beginning of more effective depopulation of the ground donor state $1s$ in QW due to stimulated interband emission related to radiative recombination of electrons from the ground donor state $1s$ and holes from the valence subband $hh1$.

In accordance with the mentioned above mechanism of terahertz radiation in QWs in conditions of interband stimulated emission, we expected to find the increase of terahertz integrated intensity at the optical pumping power corresponding to the interband lasing threshold value (this value is marked with the arrow labeled "interband lasing threshold" in figure 2) for the $1s$-$hh1$ optical transitions. But as it was mentioned earlier, the stimulated emission lines associated with the $2p_{x,y}$-$hh1$ and $e1$-A optical transitions were also revealed in the near-infrared emission spectra, simultaneously with the

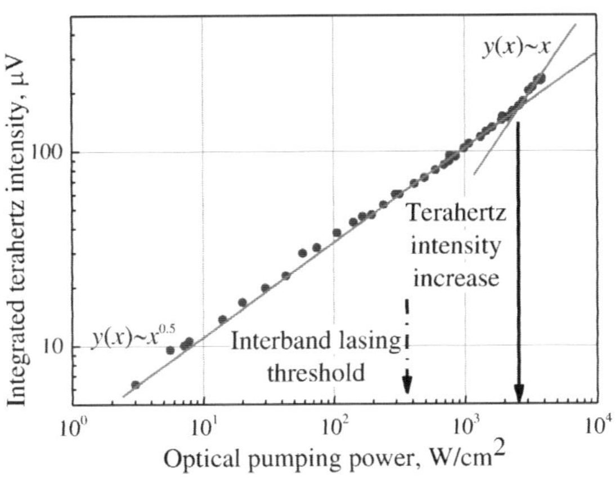

Figure 2. Dependence of the impurity-assisted integrated terahertz intensity on the optical pumping power in a doped QW laser structure.

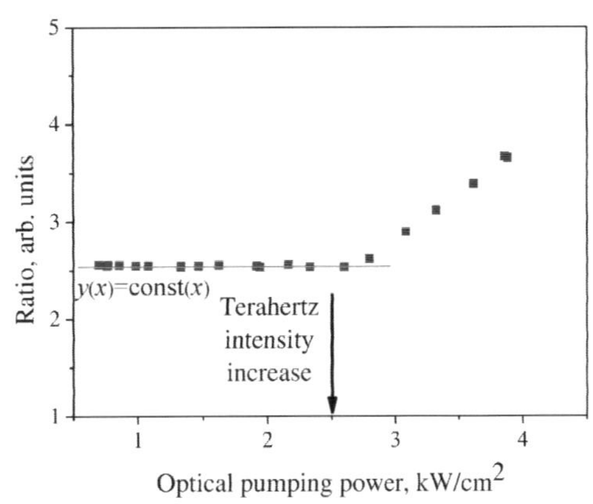

Figure 3. Dependence of the ratio of the integrated intensities of the $1s$-$hh1$ to the $2p_{x,y}$-$hh1$ stimulated emission lines on the optical pumping power.

$1s$-$hh1$ one. This fact means that all of our states $e1$, $1s$ and $2p_{x,y}$ are well enough depopulated due to interband stimulated transitions, and there are no reasons to observe the expected increase of the terahertz integrated intensity. However, the intensity of the $1s$-$hh1$ and $2p_{x,y}$-$hh1$ stimulated emission lines is much higher than the intensity of the $e1$-A stimulated emission line within the all studied pumping power range, and their comparative intensities depend on the pumping power.

The dependence of the ratio of the integrated intensities of the $1s$-$hh1$ to the $2p_{x,y}$-$hh1$ stimulated emission lines on the optical pumping power is presented in figure 3. From this dependence, one can see that the ratio between the integrated intensities of the $1s$-$hh1$ and $2p_{x,y}$-$hh1$ stimulated optical transitions is constant up to the optical pumping power of about 2.5 kW/cm^2 and at higher values of the optical pumping power this ratio begins to increase linearly. This ratio increase is connected with the fact that the integrated intensity of the $2p_{x,y}$-$hh1$ stimulated optical transitions begins to saturate, while it continues to increase for the $1s$-$hh1$ transitions. We recall that at the optical pumping power just of about 2.5 kW/cm^2, the terahertz intensity starts to increase more quickly (this pumping power value is shown with the arrows labeled "terahertz intensity increase" in both figures 2 and 3). Thus, the more intensive increase of terahertz intensity observed at the optical pumping power exceeding 2.5 kW/cm^2 (see figure 3) can be connected with more effective depopulation of the ground donor states in QWs under interband lasing conditions.

Conclusion

Luminescence of terahertz and near-infrared spectral ranges in GaAs/AlGaAs QWs doped with donor impurities under interband photoexcitation is studied under conditions of interband stimulated emission. Terahertz photoluminescence from doped QWs is connected with the optical transitions of nonequilibrium electrons with the participation of donor states in QWs. The increase of the impurity-assisted terahertz luminescence intensity in doped QWs due to fast depopulation of impurity states under conditions of impurity-assisted interband lasing is shown.

Acknowledgments

This work was supported by the Russian Foundation for Basic Research (grant 16-32-60085) and the Ministry of Education and Science of the Russian Federation (state assignments, projects 3.933.2017/4.6 and 3.5518.2017/7.8).

References

[1] Duque C A, Tiutiunnyk A, Morales Á L, Akimov A., Restrepo R L, Fomina O and Tulupenko V 2016 *Momento* **51E** 103

[2] Firsov D A, Vorobjev L E, Panevin V Yu, Sofronov A N, Balagula R M, Makhov I S, Kozlov D V and Vasil'ev A P 2015 *Semiconductors* **49**(1) 28

[3] Vinnichenko M Ya, Makhov I S, Panevin V Yu, Sofronov A N, Firsov D A, Vorobjev L E, Sadofev Yu G and Vasil'ev A P 2016 *41st International Conference on Infrared, Millimeter, and Terahertz waves* 7758782

[4] Vorob'ev L E, Firsov D A, Shalygin V A, Tulupenko V N, Shernyakov Yu M, Ledentsov N N, Ustinov V M and Alferov Zh I 1998 *JETP Letters* **67**(4) 275

[5] Fraizzoli S, Bassani F and Buczko R 1990 *Physical Review* B **41**(8) 5096

[6] Dingle R, Wiesbuch C, Störmer H L, Morkoç H and Cho A Y 1982 *Applied Physics Letters* **40** 507

[7] Andrianov A V, Zakhar'in A O, Ivanov Yu L and Kipa M S 2010 *JETP Letters* **91**(2) 96

The development of technology for growing InAs/GaSb superlattices by MOCVD

I V Fedorov[1,2], R V Levin[2,3] and V N Nevedomsky[2]

[1] Saint Petersburg National Research University of Information Technologies, Mechanics and Optics, Kronverkskiy pr., 49, lit. A, St. Petersburg 197101, Russia
[2] Ioffe Institute, 26 Polytekhnicheskaya, St. Petersburg 194021, Russia
[3] Submicron Heterostructures for Microelectronics Research and Engineering Center of the RAS, 26 Polytekhnicheskaya, St. Petersburg 194021, Russia

E-mail: kingwash@yandex.ru , lev13@yandex.ru

Abstract. This study is dedicated to developing the technology for growing InAs/GaSb superlattices (SLs) by MOCVD. The structures were studied by transmission electron microscopy (TEM) and photoluminescence (PL) methods. We concluded that hetero-interface sharpness is not affected by the pause time between growth stages for separate layers or by switching the layer direction. A possible interpretation for the spectra of SLs was suggested.

1. Introduction

Devices operating in the near-infrared (NIR) and mid-wavelength infrared (MWIR) ranges can be used in various fields, from detection of harmful substances in the atmosphere to military applications. Development and production of such devices is a promising task.

One of the materials that can be used for the creation of photovoltaic converters and light emitting devices for NIR and MWIR is InAs/GaSb SLs. The use of InAs/GaSb SLs in devices makes it possible to reduce tunneling currents and the Auger recombination rate, as well as to increase the operating temperature of devices [1-2]. Molecular beam epitaxy (MBE) is the main technique to grow InAs/GaSb SLs, which is why the majority of published studies are concerned with InAs/GaSb SLs grown by MBE. The paper [3] also deals with the incapacity to grow InAs/GaSb SLs by metal organic chemical vapor deposition (MOCVD). Attempts to create such superlattices were reported in a few articles [4-6]. Follow-ups to these studies have not been published. The aim of this study is to develop the technology for growing InAs/GaSb SLs by MOCVD. The following tasks are an inherent part of this goal: to find the best conditions for growing InAs/GaSb superlattices, and to study the structural and optical-energy characteristics of the grown structures.

2. Experimental details

All the experiments were carried out with an AIXTRON AIX-200 reactor. The structures were grown on n-GaSb (001) substrates at a temperature of 500°C. The pressure in the reactor was 76 Torr. The substrate was rotated at a rate of 100 revolutions per minute. The carrier gas was purified hydrogen with the dew point not worse than −100°C. The total flow through the reactor was 5.5 liters per minute. The sources of elements for growth were the following compounds: TMIn, TEGa, TMSb, and AsH_3. The grown structures were studied by TEM and PL.

IOP Conf. Series: Journal of Physics: Conf. Series **993** (2018) 012018 doi:10.1088/1742-6596/993/1/012018

3. Results and discussion

The influence of the pause time between superlattice growth stages on the sharpness of heterointerfaces was studied. (Pause time is the time when all flows of TMIn, TEGa, TMSb, and AsH_3 are interrupted, except for the flow of H_2.) For this purpose, a superlattice consisting of 6 alternating layers of InAs and GaSb was grown (figure 1). The growth of each layer of GaSb started with the TEGa supply, and the growth of each layer of InAs started with the AsH_3 supply. The flows of TMIn, TEGa, TMSb and AsH_3 were interrupted after each grown layer of InAs or GaSb, but the flow of H_2 was continued for a time period (t). This time was defined as the following ratio: $t = V_p / G$, where V_p is the volume of the reactor, cm^3, and G is the flow rate of hydrogen, cm^3/sec.

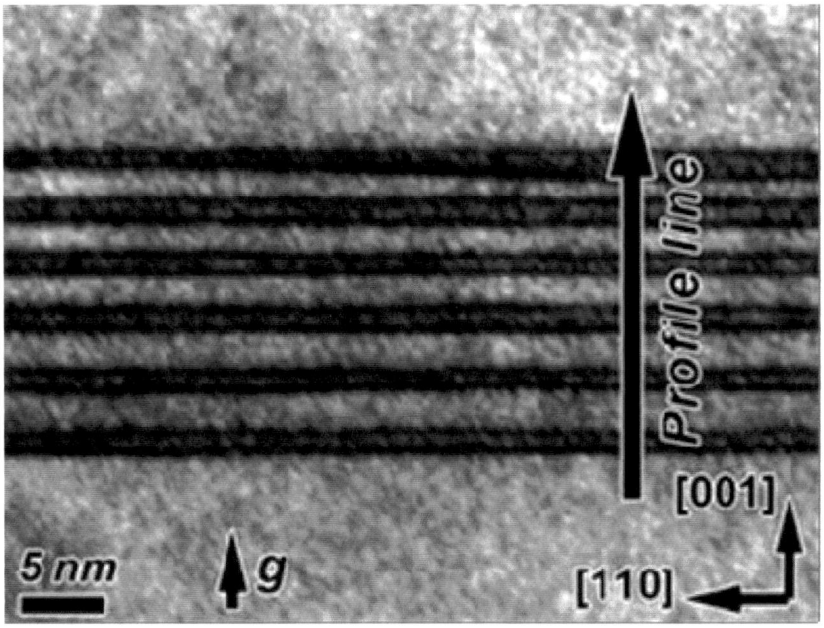

Figure 1. TEM image of a superlattice consisting of 6 alternating layers of InAs and GaSb.

The pause time between growing 2 pairs of the layers of InAs and GaSb, when the flowing of the compounds of TEGa and Arsine was stopped, was 1 minute. The exposure time after which the layers were changed was: 10 seconds for the first 2 pairs of InAs and GaSb layers, 30 seconds for the second 2 pairs, and 50 seconds for the third 2 pairs.

The results of this research suggest that all the borders of InAs/GaSb were of equal quality, regardless of the exposure time or the direction of switching (GaSb→ InAs or InAs→GaSb).

The InAs/GaSb superlattices of 5 and 10 pairs of interlacing layers of InAs and GaSb grown at a temperature of 500°C were created after finding the optimal conditions for creating them. The microstructure of the samples was studied by TEM. The obtained layers of InAs and GaSb have clearly defined borders and thicknesses on the electron micrograph (figure 2). The growth rate for InAs was 0.2 nm per sec. The growth rate for GaSb was 0.03 nm per sec. The thickness of 5-pair layers of the superlattice varied in the range of 1.4 nm to 2.0 nm for InAs and of 2.9 nm to 3.1 nm for GaSb. The thickness of 10-pair layers of the superlattice varied in the range of 1.4 nm to 1.9 nm for InAs and of 2.5 nm to 3.0 nm for GaSb.

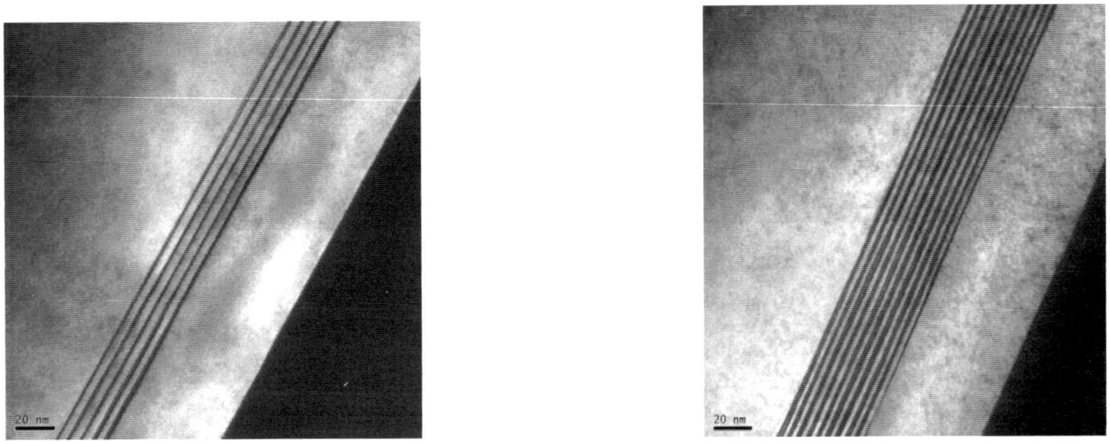

(a) (b)

Figure 2. TEM images of: (a) a superlattice consisting of 5 pairs of alternating layers of InAs and GaSb; (b) a superlattice consisting of 10 pairs of alternating layers of InAs and GaSb.

Photoluminescence measurements for the superlattice of 5 alternating layer pairs at a temperature $T = 77$K gave the following results: the superlattice showed a PL intensity maximum at 0.801 eV (this peak is related to the interband transition in GaSb layers) and a maximum at 0.725 eV (this peak is related to the GaSb substrate). The radiation of the superlattice itself was not observed, which is due to the insufficient thickness of 5 pairs of alternating InAs and GaSb layers.

The PL spectra of the superlattice of 10 alternating layer pairs of InAs and GaSb was measured at a temperature T = 77K in the range of 0.45 to 0.9 eV. They had the same peaks as those of the superlattice of 5 alternating layer pairs. At the same time, the PL spectra differ in the spectral range of 0.27 to 0.55 eV. Figure 3 shows the PL spectra of the superlattice of 10 alternating layer pairs. The peak with a maximum at 0.315 eV and a half-width of 78 meV is most probably related to the energetic transit in InAs/GaSb superlattice layers.

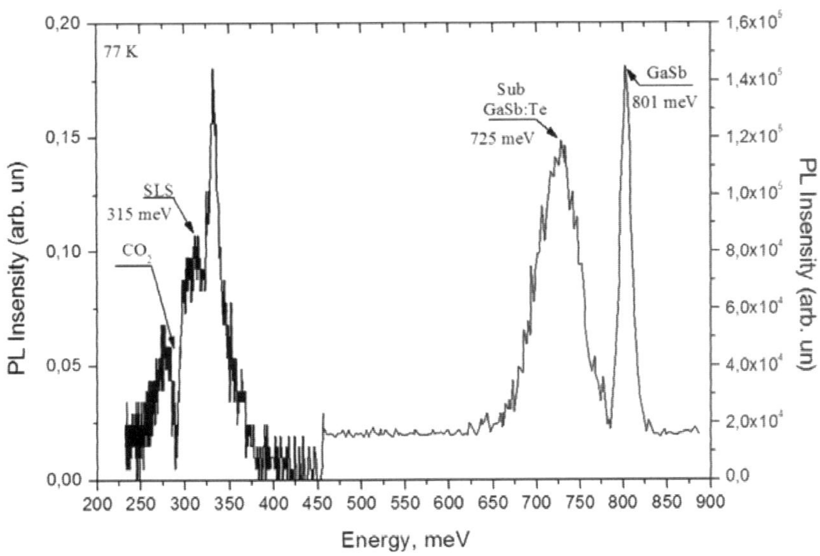

Figure 3. The PL spectra of the superlattice of 10 alternating InAs and GaSb layer pairs.

The result of this research is a development of the technology for the creation of InAs/GaSb superlattices by MOCVD at a temperature of 500°C. The following superlattices were created: the superlattice of 5 alternating InAs and GaSb layer pairs, and the superlattice of 10 alternating InAs and GaSb layer pairs. The resulting structures were examined by PL and TEM. The obtained superlattice had an InAs thickness of 1.4 nm to 2.0 nm and a GaSb thickness of 2.5 nm to 3.1 nm. The peak of the radiation related to the superlattice in the PL spectra at a temperature of 77 K was in the spectral range of 2.4 μm to 3.4 μm.

Acknowledgments

The authors thank L. V. Danilov, B. A. Matveev and M. A. Remennoii for their assistance in the research. The work was performed with partial support from the RFBR grant No. 16-08-01130-a.

References

[1] Grein C H, Young P M and Ehrenreich H 1992 *Appl. Phys. Lett.* **61**, 2905
[2] Smith D L and Mailhiot C 1987 *J. Appl. Phys.* **62** 2545
[3] Razeghi M and Nguyen B M 2014 *Rep. Prog. Phys.* **77**, 082401
[4] Booker G R, Klipstein P C, Lakrimi M, Lyapin, Mason N J, Nicholas R J, Seong T Y, Symons D M, Vaughan T A and Walker P J 1994 *Journal of Crystal Growth* **145** 778-785
[5] Zhang X B, Ryou J H and Dupuis R D 2006 *Appl. Phys. Lett.* **88**, 072104
[6] Huang Y, Ryou J H, Dupuis R D, Zuo D, Kesler B, Chuang S L, Hu H, Kim K H, Lu Y T, Hsieh K C and Zuo J M 2011 *Appl. Phys. Lett.* **99**, 011109

IOP Conf. Series: Journal of Physics: Conf. Series **993** (2018) 012019 doi:10.1088/1742-6596/993/1/012019

Particle binding over a nanofiber

I D Toftul[1,2,*], D F Kornovan[2,#] and M I Petrov[1,2,†]

[1]Department of Condensed Matter Physics, St Petersburg Academic University of the Russian Academy of Sciences, St. Petersburg 194021, Russia

[2]Department of Nanophotonics and Metamaterials, ITMO University, Saint Petersburg 199034, Russia

E-mail: *toftul.ivan@gmail.com, #d.kornovan@metalab.ifmo.ru, †m.petrov@metalab.ifmo.ru

Abstract. We propose a configuration for optical binding of nanoparticles over a thin fiber under a plane wave illumination. We have shown that the interference of the field scattered into a guided mode of a nanofiber results in the formation of optomechanically stable configurations. Our estimations of the binding force and the trapping parameter for dielectric particles show that it is potentially possible to observe particle binding over a nanofiber experimentally.

1. Introduction

Optical manipulation has been extensively studied during the last few decades [1]. Following the realization of optical trapping, a new phenomenon, termed *optical binding*, was observed when two dielectric particles were trapped and bound transversely in a large beam waist [2]. This was due to the modification of the field caused by the presence of the particles and governed by their relative positions.

Using an evanescent wave of nanofibers for an optical trapping is already a well known process [3, 4, 5, 6]. All the above-mentioned dealt with the use of external laser beams to excite the waveguide mode. In contrast to such an approach, we advance a consideration of a normal incidence of a laser beam (see figure 1). This configuration has not yet been studied.

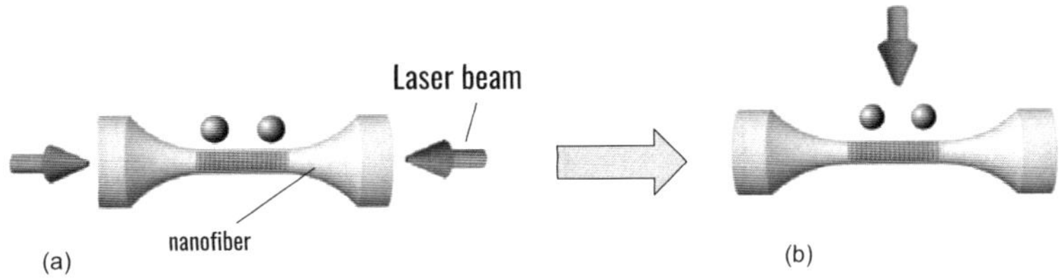

Figure 1. Configuration for stable optical binding: (a) studied before; (b) studied in this paper.

This is also an alternative configuration of optical binding (contrary to surface binding [2, 7, 8]). We suggest placing two dielectric nanoparticles close to a thin optical fiber and applying laser illumination perpendicular to the nanofiber axis (see figure 2). The electric field induces dipole moments in the particles. The particles start to radiate into free-space and the fiber. The field is scattered into the waveguide mode from the interference pattern, and nanoparticles bind into stable long-range states with forces of several pN.

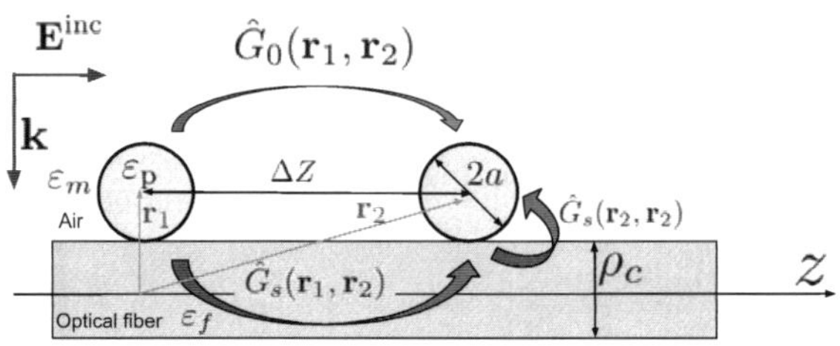

Figure 2. Two nanoparticles under plane wave illumination.

2. Theoretical description

2.1. Binding force

There are many different ways to describe such electromagnetic systems. One approach is to use the scattering matrix approach [3], and another option is to use the finite element method (FEM), which requires a lot of computation power. In this paper, we used a comprehensive method based on the Green's function.

The formulation of the problem is the following: plane wave illumination is scattered on nanoparticles, after which the scattered field propagates through a nanofiber for a longer distance without significant losses compared with surface plasmon binding [7].

We used the following approximations:

(1) Only the HE_{11} mode is considered as being propagated in a fiber. In other words, we assume that the single-mode condition is fulfilled [9]:

$$V = k\rho_c\sqrt{\varepsilon_f - \varepsilon_m} < V_c \approx 2.405, \tag{1}$$

where ρ_c is the fiber radius, $k = \omega/c$ is the free-space wavenumber of the incident wave, ε_f and ε_m are the permittivities of the fiber and the background media.

(2) The dipole approximation. We assume that the external field induces only the electric dipole moment in particles.

(3) Only the z dipole component is induced. In other words, we assume that $\mathbf{p} \parallel \mathbf{E}^{inc}$ and $|\mathbf{p}| \approx p_z$, where $\mathbf{E}^{inc}$ is the electric field of the incident wave.

(4) The effective polarizability tensor is a slow function of z, so $\partial_z \hat{\alpha}_{\text{eff}} \approx 0$.

(5) 1D motion along the z-axis. We assume that the distance from the particle centres to the fiber z-axis is constant and the polar angle φ is equal and fixed for both particles.

The main goal is to calculate the force between two nanoparticles placed over the fiber. As soon as we use the dipole approximation, the force acting on the particle in $\mathbf{r}_2$ may be written as [10]:

$$\mathbf{F}_{12} = \frac{1}{2} \sum_{i=x,y,z} \text{Re}\left\{p_{2,i}^* \nabla E_i(\mathbf{r})\right\}_{\mathbf{r}=\mathbf{r}_2}, \tag{2}$$

where $\mathbf{p}$ is the dipole moment, $\nabla = \partial_{\mathbf{r}}$ is the nabla operator and $\mathbf{E}$ is the total electric field. The problem then is formulated as a self–consistent problem of finding $\mathbf{E}$ and $\mathbf{p}$.

To find the electric field in the system, we apply the Green's function formalism. The total field in the system at the point $\mathbf{r}_2$ is the sum of the incident and scattered fields

$$\mathbf{E}(\mathbf{r}_2) = \mathbf{E}^{\text{inc}}(\mathbf{r}_2) + \mathbf{E}^{\text{sc}}(\mathbf{r}_2), \tag{3}$$

IOP Conf. Series: Journal of Physics: Conf. Series **993** (2018) 012019 doi:10.1088/1742-6596/993/1/012019

where the scattered field in $\mathbf{r}_2$ is the sum of the scattered fields from the dipole located in $\mathbf{r}_1$ and the effective dipole induced by the nanofiber in $\mathbf{r}_2$. In SI units, it may be written as [10]

$$\mathbf{E}^{sc}(\mathbf{r}_2) = \frac{k^2}{\varepsilon_0}\left(\hat{\mathbf{G}}_0(\mathbf{r}_2,\mathbf{r}_1) + \hat{\mathbf{G}}_s(\mathbf{r}_2,\mathbf{r}_1)\right)\mathbf{p}_1 + \frac{k^2}{\varepsilon_0}\hat{\mathbf{G}}_s(\mathbf{r}_2,\mathbf{r}_2)\mathbf{p}_2. \tag{4}$$

where ε_0 is the vacuum permittivity, $\hat{\mathbf{G}}_0$ is the dyadic vacuum Green's function and $\hat{\mathbf{G}}_s$ is the dyadic fiber Green's function.

The explicit forms of the Green's functions are the following. The vacuum part is:

$$\hat{\mathbf{G}}_{0,ij}(\mathbf{r},\mathbf{r}') = \frac{e^{ik_1R}}{4\pi R}\left[\left(1+\frac{ik_1R-1}{k_1^2R^2}\right)\delta_{ij} + \frac{3-3ik_1R-k_1^2R^2}{k_1^2R^2}\frac{R_iR_j}{R^2}\right], \qquad \mathbf{R} = \mathbf{r}-\mathbf{r}'. \tag{5}$$

The fiber Green's function [9, 11] is:

$$G^{11}_{s,zz}(\mathbf{r},\mathbf{r}') = -\frac{1}{2}P^{11}_{11}(\beta)\frac{k^2_{\rho_1}}{k_1^2}H_1^{(1)}(k_{\rho_1}\rho)H_1^{(1)}(k_{\rho_2}\rho)\cdot e^{i\beta|z_1-z_2|}, \tag{6}$$

where $k_{\rho_1} = \sqrt{\varepsilon_m k^2-\beta^2}$, $k_{\rho_2} = \sqrt{\varepsilon_f k^2-\beta^2}$, $\beta=\beta(\omega)$ is the fiber propagation constant, which comes from the solution of the transcendental equation; $H_n^{(1)}(z)$ is the Hankel function of the first kind. Other coefficients are given in Appendix A.

In order to find the dipole moments, we need to notice two things:

(1) An ordinary polarizability connects a dipole moment and an external field as

$$\mathbf{p} = \alpha_0\mathbf{E}, \tag{7}$$

where $\alpha_0 = \alpha_0(a,\omega,\varepsilon_p,\varepsilon_m)$ is the polarizability, which has an exact expression given by the Mie theory [12]. The explicit expression is given in Appendix B.

(2) The detailed consideration of the scattered field in form (4) allows us to notice that $\mathbf{E}^{sc}$ depends on the dipole moments:

$$\mathbf{E}^{\rm sc}(\mathbf{r}) = \mathbf{E}^{\rm sc}(\mathbf{r},\mathbf{p}_1,\mathbf{p}_2). \tag{8}$$

This means that we need to solve the following system of equations:

$$\begin{cases}\mathbf{p}_1 = \alpha_0\left(\mathbf{E}^{\rm inc}(\mathbf{r}_1) + \mathbf{E}^{\rm sc}(\mathbf{r}_1,\mathbf{p}_1,\mathbf{p}_2)\right),\\ \mathbf{p}_2 = \alpha_0\left(\mathbf{E}^{\rm inc}(\mathbf{r}_2) + \mathbf{E}^{\rm sc}(\mathbf{r}_2,\mathbf{p}_1,\mathbf{p}_2)\right).\end{cases} \tag{9}$$

which gives the solution of the self-consistent problem. In terms of the effective polarizability, we have:

$$\hat{\alpha}_i^{\rm eff} = \alpha_0\left[\hat{\mathbf{I}} - \alpha_0\frac{k^2}{\varepsilon_0}\hat{\mathbf{G}}^s_{jj} - \alpha_0\left(\frac{k^2}{\varepsilon_0}\right)^2\hat{\mathbf{G}}_{ij}\hat{\alpha}^s_j\hat{\mathbf{G}}_{ji}\right]^{-1}\left(\hat{\mathbf{I}} + \frac{k^2}{\varepsilon_0}\hat{\mathbf{G}}_{ij}\hat{\alpha}^s_j\right),$$
$$i = 1,2 \quad j = \begin{cases}2, \text{ if } i=1,\\ 1, \text{ if } i=2,\end{cases} \tag{10}$$

where we used the notation $\hat{\mathbf{G}} = \hat{\mathbf{G}}^s + \hat{\mathbf{G}}^0$, $\hat{\mathbf{G}}_{ij} \overset{\rm def}{=} \hat{\mathbf{G}}(\mathbf{r}_i,\mathbf{r}_j)$ and $\hat{\alpha}^s_i = \alpha_0\left[1-\alpha_0\frac{k^2}{\varepsilon_0}\hat{\mathbf{G}}^s_{ii}\right]^{-1}$. We have introduced an effective polarizability such that the induced dipole moment may be found as the product of polarizability and the vector magnitude of the incident wave:

$$\mathbf{p}_i = \hat{\alpha}_i^{\rm eff}\mathbf{E}_0, \qquad i=1,2. \tag{11}$$

Considering $|\mathbf{p}| \approx p_z$ and $\partial_z \hat{\alpha}_{\text{eff}} \approx 0$, we can finally write the expression for the binding force

$$F_{12,z} = \frac{1}{2}\text{Tr}\left(\hat{\alpha}^*_{\text{eff}}\hat{\alpha}_{\text{eff}}\right)|\mathbf{E}_0|^2 \frac{k_1^2}{\varepsilon_0}\text{Re}\left\{\partial_z\left[G^s_{zz}(\mathbf{r},\mathbf{r}_2) + G^0_{zz}(\mathbf{r},\mathbf{r}_2)\right]_{\mathbf{r}=\mathbf{r}_1}\right\}, \tag{12}$$

which was used for the numeric simulation afterwards.

2.2. Trapping parameter

In order to understand how stable each equilibrium point is, we need to define a dimensionless parameter. The main source of instability is the thermal motion, so we define the trapping parameter as the ratio of the potential energy of the trap U_{tr} to the energy of thermal motion

$$\gamma_{\text{tr}} \stackrel{\text{def}}{=} \frac{\text{potential energy of the trap}}{\text{energy of thermal motion}} = \frac{U_{\text{tr}}}{kT}. \tag{13}$$

The question naturally arises whether we can reason in terms of potential energy. Indeed, the binding force has two components: a gradient part and a radiation (scattering) part. The contribution of each part is defined by the real and imaginary parts of polarizability $\text{Re}\{\hat{\alpha}_{\text{eff}}\}$ and $\text{Im}\{\hat{\alpha}_{\text{eff}}\}$. Numerical computations show that the scattering part of the binding force is negligible (figure 3), so the definition of the trapping parameter according to equation (13) makes sense.

Considering a linear dependence near the equilibrium point, we can introduce an effective stiffness $k_{\text{eff}} = -\frac{\partial F_z}{\partial z}$, so $U \approx \frac{1}{2}k_{\text{eff}}(\Delta x)^2$. The period of the binding force over z is about the wavelength value, This means that we can put $\Delta x \approx \lambda/2$. Now we have an approximate analytical expression for the trapping parameter:

$$\gamma_{\text{tr}} = -\frac{\lambda^2}{8kT}\left.\frac{\partial F_z}{\partial z}\right|_{z=z_{\text{eq}}}, \tag{14}$$

where F_z is given by (12).

This dimensionless parameter is very convenient because we can distinguish two cases:

(1) $\gamma_{\text{tr}} \ll 1$ — unstable configuration.

(2) $\gamma_{\text{tr}} \gg 1$ — stable configuration.

Within this context, stability (or unstability) is understood in terms of thermal perturbation.

3. Results

In order to make some numerical estimations, we have taken the typical parameters of nanoparticles and nanofibres which potentially may be manufactured [13].

All the taken parameters are presented in table 1. The longitudinal component of the force F_z for such parameters has been calculated using (2) (see figure 3). The maximum amplitude is about several pN. As expected, the interaction does not tend to zero with a large distance between particles Δz as a consequence of a long-range interaction through the guided mode.

Let us estimate the trapping parameter γ_{tr}. At the second equilibrium point:

$$\gamma^{\text{near}}_{\text{tr}} = \frac{U_{\text{tr}}}{kT} \approx 40, \tag{15}$$

where $U_{\text{tr}} \approx \bar{F}_{\text{max}} \cdot \Delta z \approx 1$ pN $\cdot$ 200 nm and $T = 300$ K. The obtained result is very promising from the experimental point of view. At distant equilibrium points, the trapping parameter appears to be:

$$\gamma^{\text{far}}_{\text{tr}} \approx 5. \tag{16}$$

Table 1. Parameters of the model

Parameter	laser power	focus radius	appropriate intensity
Notation	P	R_{focus}	I
Value	100 mW	1 μm	3.2×10^{10} W/m^2

Parameter	particle radius	fiber radius	medium permittivity	fiber permittivity	particle permittivity	wave length
Notation	a	ρ_c	ε_m	ε_f	ε_p	λ
Value	200 nm	100 nm	1.0 (Air)	2.09 (SiO_2)	2.5 (Polystyrene)	550 nm

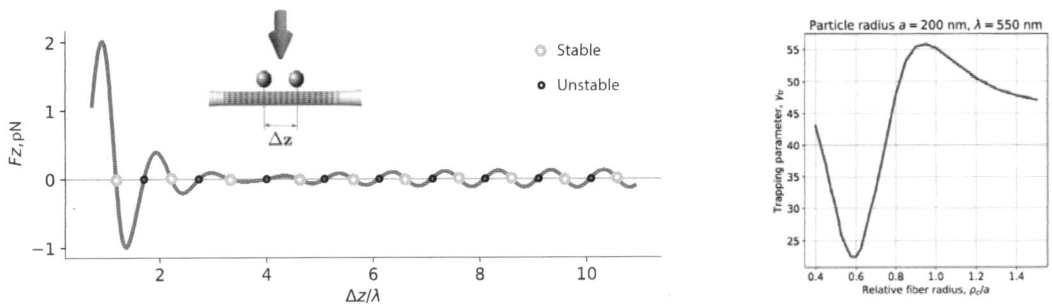

(a) (b)

Figure 3. (a) The longitudinal force for a longitudinal polarization. (b) The trapping parameter dependence on the fiber radius calculated at the second equilibrium position. The particle radius is $a = 200$ nm and the incident wavelength is $\lambda = 550$ nm. The laser power is set to 100 mW with a beam waist $w_0 = 2$ μm.

The main benefit of such system is a long–range interaction due to low losses in a nanofiber, which is now clearly visible.

The trapping parameter depends on many arguments. In order to analyze its dependence on the ratio of the fiber radius and the nanoparticle radius we have fixed the distance between nanoparticles at the second equilibrium point and varied the nanofiber radius (figure 3). The local maximum has been found at the point $\rho_c/a \approx 0.9$. This result helps to match a proper nanofiber radius to observe better binding. It should be noted that one cannot trust the results for a very thin nanofiber, as the approximations of the model will no longer be correct.

4. Conclusion

We have proposed a configuration for optical binding of nanoparticles over a thin fiber under plane wave illumination. We have shown that the interaction of nanoparticles through a single guiding mode of a nanofiber results in optical binding. Our estimations of the binding force and the trapping parameter for dielectric particles show that it is potentially possible to observe particle binding over a nanofiber experimentally.

Consideration of nanofiber modes of a higher order may increase the binding effect and allow consideration of different frequency ranges, which is a subject of future studies.

Acknowledgments
This work was supported by the Russian Foundation of Basic Research 16-32-60167, 16-37-60064. M.I.P. also acknowledges the support from the BASIS foundation.

Appendix A. The fiber Green's function

Considering the Green's function of an infinite optical fiber [9, 11], we need to split off only the waveguide part due to its significant contribution to the binding force:

$$G^{11}_{s,zz}(\mathbf{r},\mathbf{r}') = -\frac{1}{2}P^{11}_{11}(\beta)\frac{k^2_{\rho_1}}{k^2_1}H^{(1)}_1(k_{\rho_1}\rho)H^{(1)}_1(k_{\rho_2}\rho)\cdot e^{i\beta|z_1-z_2|}, \tag{A.1}$$

where

$$P^{11}_{11}(\beta) = \frac{J_1(k_{\rho_1}\rho_c)}{F(\beta)H^{(1)}_1(k_{\rho_1}\rho_c)}\Bigg[\left(\frac{1}{k^2_{\rho_2}}-\frac{1}{k^2_{\rho_1}}\right)\beta^2 - \\ -\left(\frac{(J_1(k_{\rho_2}\rho_c))'}{k_{\rho_2}J_1(k_{\rho_2}\rho_c)}-\frac{(H^{(1)}_1(k_{\rho_1}\rho_c))'}{k_{\rho_1}H^{(1)}_1(k_{\rho_1}\rho_c)}\right)\times\left(\frac{(J_1(k_{\rho_2}\rho_c))'\,k^2_2}{k_{\rho_2}J_1(k_{\rho_2}\rho_c)}-\frac{(J_1(k_{\rho_1}\rho_c))'\,k^2_1}{k_{\rho_1}J_1(k_{\rho_1}\rho_c)}\right)\rho^2_c\Bigg], \tag{A.2}$$

$$F(\beta) = -2AA'\beta^2 - 2A^2\beta + \left[(B'-C')\left(k^2_2B-k^2_1C\right)+(B-C)\left(k^2_2B'-k^2_1C'\right)\right]\rho^2_c, \tag{A.3}$$

$$A = \frac{1}{k^2_{\rho_2}}-\frac{1}{k^2_{\rho_1}}, \qquad B = \frac{(J_1(k_{\rho_2}\rho_c))'}{k_{\rho_2}J_1(k_{\rho_2}\rho_c)}, \qquad C = \frac{(H^{(1)}_1(k_{\rho_1}\rho_c))'}{k_{\rho_1}H^{(1)}_1(k_{\rho_1}\rho_c)}, \tag{A.4}$$

$$A' = 2\beta\left(\frac{1}{k^4_{\rho_2}}-\frac{1}{k^4_{\rho_1}}\right), \tag{A.5}$$

$$B' = -\frac{(J_1(k_{\rho_2}\rho_c))''\beta\rho_c}{k^2_{\rho_2}J_1(k_{\rho_2}\rho_c)}+\frac{(J_1(k_{\rho_2}\rho_c))'\beta\left[J_1(k_{\rho_2}\rho_c)/k_{\rho_2}+\rho_c(J_1(k_{\rho_2}\rho_c))'\right]}{(k_{\rho_2}J_1(k_{\rho_2}\rho_c))^2}, \tag{A.6}$$

$$C' = -\frac{(H^{(1)}_1(k_{\rho_1}\rho_c))''\beta\rho_c}{k^2_{\rho_1}H^{(1)}_1(k_{\rho_1}\rho_c)}+\frac{(H^{(1)}_1(k_{\rho_1}\rho_c))'\beta\left[H^{(1)}_1(k_{\rho_1}\rho_c)/k_{\rho_1}+\rho_c(H^{(1)}_1(k_{\rho_1}\rho_c))'\right]}{(k_{\rho_1}H^{(1)}_1(k_{\rho_1}\rho_c))^2}, \tag{A.7}$$

where $k_{\rho_1} = \sqrt{\varepsilon_m k^2-\beta^2}$, $k_{\rho_2} = \sqrt{\varepsilon_f k^2-\beta^2}$, $\beta=\beta(\omega)$ is the fiber propagation constant, which comes from the solution of the transcendental equation; $J_n(z)$ is the Bessel function of the first kind, and $H^{(1)}_n(z)$ is the Hankel function of the first kind, $(J_n(z))' \stackrel{\text{def}}{=} \frac{\partial J_n(z)}{\partial z}$.

Appendix B. Polarizability of the sphere (Mie theory results)

The exact expression for the electric polarizability of the sphere can be found using the Mie scattering theory [12]:

$$\alpha_0 = i\frac{3a^3}{2x^3}a_1(x,m), \tag{B.1}$$

where $x=\sqrt{\varepsilon_m}ka$, $m=\sqrt{\varepsilon_p}/\sqrt{\varepsilon_m}$ and

$$a_1(x,m) = \frac{m\psi_1(mx)\,(\psi_1(x))' - \psi_1(x)\,(\psi_1(mx))'}{m\psi_1(mx)\,(\xi_1(x))' - \xi_1(x)\,(\psi_1(mx))'}, \tag{B.2}$$

where

$$\psi_1(z) = zj_1(z), \qquad \xi(z) = zh^{(1)}_1(z), \qquad h^{(1)}_1(z) = j_1(z)+iy_1(z), \tag{B.3}$$

where j_1 and y_1 are the spherical Bessel functions of the first and second kind correspondingly.

IOP Conf. Series: Journal of Physics: Conf. Series **993** (2018) 012019 doi:10.1088/1742-6596/993/1/012019

References

[1] Gao D, et al 2017 Optical manipulation from the microscale to the nanoscale: fundamentals, advances and prospects *Light: Science and Applications* **6** 9

[2] Burns M M, Fournier J M and Golovchenko J A 1989 Optical binding *Phys. Rev. Lett.* **63** 12

[3] Maimaiti Aili, et al 2016 Nonlinear force dependence on optically bound micro-particle arrays in the evanescent fields of fundamental and higher order microfibre modes *Scientific reports* **6** 30131

[4] Gusachenko I, Frawley M C, Truong V G and Chormaic S N 2014 Optical nanofiber integrated into an optical tweezers for particle manipulation and in-situ fiber probing *arXiv preprint arXiv:1401.1550*

[5] Frawley M C, Gusachenko I, Truong V G, Sergides M and Chormaic S N 2014 Selective particle trapping and optical binding in the evanescent field of an optical nanofiber *Optics express* **22** 13 16322–16334

[6] Daly M, Truong V G and Chormaic S N 2016 Evanescent field trapping of nanoparticles using nanostructured ultrathin optical fibers *Optics express* **24** 13 14470–14482

[7] Kostina N, Ivinskaya A, Sukhov S, Bogdanov A, Toftul I, Nieto-Vesperinas M, Ginzburg P, Petrov M and Shalin A 2017 Optical binding via surface plasmon polariton interference *arXiv preprint arXiv:1708.05471*

[8] Karásek V, Čižmár T, Brzobohatý O, Zemánek P, Garcés-Chávez V and Dholakia K 2008 Long-range one-dimensional longitudinal optical binding *Phys. Rev. Lett.*, **101** 14

[9] Chew W C 1995 *Waves and fields in inhomogeneous media* vol 522 (New York: IEEE press)

[10] Novotny L and Bert H 2012 *Principles of nano-optics* (Cambridge: Cambridge university press)

[11] Tai CT 1994 *Dyadic Green functions in electromagnetic theory* (New York: IEEE press)

[12] Doyle W T 1989 Optical properties of a suspension of metal spheres *Phys. Rev.* B **39** 14 9852

[13] Wu X and Limin T 2013 Optical microfibers and nanofibers *Nanophotonics* **2** 5-6 407-428

Ring resonator optical modes in InGaN/GaN structures grown on micro-cone-patterned sapphire substrates

D R Kazanov[1], G Pozina[2], V N Jmerik[1] and T V Shubina[1]

[1] Ioffe Institute, 194021 St. Petersburg, Russia

[2] Department of Physics, Chemistry and Biology (IFM), Linköping University, S-581 83 Linköping, Sweden

E-mail: kazanovdr@gmail.com

Abstract. Molecular beam epitaxy (MBE) of III-nitride compounds on specially prepared cone-shaped patterned substrates is being actively developed nowadays, especially for nanophotonic applications. This type of substrates enables the successful growth of hexagonal nanorods (NRs). The insertion of an active quantum-sized region of InGaN inside a GaN NR allows us to enhance the rate of optical transitions by coupling them with resonant optical modes in the NR. However, we have observed the enhancement of emission not only from the NR but also around the circumference region of the cone-shaped base. We have studied this specific feature and demonstrated its impact on the output signal.

1. Introduction

Microresonators are frequently used for amplification of the radiative transitions inside of them. The coupling of optical transitions with resonator modes results in a significant enhancement of emission. This effect is proportional to the Purcell factor F_P which mainly depends on the quality (Q) factor, the transition energy, and the volume of the mode [1]. Therefore, the creation

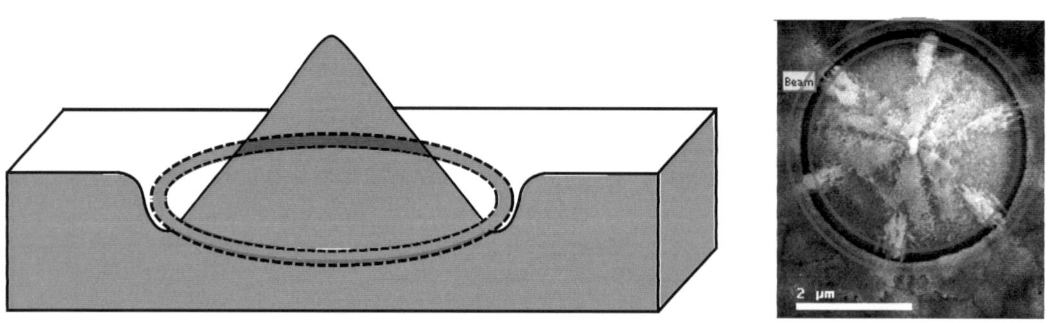

Figure 1. Side view sketch of the cone with a formed ring-like microresonator in the dip of its circumference (left); SEM image illustrating the top view of a cone with the ring-like resonator situated in the darkest region (red lines are shown here for clarity) (right).

of the high quality structures is one of the ways to obtain a larger Q. This can be achieved by using the MBE growth technology. Recently, we have demonstrated that there is an approach to fabricate InN and GaN single-crystal microresonators on patterned cone-shaped substrates [2, 3]. This approach also makes it possible to obtain standardized NR microresonators with a quantum-sized active region inside [4]. However, the special form of a substrate causes the simultaneous appearance of another type of microresonator - the ring one. In this paper, we show how an InGaN/GaN quantum well (QW) fabricated on a cone-shaped substrate couples with a ring microresonator and how it influences the output radiation in this area. We also observed cathodoluminescence of the ring microresonator, which confirms the enhancement of the optical transitions particularly in the circumference of the cones.

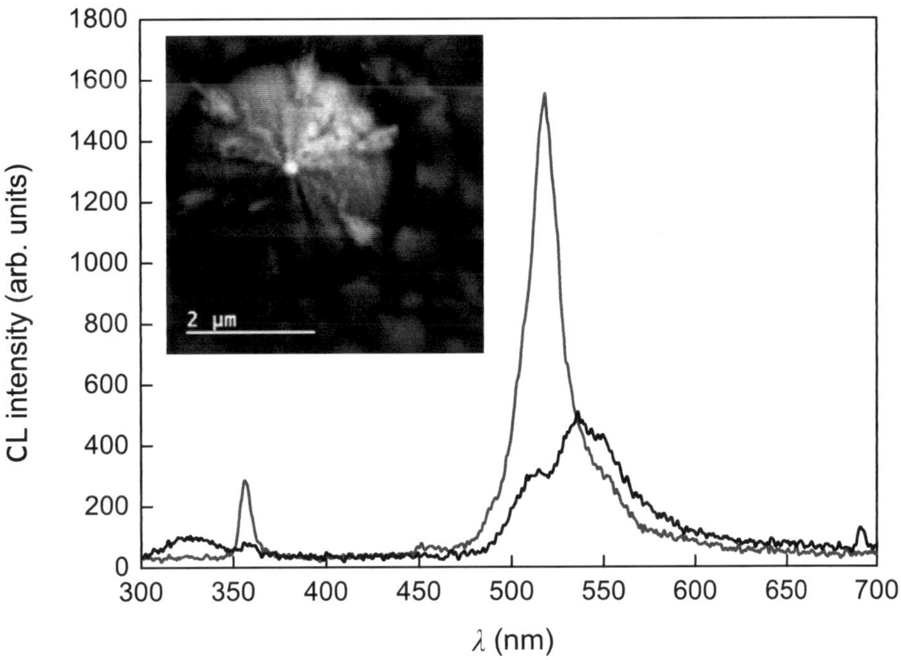

Figure 2. μ-CL from the surface near the circumference of the cone (blue line) and far (black line) from it; the inset shows a SEM image illustrating inhomogeneous surface growth on and around the cone.

2. MBE growth and experimental characterization

Here, we present the formation of ring-like resonators in a single growing process of GaN NRs on a cone-shaped patterned c-sapphire substrate with regularly located cones with a diameter and a height of 3.2 and 1.7 μm, respectively. The structures were grown by using a two-stage process. In the first stage, a nucleation layer of GaN was grown by migration enhanced epitaxy. In the second stage, strongly N-rich conditions were used to stimulate the growth of NRs. Finally, a 3-nm-thick InGaN/GaN single QW was formed by standard plasma-assisted MBE at a reduced growth temperature. The details of the MBE growth details demonstrating the successful formation of the NRs on the cone tops are described in [5]. An interesting feature we observed by using scanning electron microscopy (SEM) and micro-cathodoluminescence (μ-CL) in the circumference region around the cones was a slight dip next to them (figure 1, right part). It is the hole visible in the darkest region in a SEM top view image of a single cone and highlighted here by two red circles for clarity. The existence of such a hole makes it possible for

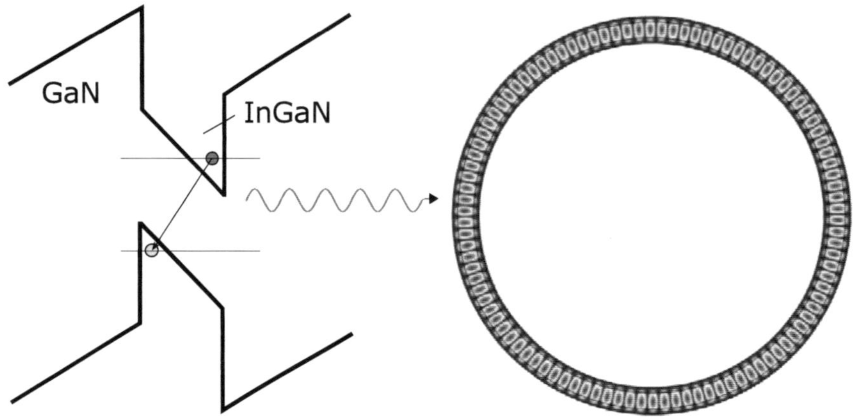

Figure 3. Scheme of the optical transitions in a GaN/InGaN QW which couples with a microresonator(left); Modulus of the electric field distribution of the coupled light inside the ring-like microresonator (right).

the formation of the ring-like microresonators to form. The scheme of a side view of the cone where a ring is formed in a dip is shown in figure 1 (left part).

It is clear that an InGaN QW is located in different parts of the sample: inside GaN NRs, on the circumference of the cones and on the conventional 3D unperturbed layer between the cones. The formation of the microresonator would reveal itself by the enhancement of the output signal in the near-cone area. Therefore, we have performed the μ-CL experiments to validate that. Figure 2 shows two different CL spectra for two different spots: on the circumference (blue line) and far from it (black line). As we see, the CL intensity is much higher for the near-cone region. It is presumably due to the coupling of the QW emission to the mode of a ring-like resonator.

3. Theoretical modelling

In order to understand the μ-CL signal enhancement, we have calculated eigenfrequencies of the formed 3 nm QW in GaN/InGaN for electrons and heavy holes. We have used the effective-mass approximation taking into account a large internal electric field, which is a common thing for piezoelectric materials such as III-nitride with C_{6v} point group symmetry. Moreover, both GaN and InGaN materials are known for their internal spontaneous polarization [6]. The left part of figure 3 shows the scheme of a band profile for electrons and heavy holes (hh). The horizontal lines show the confined energies for both electrons and hh in $In_{0.2}Ga_{0.8}N$ QW of 3 nm width. Radiative optical transitions are reduced due to the spatial separation. However, narrow QWs allow to have long tails of wave functions inside the barriers and, hence, a higher probability of optical transitions. The calculated wavelength of such an optical transition is at $\lambda_{opt} = 525$ nm which is in close agreement with the maximum peak intensity of the measured μ-CL spectra.

To model the real ring-like resonator and its eigenmodes on frequencies close to optical transitions, we have taken the real size and shape of a ring microresonator, assuming $r_1 = 1.65\ \mu$m and external one $r_2 = 1.8\ \mu$m for the internal and external radii, respectively, and solved the Maxwell's equations using the perfectly-matched-layer approximation [7] at the boundaries. Such an approach is useful for modelling the electric field which can leak into nearby layers depending on the imaginary part of the refractive index. Figure 3 demonstrates the modulus of the electric field distribution in the ring-like resonator at a wavelength of 523 nm, which corresponds to the eigenmode of such a system. It turns out that this mode is a whispering gallery mode (WGM) with the mode order $m \sim 50$. A simple approximation shows that the wavelengths inside the ring should satisfy the expression $\lambda \sim \pi n(r_1 + r_2)/m$, where $n \sim 2.5$ is the refractive index. This estimation shows reasonable consistency between the experiment and modelling. The WGMs are known for their high quality factor and coupling with light [8]. Considering the imaginary part of the refractive index, we find the quality factor of $Q \sim 200$. Consequently, the coupling of the existing WGM in a ring-like resonator with a QW mode explains more intensive μ-CL on the circumference of the cones in contrast to a weaker intensity of the unperturbed layer emission far from the cone.

4. Summary

In this paper, we have discovered a specific feature of the MBE growth on the cone-shaped substrates. The unique ring-like resonators were self-formed on the c-substrates, supporting high-order optical modes. To prove that, we have calculated the optical transitions in QWs of InGaN/GaN taking into account the internal electric fields, and we have also simulated the eigenmodes of the ring-like microresonators. It turns out that the modes that are close to the optical transitions frequencies were WGMs of a high order. They cause the enhancement of emission. μ-CL experiments showed significant enhancement of the signal in the near-cone region, which proves the formation of the ring-like microresonators. Two microresonators fabricated simultaneously in a single growth process (the nanorod resonator and the ring one that are in close proximity to each other) is a unique feature that may be used for different applications, e.g., as an alternative way of indirect pumping by the energy transfer from one microresonator to another.

Acknowledgments

This study is supported by the the Russian Science Foundation (Project 14-22-00107).

References

[1] Purcell E M 1946 *Phys. Rev. B* **69** 681
[2] Shubina T V, Pozina G, Jmerik V N, Davydov V Yu, Hemmingsson C, Andrianov A V, Kazanov D R and Ivanov S V 2015 *Sci. Rep.* **5** 17970
[3] Shubina T V, Jmerik V N, Davydov V Yu, Kazanov D R, Smirnov A N, Nechaev D V, Kuznetsova N, Pozina G, Hemmingsson C and Ivanov S V 2016 *Phys. Status Solidi B* **253** 845-852
[4] Kazanov D R, Kuznetsova N, Zotolukhin D S, Jmerik V N, Davydov V Yu, Smirnov A N and Shubina T V 2016 *J. Phys.: Conf. Series* **741** 012126
[5] Jmerik V N, Kuznetsova N V, Nechaev D V, Shubina T V, Kirilenko D A, Troshkov S I, Davydov V Yu, Smirnov A N and Ivanov S V 2017 *J. Crystal Growth* **477** 207-211
[6] Bernardini F, Fiorentini V and Vanderbilt D 2001 *Phys. Rev. B* **63** 193201
[7] Berenger J-P 1994 *J. Comp. Phys.* **114** 185-200
[8] Yufeng L, Lungang F, Xilin S, Qiang L, Feng Y, Ge Y and Jung H 2017 *Opt. Express* **25**(15) 18073

Misfit stress relaxation in composite core-shell nanowires with parallelepiped cores using rectangular prismatic dislocation loops

S A Krasnitckii[1,2], D R Kolomoetc[1], A M Smirnov[2] and M Yu Gutkin[1-3]

[1]Peter the Great St. Petersburg Polytechnic University, Polytekhnicheskaya 29, St. Petersburg, 195251 Russia
[2]ITMO University, Kronverkskii pr. 49, St. Petersburg, 197101 Russia
[3]Institute of Problems in Mechanical Engineering, Russian Academy of Sciences, Bolshoi 61, Vasil. Ostrov, St. Petersburg, 199178 Russia

Email: smirnov.mech@gmail.com

Abstract. The misfit stress relaxation via generation of rectangular prismatic dislocation loops at the interface in core-shell nanowires is considered. The core has the shape of a long parallelepiped of a square cross-section. The energy change caused by loop generation in such nanowires is calculated. Critical conditions for the onset of such loops are calculated and analyzed.

1. Motivation

Thanks to their excellent physical properties, core-shell nanowires (NWs) are widely used in novel devices for optoelectronics, photonics, and spintronics [1-4]. The stability of physical properties depends on different structural and material factors including lattice mismatch, a difference in thermal coefficients of core and shell materials, and the presence of defects at the core/shell interfaces [5-10]. Currently, there are a number of theoretical models that describe how defects are generated, in particular, misfit dislocations [11-13]. The disadvantage of all these models is the assumption of axial symmetry of cores in core-shell NWs, although in reality the cores are shaped like polyhedral prisms with flat surfaces [14-16]. This approach does not allow to take into account the effects associated with the stress concentration at the edges of the interfaces. Moreover, it cannot be used in theoretical consideration of dislocations gliding along flat interfaces.

The simplest geometric model of an NW with flat core-shell interfaces is an elastic cylinder with a long parallelepiped inclusion. Recently, we have obtained a solution for the boundary value problem in the theory of elasticity for an elastic cylinder with an eccentric inclusion of a rectangular cross section subjected to a dilatation eigenstrain [17]. The aim of the present paper is to develop a theoretical model describing stress relaxation by the generation of rectangular prismatic dislocation loops (PDLs) at the interface in core-shell NWs with cores in the form of a long parallelepiped.

2. Model

Let us consider a composite NW as an elastic core-shell cylinder of radius R with a core in the form of a long parallelepiped with a square cross section ($2a\times2a$). The mismatch of core and shell lattices is characterized by the misfit parameter $f = 2(a_c - a_s) / (a_c - a_s)$, where a_c and a_s are the lattice parameters of the core and shell materials, respectively. We assume that the core and shell materials have the same elastic modules. We also suppose that rectangular PDLs with sizes $2c\times2h$ can appear in different regions of the core-shell interface (figure 1), in which cases they are denoted as follows: PDL-1 is generated from the middle of the interface and expands to the bulk of the core, PDL-2 is generated from the edge of the interface and also expands to the bulk of the core, and PDL-3 is generated from the middle of the interface and expands to the free surface of the NW.

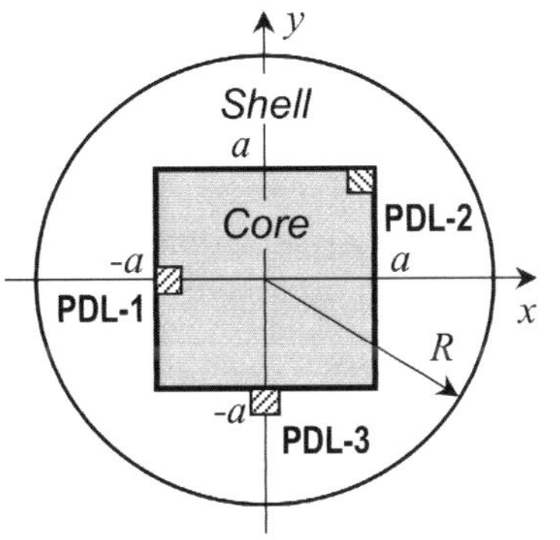

Figure 1. Generation of rectangular PDLs in a cross section of a core-shell NW with a parallelepiped core. We consider three possible cases for the PDL generation at the interface: (1) from the middle of the flat interface into the core, PDL-1; (2) from the core edge into the core, PDL-2; and (3) from the middle of the flat interface to the free surface, PDL-3.

According to the energy approach [13], the generation of PDLs is favorable when the change of the total energy is negative:

$$\Delta W = W_2 - W_1 = W_c + W_{el} + W_{int} < 0, \tag{1}$$

where W_1 is the strain energy of the initial coherent state of the NW with no PDLs, W_2 is the total energy of the NW after the onset of the PDLs, W_c is the energy of the PDL core, W_{el} is the strain energy of the PDL, and W_{int} is the energy of interaction between a PDL and the misfit stress in the NW.

Following the approach in [13], we assume that the PDL sizes c and h are much smaller than the NW radius R, while the shell is quite thick ($a \le 0.6R$). Then, in a first approximation, the strain energy W_{el} can be expressed as in the case of an infinite medium [18]:

$$W_{el} \approx Db^2\left(2c\ln\frac{4c}{r_0} + 2h\ln\frac{4h}{r_0} - c\ln\frac{K+c}{K-c} - h\ln\frac{K+h}{K-h} - 4(c+h-K)\right), \tag{2}$$

where $D = G/[2\pi(1-\nu)]$, G is the shear modulus, ν is the Poisson ratio, b is the Burgers vector magnitude for the PDL, r_0 is the core radius of the PDL, and $K = \sqrt{c^2+h^2}$.

The core energy of a rectangular PDL is approximated by [19] as:

$$W_c \approx 4(h+c)\frac{Db^2}{2} \tag{3}$$

The interaction energy W_{int} can be found as the work spent to create the PDL-k (k = 1, 2, 3) in the axial misfit stress component ${}^{\rm cyl}\sigma_{zz}$:

$$W_{int1} = -b\int_{a-2c}^{a} dx \int_{a-2h}^{a} {}^{\rm cyl}\sigma_{zz}dy, \quad W_{int2} = -b\int_{a-2c}^{a} dx \int_{-h}^{h} {}^{\rm cyl}\sigma_{zz}dy, \quad W_{int3} = -b\int_{a}^{a+2c} dx \int_{-h}^{h} {}^{\rm cyl}\sigma_{zz}dy. \tag{4}$$

The misfit stress ${}^{\rm cyl}\sigma_{zz}$ was found in a closed analytical form in [17]. In the case shown in figure 1, it can be written as follows:

$${}^{\rm cyl}\sigma_{zz} = C\left[\pi\ \mathrm{sign}\left(\frac{x-x_0}{y-y_0}\right) + 8\frac{R^2}{r^4}\left(2xy\ \ln\sqrt{R^4 - 2R^2(xx_0+yy_0) + (x^2+y^2)(x_0^2+y_0^2)} + \right.\right.$$

$$+(x^2+y^2)\,\mathrm{arctg}\left(\frac{x_0 y - x y_0}{R^2 - x_0 y - x y_0}\right)\Bigg)\Bigg]\Bigg|_{x_0=-a}^{x_0=a}\Bigg|_{y_0=-a}^{y_0=a}, \qquad (5)$$

where $C = f\,G(1+\nu)/[2\pi(1-\nu)]$. Using Eq. (5), the interaction energies (4) were calculated numerically. Thus, we determined all the terms appearing in equation (1).

3. Results

To study the energy change of an NW due to the generation of rectangular PDLs at the interface, we used the energy maps built in the phase space of the normalized PDL sizes of $2a/b$ and $2c/b$. Following [13], we considered two cases of Si/Ge ($f = 0.042$) and InAs/ZnS ($f = 0.107$) core-shell NWs for $R = 100$ nm and $a = 0.5R = 50$ nm (figure 2). As figure 2(a) demonstrates, the generation of PDL-1 and PDL-2 in Si/Ge core-shell NWs is energetically unfavorable. In contrast, the generation of these PDLs in the InAs/ZnS core-shell NWs (figure 2(b)) is energetically favorable even for very

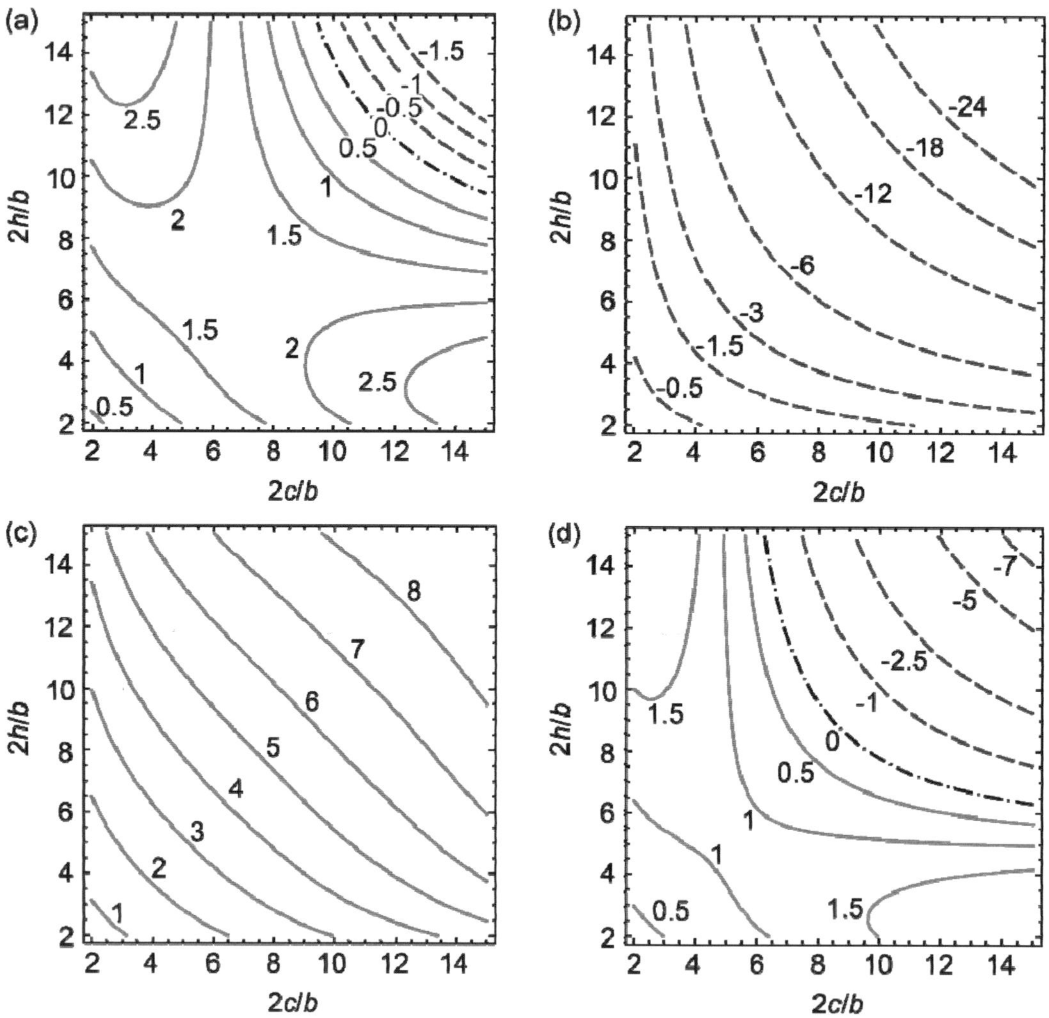

Figure 2. Maps of the energy change ΔW in the space of normalized sizes of PDLs $2a/b$ and $2c/b$ for Si/Ge (a,c) and InAs/ZnS (b,d) core-shell NWs for $R = 100$ nm and $a = 0.5R = 50$ nm. (a,b) is the case of either PDL-1 or PDL-2; (c,d) is the case of PDL-3. The energy values are given in units of Gb^3.

small PDL sizes. The generation of PDL-3 is energetically unfavorable for both the Si/Ge (figure 2(c)) and InAs/ZnS (figure 2(d)) core-shell NWs.

The misfit parameter f is in direct proportion to the interaction energy (5). Therefore, it is easily extracted from the equation $\Delta W = 0$, which determines the critical conditions needed for the generation of PDLs. This generation becomes energetically favorable ($\Delta W < 0$) if f exceeds a critical value f_c, which depends on the core and shell sizes, as well as on the type and sizes of the PDL. Figure 3 shows these dependences for different types of PDLs, which are calculated from the equation:

$$f_c = \frac{W_{el} + W_c}{W^*_{int}}, \tag{6}$$

where $W^*_{int} = -W_{int} / f$.

It is seen from figure 3 that generation of PDL-2 requires the smaller values of the critical misfit f_c, which means that this type of PDL is preferable over the other two. It is also seen that f_c increases with a for PDL-1 and PDL-2 at a fixed value of R. Indeed, an increase in the core size leads to a decrease in the axial misfit stress in the core and, therefore, in the term W^*_{int} appearing in the balance (6). As a result, the critical misfit f_c must increase. In contrast, f_c decreases with a for PDL-3 at a fixed value of R. This is so because increasing the core leads to increasing the axial misfit stress in the shell and, therefore, to increasing the term W^*_{int} . It is also worth noting that f_c decreases with an increase in R at a constant a for PDL-1 and PDL-2, and, conversely, for PDL-3, due to similar reasons.

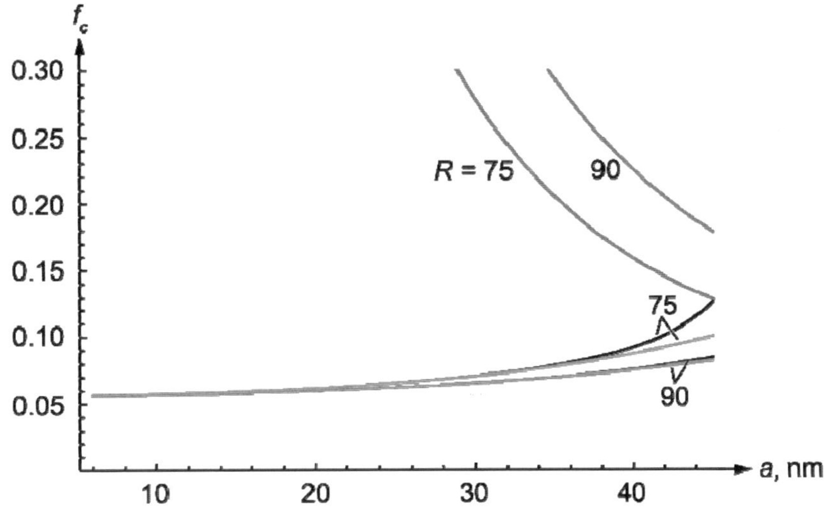

Figure 3. Dependence of the critical misfit f_c on the core size a for different values of radius R (in nanometers) at $c/h = 0.2$, $h/b = 10$, and $b = 0.3$ nm. Black, magenta and red curves correspond to PDLs of types 1, 2 and 3, respectively.

4. Conclusions

A theoretical model is suggested which describes misfit stress relaxation in core-shell NWs with symmetric parallelepiped cores by the generation of rectangular PDLs at the core-shell interface. Three different types of PDLs are considered which are generated either in the middle or on the edge of the flat interface and expand either to the bulk of the core or to the free surface of the NW. It is shown that PDLs generated on the core edge and expanding into the bulk of the core are the most preferable PDLs for the NWs. For these PDLs, the critical misfit f_c increases with the core size a and decreases with the NW radius R.

Acknowledgments
S. A. Krasnitckii and A. M. Smirnov acknowledge the Russian Foundation of Basic Research (RFBR project No. 16-32-00521) for the support in providing the analytical and numerical calculations of the critical conditions for the PDL generation in core-shell NWs. M. Yu. Gutkin acknowledges the Ministry of Education and Science of the Russian Federation (Grant No. 3.3194.2017/PCh) for the support in the invention of the theoretical model of misfit stress relaxation in the core-shell NW by the generation of PDLs.

References

[1] Dimakis E, Jahn U, Ramsteiner M, Tahraoui A, Grandal J, Kong X, Marquardt O, Trampert A, Riechert H and Geelhaar L 2014 *Nano Lett.* **14** 2604
[2] Qian F, Gradecak S, Li Y, Wen C-Y and Lieber C M 2005 *Nano Lett.* **5** 2287
[3] Ai G, Li H, Liu S, Mo R and Zhong J *2015 Adv. Funct. Mater.* **25** 5706
[4] Tomioka K Yoshimura M and Fukui T 2012 *Nature* **488** 189
[5] Sneed B T, Young A P and Tsung C 2015 *Nanoscale* **7** 12248
[6] Muralidharan N, Carter R, Oakes L, Cohn A P and Pint C L 2016 *Scientific Reports* **6** 27542
[7] Lin Y, Kim D, Li Z, Nguyen B, Li N, Zhang S , Yoo J 2017 *Nanoscale* **9** 1213
[8] Dayeh S A, Tang W, Boioli F, Kavanagh K L, Zheng H, Wang J, Mack N H, Swadener G, Huang J Y, Miglio L, Tu K and Picraux S T 2013 *Nano Lett.* **13** 1869
[9] Biermanns A, Rieger T, Bussone G, Pietsch U, Grutzmacher D and Lepsa M L *2013 Appl. Phys. Lett.* **102** 043109
[10] Salehzadeh O, Kavanagh K L and Watkins S P 2013 *J. Appl. Phys.* **113** 134309
[11] Gutkin M Y, Ovid'ko I A and Sheinerman A G 2000 J. Phys.: *Condens. Matter* **12** (25) 5391
[12] Liang Y, Nix W D, Griffin P B and Plummer J D 2005 *J. Appl. Phys.* **97** 043519
[13] Gutkin M Y and Smirnov A M 2015 Acta Materialia **88** 91
[14] Jenichen B Hilse M Herfort J and Trampert A 2015 *J. of Crystal Growth.* **427** 21
[15] Jenichen B Hilse M Herfort J and Trampert A *2015 J. of Crystal Growth.* **410** 1-Aksoy B, Kalay Y E, and Unalan H E 2014 *Journal of Crystal Growth* **392** 20
[16] Krasnitckii S A, Smirnov A M and Gutkin M Y 2016 *J. Phys.: Conf. Ser.* **690** 012022
[17] Krasnitckii S A, Kolomoets D R, Smirnov A M and Gutkin M Y 2017 *J. Phys.: Conf. Ser.* **816** 012043
[18] Gutkin M Y, Ovid'ko I A and Sheinerman A G 2003 J. Phys.: *Condens. Matter* **15** (21) 3539
[19] Hirth J P, Lothe J 1982 Theory of Dislocations *Wiley, New York*

IOP Conf. Series: Journal of Physics: Conf. Series **993** (2018) 012022 doi:10.1088/1742-6596/993/1/012022

Spontaneous parametric downconversion of light by a dielectric nanoparticle

N A Olekhno[1], M I Petrov[1,2] and I V Iorsh[1]

[1]Department of Nanophotonics and Metamaterials, ITMO University, 199034 St. Petersburg, Russia

[2]Department of Physics of Condensed Matter, St Petersburg Academic University, 194021 St. Petersburg, Russia

E-mail: nikita.olekhno@metalab.ifmo.ru

Abstract. In the present paper, we theoretically consider the generation of entangled photon pairs in the process of spontaneous parametric downconversion of light by a nonlinear dielectric nanoparticle. This nanoparticle is the only source of nonlinearity in the considered model. We assume that the particle is small compared to the wavelength of light so that the process is non-resonant. An intuitive theoretical description is developed for the case when the nonlinear susceptibility tensor of the material of a nanoparticle has a form of the Levi-Civita symbol. In this case, a two-photon amplitude which characterizes a spatial distribution of entangled photon pairs can be interpreted as a result of interference between two induced dipoles, both being orthogonal to the direction of polarization of the pump photon. A generation of entangled surface plasmon-polaritons by a dielectric nanoparticle located near the metal-dielectric interface is also considered.

1. Introduction

Entangled states are a subject of primary interest in the area of quantum optics. The wavefunction of these many-particle states can not be represented as a product of the one-particle wavefunctions. Thus, there are correlations between the particles. Entangled photon pairs are an example of such states. They are widely used in experiments testing the fundamentals of quantum mechanics, as well as for applications such as quantum cryptography and sensing. The main source of entangled photon pairs is the spontaneous parametric downconversion of light (SPDC) [1]. In this process, a medium with non-zero second-order nonlinear susceptibility $\chi^{(2)}$ absorbs a pump photon and produces two photons, namely idler and signal ones, with half the frequency.

Nowadays, the generation of quantum states with nanosystems attracts considerable attention in the areas of nanophotonics and quantum optics. For example, the generation of entangled photon pairs in semiconductor waveguides of subwavelength cross sections [2] as well as by quantum dots [3] has been considered. Also, the generation of entangled photon and plasmon-polariton pairs in hyperbolic metamaterials has been studied theoretically in [4]. However, SPDC has a very low efficiency since it is a spontaneous nonlinear process. This considerably limits the possibility of observing entangled pair creation in solitary nanostructures or in small groups of nanoscale objects, which are of special interest for the development of on-chip nanophotonic devices.

Various strategies have been considered to enhance SPDC efficiency in nanostructures. The first is the engineering of quasi-phase-matching in waveguides. Another strategy is the use of the resonant processes in nanoparticles and nanoantennas. In order to study this possibility, the production of photon pairs by a quantum dot located near a plasmonic nanoparticle [5] or a bimodal plasmonic nanoantenna [6] has been considered. However, plasmonic systems do not have an intrinsic nonlinearity due to the centrosymmetric lattice structure of noble metals. Thus, they only act as electric field amplifiers. On the other hand, all-dielectric nanophotonics is one of the most rapidly developing areas of modern nanostructure optics. Dielectric structures demonstrate electric as well as magnetic resonances and very low losses that are unattainable in metals. In addition, they possess considerable $\chi^{(2)}$ nonlinearity and, thus, are widely used for second harmonic generation. For example, in a series of recent papers [7, 8, 9] a resonant enhancement of second harmonic generation assisted by Mie resonances in dielectric nanoparticles has been reported. Second harmonic generation is the opposite process to SPDC. Nevertheless, spontaneous parametric downconversion in dielectric nanostructures, such as nanoparticles or nanoantennas, has been the subject of a very limited number of studies. Namely, photon pair creation in a semiconductor microresonator has been considered numerically in [10].

In the present paper, we consider spontaneous parametric downconversion by a dielectric nanoparticle within the Rayleigh approximation, when the following condition is satisfied:

$$\frac{\lambda}{n} \gg d, \tag{1}$$

where λ is the wavelength of the incident radiation, n is the index of the material of a nanoparticle and d is the size of a nanoparticle. Thus, a nanoparticle can be considered as a point electric dipole. This simple non-resonant model is of basic interest for the following study of SPDC in dielectric nanostructures. This model provides an analytical solution in the case of a nanoparticle located in vacuum or in the dielectric bulk as well as a semi-analytical solution in the case of a nanoparticle located near a metal-dielectric interface. To our knowledge, such a basic model has not yet been considered.

2. Theoretical description of nonlinear generation

We start with a general theoretical framework of paper [4]. Within this approach, spatial correlations between photons of the generated photon pair are described by the two-photon amplitude

$$T_{is}(\mathbf{r}_i, \mathbf{r}_s) = \int \mathbf{d}_{i,\alpha} G_{\alpha\beta}(\mathbf{r}_i, \mathbf{r}_0, \omega_i) \Gamma_{\beta\gamma}(\mathbf{r}_0) G_{\gamma\delta}(\mathbf{r}_0, \mathbf{r}_s, \omega_s) \mathbf{d}_{s,\delta} d\mathbf{r}_0, \tag{2}$$

where the subscript i denotes the first (idler) photon with a frequency ω_i, and the subscript s denotes the second (signal) one with a frequency ω_s. The dipole moments $\mathbf{d}_i$ and $\mathbf{d}_s$ of idler and signal detectors define the polarizations of the detected photons. In the equation above, $\hat{G}(\mathbf{r}, \mathbf{r}_0, \omega)$ is the matrix of the dyadic Green's function of the generating system, and $\hat{\Gamma}(\mathbf{r})$ is a generation matrix. The latter in the case of the SPDC process is defined as

$$\Gamma_{ij}(\mathbf{r}) = \chi^{(2)}_{ijk}(\mathbf{r}) E^{pump}_k(\mathbf{r}) \mathrm{e}^{-i\omega_{pump} t}, \tag{3}$$

where $\chi^{(2)}_{ijk}$ is the second-order nonlinear susceptibility tensor and E^{pump} at frequency ω_{pump} is the pump field which causes nonlinear generation. The integration in Eq.(2) is performed over the volume of the nonlinear system which generates photon pairs. The probability of simultaneous detection of idler and signal photons of given polarizations at the given positions $\mathbf{r}_i$ and $\mathbf{r}_s$ is related to the two-photon amplitude as [4]:

$$W(\mathbf{r}_i, \mathbf{r}_s) = (2\pi/\hbar)\delta(\hbar\omega i + \hbar\omega_s - \hbar\omega_{pump}) |T_{is}(\mathbf{r}_i, \mathbf{r}_s)|^2. \tag{4}$$

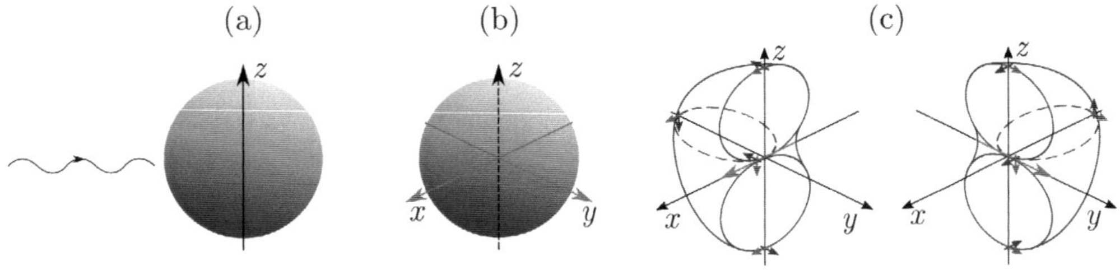

Figure 1. Sketch of photon pair generation in the process of SPDC by a nonlinear dielectric nanoparticle. (a): Initial polarization of a nanoparticle. Incident photon is z-polarized. (b): Induced dipoles. (c): Field pattern of a radiating dipole. Electric field vectors are shown with red, and magnetic field vectors are shown with blue.

3. SPDC by a point dipole

When a pump photon interacts with a nanoparticle, the latter is homogeneously polarizes (figure 1a). This polarization field acts as the actual electric field responsible for nonlinear generation in the nanoparticle. Let us define a coordinate system with the z-axis being directed along the polarization. Then, the pump field can be expressed as

$$\mathbf{E}_{pump}(\mathbf{r}) = \chi E \delta(\mathbf{r}) \left| e_z \right\rangle . \tag{5}$$

In the equation above, $|e_z\rangle$ is the unit vector in the z-direction, $\chi = \chi^{(1)}$ is the scalar linear susceptibility of the nanoparticle, and E is the electric field of the pump photon. In the case of a spherical particle, susceptibility is related to the permittivity of the nanoparticle ε by the Lorentz-Lorenz equation:

$$\chi(\omega) = \frac{3}{4\pi} \frac{\varepsilon(\omega) - 1}{\varepsilon(\omega) + 2}. \tag{6}$$

We consider the second-order nonlinear susceptibility tensor in the form of the Levi-Civita symbol

$$\chi^{(2)}_{ijk} = \epsilon_{ijk} \tag{7}$$

as an example. This case corresponds to a crystal of cubic O-class symmetry (see [11]). Let us recall that the considered tensor has 6 nonzero components $\epsilon_{xyz} = \epsilon_{zxy} = \epsilon_{yzx} = 1$, $\epsilon_{yxz} = \epsilon_{zyx} = \epsilon_{xzy} = -1$. Then, the generation matrix $\hat{\Gamma}$ (3) can be expressed as follows:

$$\Gamma_{ij} = \epsilon_{ijk} \mathbf{E}_k = \epsilon_{ijz} E_z = E(\left| e_x \right\rangle \left\langle e_y \right| - \left| e_y \right\rangle \left\langle e_x \right|), \tag{8}$$

with $|e_x\rangle$ and $|e_y\rangle$ being unit vectors in the directions x and y, correspondingly. We consider the dyadic Green's function of a point electric dipole in the following form:

$$G_{\alpha\beta}(\mathbf{r}, \mathbf{r}', \omega) = \frac{\exp(ik_0 R)}{R} k_0^2 \Big[\Big(\delta_{\alpha\beta} - \frac{R_\alpha R_\beta}{R^2} \Big) - \frac{1 - ik_0 R}{k_0^2 R^2} \Big(\delta_{\alpha\beta} - 3 \frac{R_\alpha R_\beta}{R^2} \Big) \Big], \tag{9}$$

where $\mathbf{R} = \mathbf{r} - \mathbf{r}'$, $R = |\mathbf{R}|$, $k_0 = \omega / c$.

We start with the simplest case when both of the detectors and the pump field are polarized in the z-direction. Then, the following expression for the two-photon amplitude can be obtained:

$$|T^{zz}_{is}(\mathbf{r}_i, \mathbf{r}_s)|^2 = \frac{z_i^2 z_s^2 |[\mathbf{r}_i \times \mathbf{r}_s]_z|^2}{r_i^6 r_s^6} \Big(\frac{9}{r_i^4} + \frac{3}{r_i^2} \big(\frac{\omega_i}{c}\big)^2 + \big(\frac{\omega_i}{c}\big)^4 \Big) \Big(\frac{9}{r_s^4} + \frac{3}{r_s^2} \big(\frac{\omega_s}{c}\big)^2 + \big(\frac{\omega_s}{c}\big)^4 \Big). \tag{10}$$

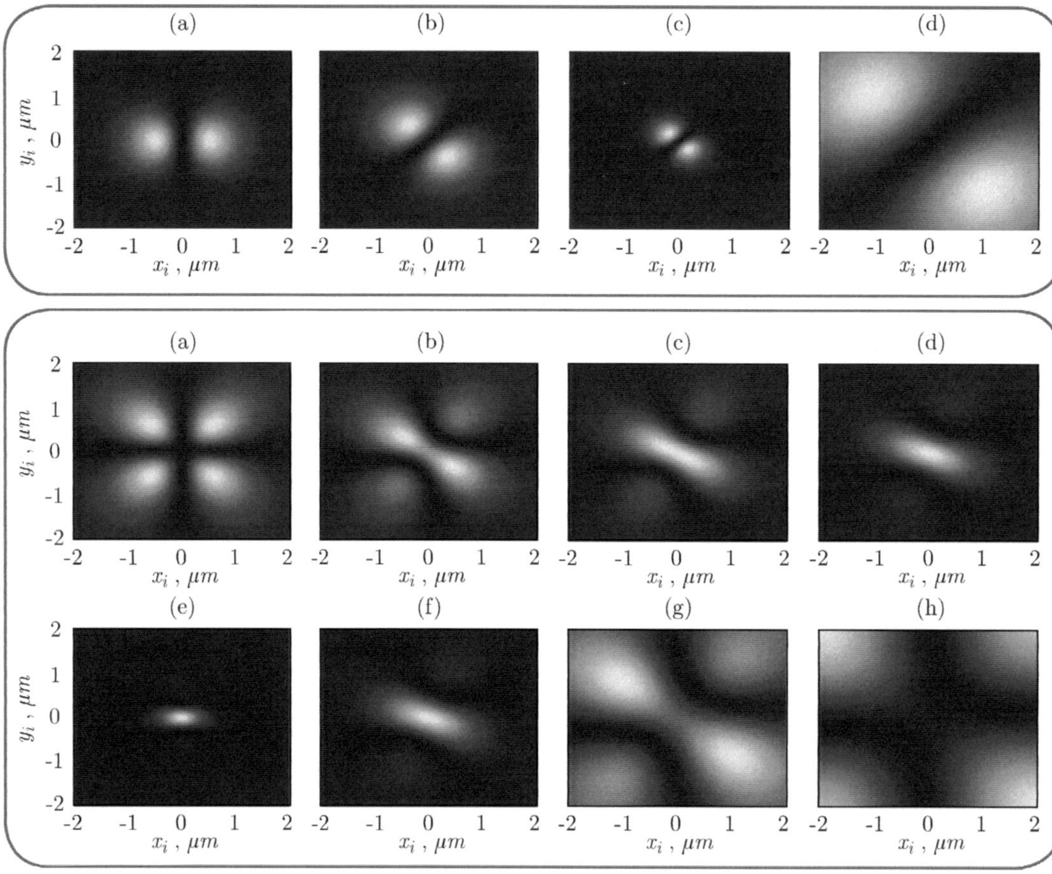

Figure 2. Simultaneous detection probability $W(\mathbf{r}_i, \mathbf{r}_s)$ as a function of the position of the idler detector x_i, y_i for a fixed position of the signal detector. Pump polarization is aligned along the z axis at the position $x_0 = y_0 = z_0 = 0$ nm, and $\omega_i = \omega_s = 3 \cdot 10^{11} s^{-1}$.
First panel: Idler and signal detectors are polarized in the z direction. (a) $x_s = 500$ nm, $y_s = 0$ nm, $z_i = z_s = 1$ μm. (b): $x_s = y_s = 500$ nm, $z_i = z_s = 1$ μm. (c): $x_s = y_s = 500$ nm, $z_i = 1$ μm, $z_s = 0.5$ μm. (d): $x_s = y_s = 500$ nm, $z_i = 1$ μm, $z_s = 3$ μm.
Second panel: Idler and signal detectors are polarized in the x direction. Top row: $x_s = 500$ nm, $z_i = z_s = 1$ μm. (a): $y_s = 0$ nm. (b): $y_s = 100$ nm. (c): $y_s = 200$ nm. (d): $y_s = 500$ nm. Bottom row: $x_s = y_s = 500$ nm, $z_i = 1$ μm. (e): $z_s = 0.5$ μm. (f): $z_s = 1$ μm. (g): $z_s = 2$ μm. (h): $z_s = 4$ μm.

Hereafter, we define the two-photon amplitude as divided by the factor $|\mathbf{d}_i\mathbf{d}_s E_{pump}|$. It is seen that the detection probability (4) vanishes if the z-component of the cross product $[\mathbf{r}_i \times \mathbf{r}_s]$ is zero, i.e. if both the signal and idler detectors are located in the plane that is orthogonal to the xy plane. This is clearly seen in figure 2. At the same time, the amplitude at the given distances between detectors and a point dipole is maximal if the idler and signal detectors are located in the perpendicular directions. Thus, we can say that the photons emitted in the perpendicular directions are correlated, whereas the photons emitted in the parallel directions are anticorrelated.

However, there are several possible alignments of detector polarizations with respect to the pump field in the considered problem. Namely, if the pump is polarized along the z-axis, after obviously equivalent variants have been excluded, the following combinations remain: $\mathbf{d}_i$, $\mathbf{d}_s || z$;

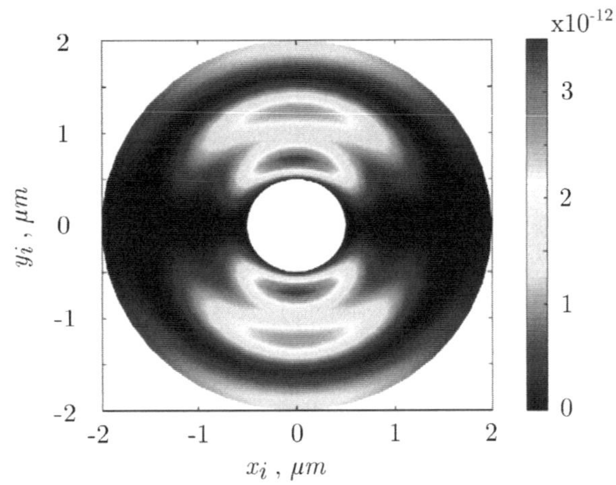

Figure 3. Simultaneous detection probability $W(\mathbf{r}_i, \mathbf{r}_s)$ calculated numerically as a function of the in-plane position of the idler detector x_i, y_i for $\omega_i = \omega_s = 1.55$ eV. The position of a nonlinear dipole is $x_0 = y_0 = 0$ nm, $z_0 = 10$ nm. The vertical position of the idler detector is $z_i = 10$ nm. The position of the signal detector is $x_s = 200$ nm, $y_s = 0$ nm, $z_s = 10$ nm.

$\mathbf{d}_i$, $\mathbf{d}_s||x$; $\mathbf{d}_i||z$, $\mathbf{d}_s||x$ and $\mathbf{d}_i||x$, $\mathbf{d}_s||y$. The expressions for the two-photon amplitude look more complicated in the last three cases due to the absence of the axial symmetry. In figure 2, the first and second situations are presented as well. As a result, it is seen that one can sufficiently tailor the spatial distribution of entangled photon pairs by selecting polarizations of the detectors. Generalization of the obtained results to the case of materials with the second-order nonlinear susceptibility tensor $\chi^{(2)}_{xyz} = \chi^{(2)}_{zxy} = \chi^{(2)}_{yzx} = \chi^{(2)}_{yxz} = \chi^{(2)}_{zyx} = \chi^{(2)}_{xzy}$ (such as GaAs) is straightforward.

4. SPDC by a point dipole located near a metal-dielectric interface

In the present section, we consider SPDC by a dielectric nanoparticle located near the planar metal-dielectric interface. There is a principal difference between this model and the previous one. Indeed, surface plasmon-polaritons as well as photons are generated in such a system. Entanglement of plasmons was observed for the first time in [12]. Since then, the generation of entangled plasmons in nanostructures attracts considerable attention both experimentally and theoretically [4].

As in the previous section, we assume that the second order susceptibility tensor has a form $\chi^{(2)}_{ijk} = \epsilon_{ijk}$, and the condition $\lambda/n \gg d$ is valid. Thus, a particle can be described as a point electric dipole. The Green's function of the point dipole located near the metal-dielectric interface consists of the free space Green's function G_0 and the reflected part G_{ref}:

$$G(\mathbf{r}, \mathbf{r}_0, \omega) = G_0(\mathbf{r}, \mathbf{r}_0, \omega) + G_{ref}(\mathbf{r}, \mathbf{r}_0, \omega). \tag{11}$$

We consider the simplest case, when a pump photon is polarized in the z-direction along with the signal and idler detectors. In this case, the reflected part of the Green's function G_{ref} includes only p-polarized waves and can be expressed as follows [13]:

$$G_{ref}(\rho, \varphi, z) = \frac{ik_0^3}{2} \int_0^\infty \hat{M}(s, \rho, \varphi) \exp(is_{1z}z) ds. \tag{12}$$

In the relation above, we use the polar representation with $\rho = \sqrt{(x - x_0)^2 + (y - y_0)^2}$, $\varphi = \mathrm{arctg}((y - y_0)/(x - x_0))$ and $z = z_{i,s} + z_0$. The normalized wavenumber $s = \sqrt{k_x^2 + k_y^2}/k_0$, and $s_{1z} = k_1 z/k_0$. The index 1 denotes an upper half-space filled with a dielectric. Elements of the matrix $\hat{M}$ that are relevant for us have the form $M_{xz} = -M_{zx} = -2\pi i s^2 r_p(\omega) J_1(s\rho)\cos(\varphi)$ and $M_{yz} = -M_{zy} = -2\pi i s^2 r_p(\omega) J_1(s\rho)\sin(\varphi)$, with $J_1(s\rho)$ being the Bessel function of the first kind and of the first order and r_p being the Fresnel reflection coefficient for p-polarized waves. The resulting probability of simultaneous detection $W(\mathbf{r}_i, \mathbf{r}_s)$ is shown in figure 3. The permittivity of the dielectric is $\varepsilon_d = 1$ and the permittivity of metal is $\varepsilon(\omega) = 1 - \omega_p^2/(\omega^2 + i\gamma\omega)$, where the plasma frequency of metal $\omega_p = 5$ eV and $\gamma = 0.01\omega_p$.

We can interpret the obtained results in the same way as in the previous case. Indeed, surface plasmons are generated by a superposition of two dipoles arranged in the x and y directions, correspondingly. Thus, one dipole does not emit any plasmons in the y-direction, whereas the other one does not emit in the x-direction. As a result, the two-photon amplitude shows mirror symmetry with respect to the plane that includes a line connecting a dipole and one of the detectors and is orthogonal to the interface. At this plane $|T_{is}(\mathbf{r}_i, \mathbf{r}_s)|^2 = 0$ (see figure 3). Hence, surface plasmon-polaritons emitted by a nonlinear particle in a non-resonant SPDC process are correlated in the orthogonal angular directions and anticorrelated in the parallel directions, if nonlinearity of the nanoparticle has a form $\chi^{(2)}_{ijk} = \epsilon_{ijk}$ and all of the excitations in the system are polarized in the z-direction.

5. Conclusion

In the present paper, we have considered spontaneous parametric downconversion of light by a dielectric nanoparticle within a theoretical framework of paper [4]. The particle is described as a point electric dipole. Such an approximation is valid for particles of a size much smaller than the production of the wavelength of incident light and the refractive index of a nanoparticle. Even this simple model demonstrates rich behavior, which depends on the polarizations of signal and idler photons. The obtained results can be interpreted in terms of the interference of the fields produced by two induced dipoles, if the nonlinear susceptibility tensor of the nanoparticle is proportional to the Levi-Civita symbol, i.e. for materials of the O symmetry group. The same is true in the case when a dipole is placed near a planar metal-dielectric interface, and surface plasmon-polaritons are excited. This interpretation is supported by the results of numerical calculations.

Acknowledgments

We are grateful to A.N. Poddubny, A.A. Sukhorukov, K.S. Frizyuk and A.A. Nikolaeva for fruitful discussions. The work was supported by the Russian Foundation for Basic Research (proj. # 18-02-01206, 16-32-60167 and 16-32-60123).

References

[1] Klyshko D N, Penin A N and Polkovnikov B F 1970 *JETP Lett.* **11** 5–8
[2] Solntsev A S, Sukhorukov A A 2017 *Reviews in Physics* **2** 19–31
[3] Orieux A, Versteegh M A M, Jons K D and Ducci S 2017 *Rep. Prog. Phys.* **80** 076001
[4] Poddubny A N, Iorsh I V and Sukhorukov A A 2016 *Phys. Rev. Lett.* **117** 123901
[5] Poddubny A N, Ginzburg P, Belov P A, Zayats A V and Kivshar Y S 2012 *Phys. Rev.* A **86** 033826
[6] Straubel J, Sarniak R, Rockstuhl C and Slowik K 2017 *Phys. Rev.* B **95** 085421
[7] Carletti L, Locatelli A, Stepanenko O, Leo G and De Angelis C 2015 *Optics Express* **23** 26544–50
[8] Timpu F, Sergeyev A, Hendricks N and Grange R 2017 *ACS Photonics* **4** 76–84
[9] Makarov S V et al 2017 *Nano Letters* **17** 3047–53
[10] Solntsev A, Carletti L, Xu L, Poddubny A, De Angelis C, Leo G, Kivshar Y, Neshev D N and Sukhorukov A A 2016 *2016 Photonics and Fiber Technology Congress* paper NT3A.4
[11] Boyd R W 2008 *Nonlinear Optics (3rd ed.)* (San Diego: Academic Press)
[12] Altewischer E, van Exter M P and Woerdman J P 2002 *Nature* **418** 304–6
[13] Sinev I S et al 2017 *Laser Photonics Rev.* **11** 1700168

IOP Conf. Series: Journal of Physics: Conf. Series **993** (2018) 012023 doi:10.1088/1742-6596/993/1/012023

Optical properties of blue-green single-photon sources based on self-assembled CdSe quantum dots

M V Rakhlin[1], K G Belyaev[1], S V Sorokin[1], I S Mukhin[2,3], S V Ivanov[1] and A A Toropov[1]

[1]Ioffe Institute, 26 Polytekhnicheskaya str., St. Petersburg, 194021, Russia
[2]St. Petersburg Academic University RAS, 8/3 Khlopina str., St. Petersburg, 194021, Russia
[3]ITMO University, 49 Kronverksky pr., St. Petersburg, 197101, Russia

E-mail: maximrakhlin@mail.ru

Abstract. We report on single photon emission with a wavelength below 500 nm from single CdSe quantum dots (QDs) grown by migration-enhanced epitaxy providing a reduced QD lateral density below 10^{10} cm^{-2}. The QD photoluminescence was observed at the temperature of 8 K in 200-nm-wide mesa-structures made of CdSe QD heterostructures. The antibunching effect under cw excitation with $g^{(2)}(0) \sim 0.2$ was demonstrated.

1. Introduction

Reliable generation of single photons on demand is essential for quantum cryptography and quantum computing [1]. Using visible single photons instead of the infrared ones is important because of the availability of efficient high-speed single-photon detectors in this wavelength range. Single self-organized quantum dots (QDs) based on wide-gap II-VI CdSe and CdTe compounds grown by epitaxial techniques (either by conventional molecular beam epitaxy, MBE [2,3], or by migration enhanced epitaxy, MEE [4,5]), are considered to be promising candidates for creation of single-photon or even entangled-photon sources of visible light because of their high quantum efficiency, fast radiative decay time and ability to operate at elevated temperatures. They are also less susceptible to blinking, which is an essential drawback of colloidal QDs. However, there are only a few reports on the single-photon emission in epitaxial QDs at room temperature, which is mainly hampered by spectral overlapping of the emission bands of different QDs due to their large density [6,7]. In particular, the conventional approaches to distinguishing the single-QD emission lines (fabrication of etched mesa-structures [8] or nanoapertures in a non-transparent mask [9]) require a relatively low QD density of at least 10^{10} cm^{-2}. Note that CdSe QDs with a wider band gap are more preferable for high-temperature operation than the CdTe ones due to the larger binding energies of excitons, biexcitons, and trions, providing a reduced overlap of the respective PL lines originating from individual QDs [10]. Moreover, the emission of CdSe QDs can be obtained in an extended spectral range including blue light (450-590 nm) [11], whereas the emission of CdTe QDs is limited to the spectral range above 540 nm.

In this paper, we apply micro-photoluminescence (μ-PL) spectroscopy to compare emission properties of individual CdSe QDs grown by either MBE or MEE techniques. The data of μ-PL spectroscopy allow us to estimate the QD density, which turns out to be sufficient for spatial and spectral selection of single quantum emitters. The correlation function $g^{(2)}(\tau)$ was measured and analyzed in order to confirm single photon statistics.

2. Experiment

We have studied emission properties of two samples (labeled as A and B) grown on GaAs (001) substrates. The structures included the following layers sequentially from the substrate: a 10-nm-thick ZnSe buffer layer, a 100-nm-thick ZnSe/ZnSSe short-period superlattice (SL), CdSe QDs surrounded on both sides by 50-nm-thick ZnMgSSe, 1.5-nm-thick ZnSSe and ZnSe layers for better carrier confinement, and a top layer of 3-nm-thick ZnSe. CdSe QDs were fabricated by MEE in sample A, while they were grown by conventional MBE in sample B.

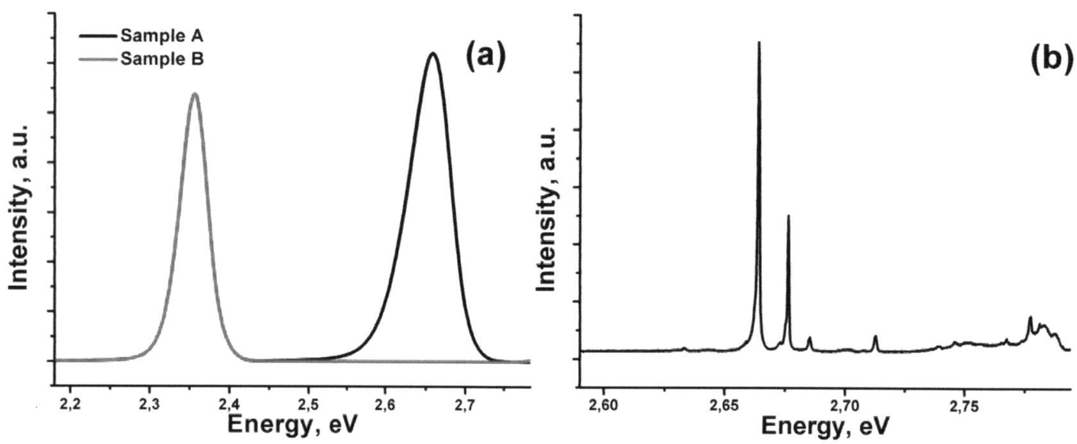

Figure 1. (a) Normalized low temperature (77 K) spatially integrated PL spectra of the structures fabricated using the MEE (sample A) and MBE (sample B) growth modes. (b) μ-PL spectra measured at 8 K in sample A, using a 200-nm mesa.

The emission properties of QDs were investigated using cylindrical mesa-structures of 200 nm in diameter, which were fabricated using a combination of electron lithography and plasma etching. Scanning electron microscopy studies have confirmed the formation of cylindrical mesas with vertical sidewalls. The sample was mounted in a He-flow cryostat with an Attocube XYZ piezo driver inside, which allowed us to optimize and precisely maintain the positioning of the chosen mesa with respect to a laser spot during an extended period (several hours). The μ-PL measurements were carried out under optical excitation by a cw laser line (405 nm). The laser power density was 4 W/cm^2. Photon correlation measurements were performed in a Hanbury-Brown–Twiss detection scheme exploiting two single-photon avalanche diodes (from Micro Photon Devices) possessing a full width at half maximum (FWHM) of photon timing resolution of about 40 ps.

3. Results and discussion

The cw PL spectra (integrated over the area of ~200 μm^2) of samples A and B, measured at 77 K, are shown in figure 1(a). The emission line in sample B is markedly red-shifted relative to that in sample A, in spite of the same amount of deposited CdSe. This shift is a sign of larger QD sizes in sample B. Besides, samples A and B have more or less similar FWHM of the emission line (45 meV and 60 meV, respectively), which indicates approximately the same QD size dispersion leading to the dispersion of the emission energies of individual QDs.

The μ-PL spectra of sample A, measured at 8 K, reveal a number of relatively narrow lines assigned to the emission of excitons, biexcitons, and other electron-hole complexes from individual QDs (figure 1(b)). For the 200-nm mesas, the ballpark number of narrow excitonic lines in μ-PL spectra is estimated to be in the range of 2 - 7. The recorded number of lines in sample A corresponds to the QDs density as small as 10^{10} cm^{-2}. The obtained density of QDs is perfectly suited for creation of single-photon sources based on QD emission. In contrast, the μ-PL spectra measured in sample B do not show pronounced narrow lines at all, indicating a large density of QDs.

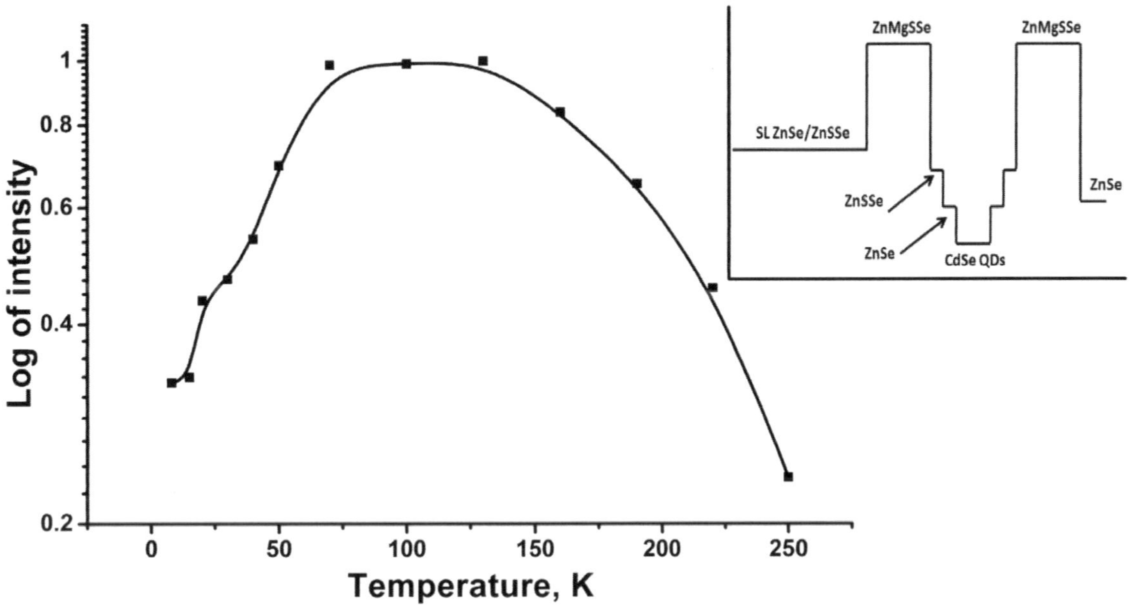

Figure 2. Spectrally integrated PL intensity vs temperature in sample A. The intensity is normalized to the maximum value. The inset shows the band structure of sample A.

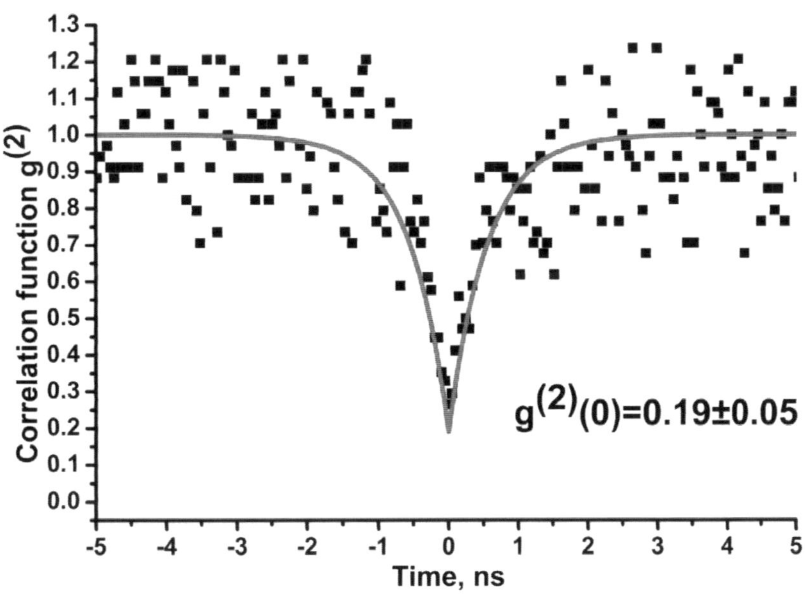

Figure 3. Normalized autocorrelation $g^{(2)}$ function of the single-photon emission measured in sample A at $T = 8$ K using the 200-nm mesa.

The spectrally integrated PL intensity is plotted in figure 2 for the temperature range between 8 and 250 K. The integrated intensity is normalized to the maximum value. The carriers are trapped in a ZnSe/ZnSSe SL at low temperatures and, due to the lack of energy, they cannot pass through a sufficiently thick layer of ZnMgSSe (inset in figure 2). So, a rise in the PL intensity is observed with increasing temperature from 8 to 130 K. The decrease in the PL intensity, observed between 160 to 250 K, is due to carrier delocalization and migration towards centers of nonradiative recombination. In total, the integrated PL intensity drops by less than a factor of 1.5.

Sample A fabricated by the MEE growth mode demonstrates better optical characteristics with respect to quantum light generation. Indeed, the distinct antibunching behavior has been recorded using the 200-nm mesas fabricated in this sample. Figure 3 shows the autocorrelation function $g^{(2)}(\tau)$ of a single QD, registered at T= 8 K. This function is obtained by fitting the experimental data to the equation:

$$f(\tau) = a - b\exp\left(-\frac{|\tau|}{c}\right) \qquad (1)$$

and normalized according to the formula $C_N(t) = c(t)/(N_1 N_2 \omega T)$, where $c(t)$ is the raw coincidence number, $N_{1,2}$ are single counter rates, ω is the width of the time bin, and T is the total acquisition time [12]. The fitted value c = 510 ps gives an estimation of the intrinsic width of the correlation function dip, which is essentially due to the lifetime of the involved QD exciton. For the measured correlation function, the fitting gives the value of $g^{(2)}(0) = 0.19 \pm 0.05$ that evidences the dominance of single-photon emission.

4. Conclusions

We have presented comparative studies of self-assembled CdSe QDs grown either by MEE or MBE. Only the approach involving the MEE growth mode allowed us to obtain a low enough QD density of 10^{10} cm^{-2} and to extend the available spectral range of the single-photon emission below 500 nm. The low value of the second-order correlation function measured in this sample, as well as rather high carrier confinement, provide an opportunity to use similar CdSe QDs in various quantum systems working at high temperatures.

Acknowledgements

The authors gratefully acknowledge the financial support of the Russian Science Foundation (Project #14-22-00107).

References

[1] Knill E, Ladlamme R and Milburn G 2001 *Nature* **409** 46
[2] Bacher G, Weigand R, Seufurt J, Kulakovskii V D, Gippius N A and Forchel A 1999 *Phys. Rev. Lett.* **83** 21
[3] Sorokin S V, Sedova I V, Gronin S V, Klimko G V, Belyaev K G, Rakhlin M V, Mukhin I S, Toropov A A and Ivanov S V 2017 *J.of Cryst. Growth* **477** 127
[4] Sedova I V, Sorokin S V, Sitnikova A A, Nekrutina O V, Reznitsky A N and Ivanov S V 2003 *Inst. Phys. Conf.* **174** Edited by Ilegems M, Weimann G and Wagner J 161
[5] Marsal L, Besombes L, Tinjod F, Kheng K, Wasiela A, Gilles B, Rouviere J-L and Mariette H 2002 *J. of Appl. Phys.* **91** 4936
[6] Arians R, Gust A, Kummell T, Kruse C, Zaitsev S, Bacher G and Hommel D 2008 *Appl. Phys. Lett.* **93** 173506
[7] Fedorych O, Kruse C, Ruban A, Hommel D, Bacher G and Kummel T 2012 *Appl. Phys. Lett.* **100** 061114

[8] Sebald K, Michler P, Passow T, Hommel D, Bacher G and Forchel A 2002 *Appl. Phys. Lett.* **81** 2920
[9] Rakhlin M V, Belyaev K G, Sedova I V, Sorokin S V, Gronin S V, Usikova A A, Sitnikova A A, Brunkov P N, Ivanov S V and Toropov A A 2016 *Phys. Status Solidi C* **13** 514
[10] Patton B, Langbein W and Woggon U 2003, *Phys. Rev. B* 68 125316
[11] Ivanov S V, Toropov A A, Sorokin S V, Shubina T V, Sedova I V, Sitnikova A A, Kop'ev P S, Alferov Zh I, Lugauer H-J, Reuscher G, Keim M, Fischer F, Waag A and Landwehr G 1999 *Appl. Phys. Lett.* **74** 4
[12] Brouri R, Beveratos A, Poizat J-P and Grangier P 2000 *Opt. Lett.* **25** 1294

Synthesis of optosensitive structures based on zinc oxide

A A Ryabko, A I Maximov, V A Moshnikov and S S Nalimova

St. Petersburg Electrotechnical University "LETI", 5 Prof. Popova st., St. Petersburg 197376, Russia

E-mail: a.a.ryabko93@yandex.ru

Abstract. In the research, light-sensitive structures based on a heterostructure of ZnO-nanorods/PbS colloidal quantum dots (CQDs) were fabricated. The paper describes technological features of obtaining coatings by the low-temperature hydrothermal method and fabrication of a ZnO-nanorods/PbS CQDs heterostructure. It is shown that this structure is capable of converting light into electrical energy. These structures are promising for use as gas sensors functioning at temperatures close to room temperature.

1. Introduction

Nanostructured materials based on zinc oxide, due to their large specific surface and semiconductor properties, find applications in photocatalysis [1], gas sensors [2, 3], photovoltaic cells [4–7], and piezoelectric nanogenerators [8]. The use of zinc oxide in the form of filamentary nanocrystals is of special interest due to their high crystalline qualities and high charge carrier mobility.

Presently, gas sensors based on metal oxides generally operate at high temperatures (about 400 °C). The need for heating oxide gas sensors leads to an increase in energy consumption, which is unacceptable for stand-alone and mobile devices. A new approach to solve this problem is the replacement of heating by light irradiation [9]. The sensitization of zinc oxide nanostructures by colloidal quantum dots (QCDs) is promising for the use of optical and infrared radiation.

The appearance of an electric current upon irradiation of the ZnO/QCDs structure can be a sign of the effective injection of charge carriers into the conduction band of zinc oxide, that is, the operation of this structure as a photovoltaic cell. In this paper, heterostructures of ZnO nanorods/PbS QCDs were fabricated. The colloidal quantum dots of lead sulphide were chosen due to their widely tuned band gap (0.71–1.28 eV) [10]. PbS QCDs with a band gap of 1.35 eV were also studied in this paper.

2. Experiment

The fabrication of ZnO-nanorod/PbS QCD heterostructures can be divided into three consecutive stages: the formation of ZnO seed layers, the synthesis of coatings based on ZnO nanorods by the low-temperature hydrothermal method, and the deposition of PbS colloid quantum dots. In this work, the seed layers were formed by the ultrasonic spray-pyrolysis method. The main advantage of this method is the simplicity of the equipment, which determines the low cost of the technology. The spray pyrolysis method is based on thermal decomposition of solution aerosols. The use of ultrasonic spraying, as a rule, leads to the creation of aerosols from smaller droplets, which ultimately leads to more uniform coatings. Thus, an aqueous solution of zinc acetate $Zn(CH_3COO)_2{\cdot}2H_2O$ with a concentration of 0.05 M was used as a spray solution. The temperature of the heating element was maintained at 380 °C. At present, the low-temperature hydrothermal method has become one of the

IOP Conf. Series: Journal of Physics: Conf. Series **993** (2018) 012024 doi:10.1088/1742-6596/993/1/012024

main methods for the synthesis of nanorods, since the use of alkaline reagents allows synthesis at temperatures below 100 °C and does not require complicated equipment. An aqueous solution with equimolar concentrations (0.025 M) of hexamethylenetetramine (HMTA) and zinc nitrate was used as a solution for the synthesis of ZnO nanorods. To form the ZnO–nanorods/PbS QCD structure, a dispersion of colloidal quantum dots was introduced onto ZnO nanorods by spin coating at 2500 rpm for 20 seconds.

3. Results and discussion

The morphology of a typical ZnO seed layer obtained by ultrasonic spray pyrolysis is shown in figure 1.

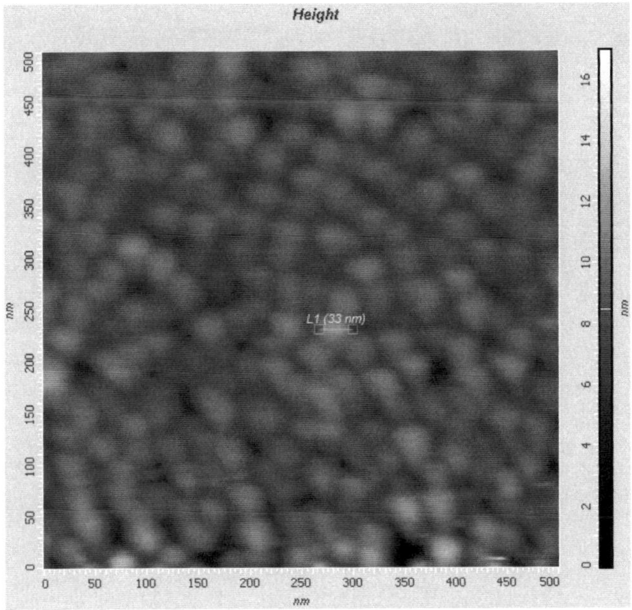

Figure 1. AFM image of a seed layer obtained by ultrasonic spray pyrolysis (on a silicon substrate).

The image shows that a uniform seed layer of zinc oxide consisting of ~ 30 nm crystallites with a small spread in height is formed.

Zinc nitrate serves as a source of Zn^{2+} ions in the low-temperature hydrothermal method. HMTA, first of all, serves as a source of hydroxyl group OH^-, and secondly, it acts as a buffer, slowly releasing OH^-. The rate of hydrolysis of HMTA decreases with increasing pH and vice versa. However, nucleation in the bulk of the solution is observed in the synthesis process, which leads to the contamination of the sample surface, rapid depletion of the solution and, accordingly, a relatively low growth rate of nanorods. To suppress nucleation in the volume of the solution, ammonium hydroxide and polyethyleneimine were used. Ammonium hydroxide, forming complexes with zinc ions, reduces the degree of supersaturation in zinc, and polyethyleneimine molecules adsorbed on the nuclei due to the steric effect prevent the achievement of the critical size of ZnO nuclei, at which further growth becomes thermodynamically stable. So polyethyleneimine (0.005 M) and ammonium hydroxide (0.66 M) were added to the growth solution with equimolar concentrations (0.025 M) of HMTA and zinc nitrate. In this case, suppression of bulk nucleation and an increase in the growth rate of ZnO nanorods (figure 2 (b)) were observed.

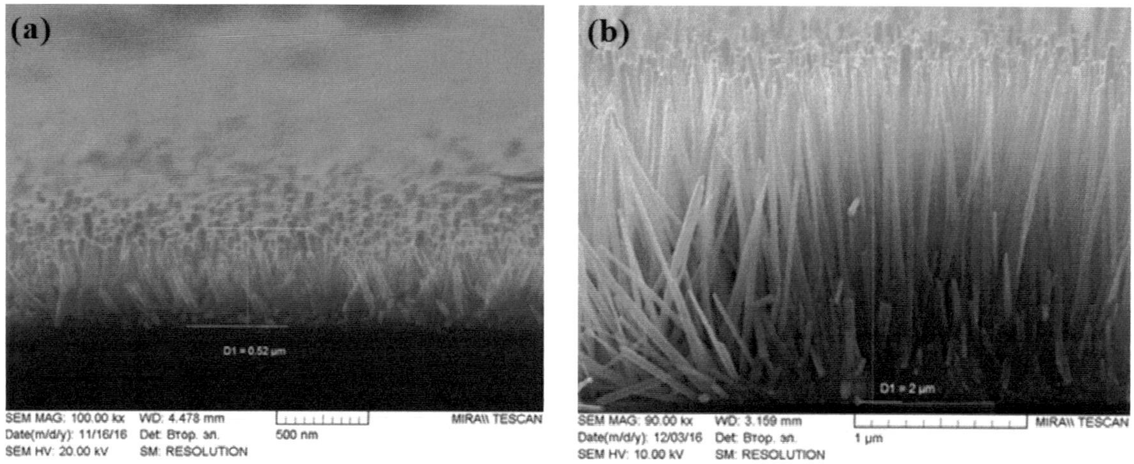

Figure 2. SEM images of: a sample without suppression of bulk nucleation (a); a sample with suppression of bulk nucleation (b). The duration of the synthesis was 60 minutes.

Figure 2 shows that suppression of bulk nucleation leads to an increase in the length of ZnO nanocrystals by approximately 4 times. The shape of the nanorods also changes, which is due to the gradual depletion of reagents in the solution. The suppression of bulk nucleation also avoids the contamination of the growing ZnO nanorod arrays by agglomerates appearing in the solution.

The optical absorption spectrum of colloidal quantum dots is shown in figure 3.

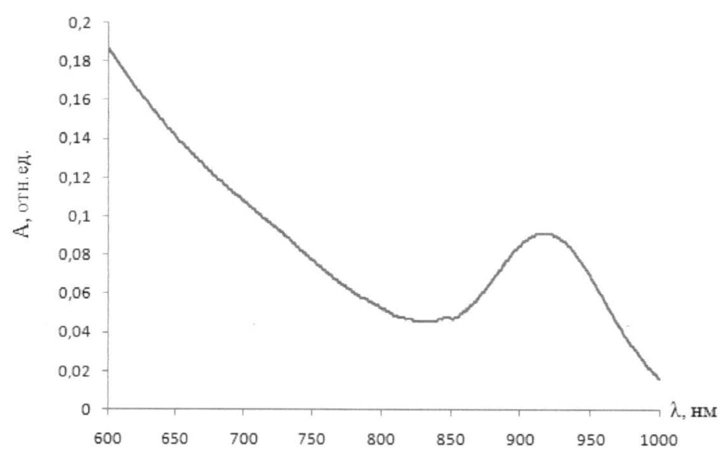

Figure 3.Optical absorption spectrum of colloidal quantum dots.

The exciton peak of absorption of the PbS QCDs is ~ 920 nm, which corresponds to a band gap of 1.35 eV. To improve transport in the QCDs layer, long molecules of oleic acid were replaced with short ligands, which were tetrabutylammonium iodide (TBAI), because PbS QCDs were stabilized with oleic acid and dispersed in toluene. The solution of tetrabutylammonium iodide in methanol was introduced to the PbS layer by spin coating, followed by washing in methanol. With a different number of deposition cycles of colloidal quantum dots, the values of the short-circuit current changed. However, the values of the open-circuit voltage V_{oc} remained almost unchanged (~ 50 mV). The SEM image of the sample with the highest value of the short-circuit current is shown in figure 4.

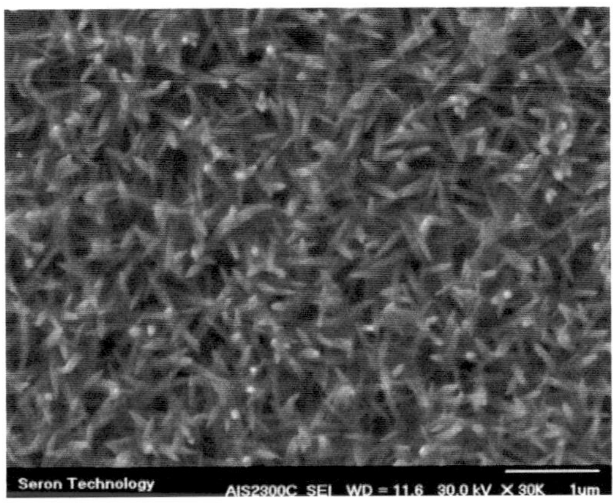

Figure 4. SEM image of the measured structure.

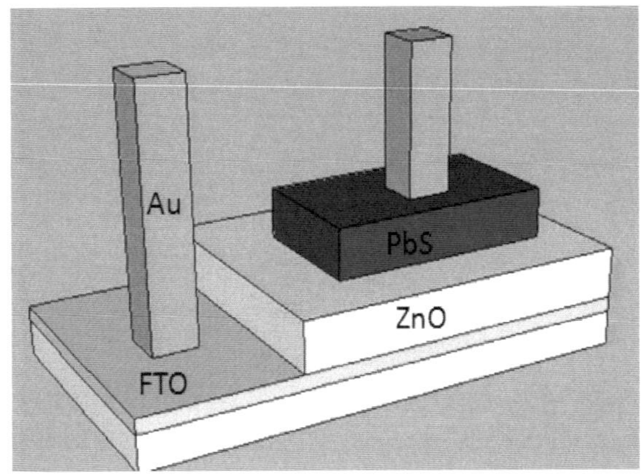

Figure 5. Schematic representation of the measured structure with clamping contacts.

Figure 4 shows that there is a second phase on the ZnO nanorods — the PbS QCDs. Measurement of the voltage-current characteristics of the heterostructure was carried out using clamping gold contacts (figure 5). One contact was fed to the layer of colloidal PbS quantum dots, and the second to the FTO coating.

The energy parameters of the measured structure are shown in figure 6. The electronic affinity for the FTO was adopted at 4.4 eV [11]. The electron affinity for ZnO in the direction (000-1), (0001) usually lies in the range 4.44 - 4.2 eV [12]. The value of the work function was adopted at 4.3 eV [13]. A value of the zinc oxide band gap was taken at 3.4 eV. The electron affinity for PbS usually takes the value of ~ 4.3 eV [7]. The position of the Fermi level is influenced by the type of ligands. So in [7] the position of the Fermi level was at 0.51 eV below the bottom of the conduction band in the case of using TBAI.

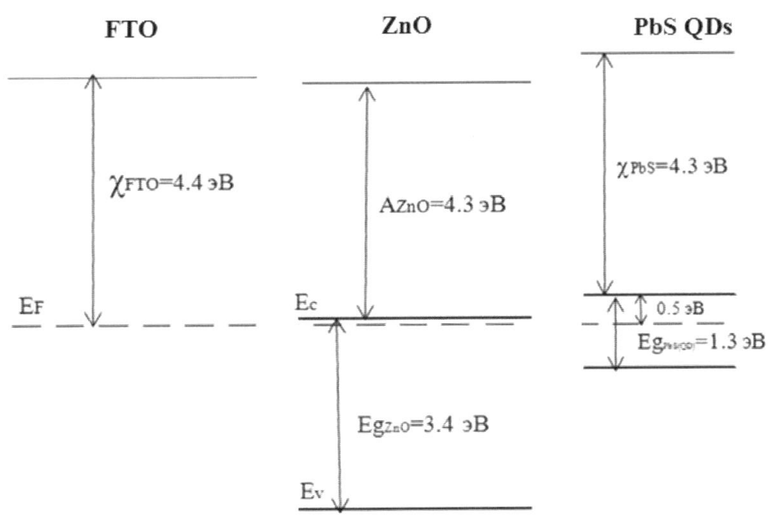

Figure 6. Energy parameters of the ZnO/PbS heterostructure materials.

The current-voltage characteristic of the sample showed that there is a nonzero current at zero voltage both under illumination and without illumination (figure 7).

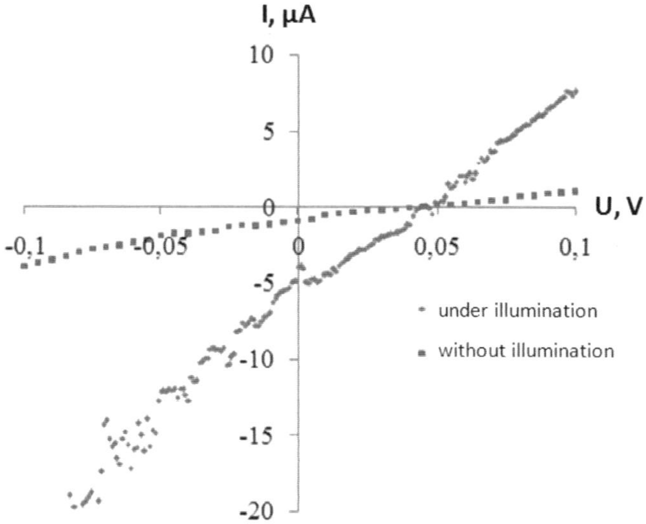

Figure 7. Voltage-current characteristic of the heterostructure.

Generation of an electric current without illumination can be related to the background of infrared radiation. During illumination, the short-circuit current increased, while voltage remained unchanged at V_{oc} ~ 50 mV. Thus, the structure of ZnO-nanorods/PbS-colloidal quantum dots is sensitive to radiation in the optical and infrared regions of the spectrum.

4. Conclusion

The use of hydrothermal synthesis makes it possible to form dense arrays of zinc oxide nanowires that are promising for creating highly sensitive gas sensors. The heterojunction of ZnO nanorods / PbS-QCDs is designed to activate the gas sensitive layers operating at temperatures close to room temperature by irradiating the structure and injecting the charge carriers into the zinc oxide conduction band. Irradiation of the structure leads to an increase in the electron concentration in zinc oxide crystals and to an increase in its catalytic activity.

Acknowledgments

The investigation was made possible by the Russian Science Foundation (project № 17-79-20239).

References

[1] Bozhinova A S, Kaneva N V, Kononova I E, Nalimova S S, Syuleiman Sh A, Papazova K I, Dimitrov D Ts, Moshnikov V A and Terukov E I 2013 *Semiconductors* **47** 1636–40
[2] Karpova S S, Moshnikov V A, Maksimov A I, Mjakin S V and Kazantseva N E 2013 *Semiconductors* **47** 1026–30
[3] Karpova S S, Moshnikov V A, Mjakin S V and Kolovangina E S 2013 *Semiconductors* **47** 392–5
[4] Bobkov A A, Maximov A I, Moshnikov V A, Somov P A and Terukov E I 2015 *Semiconductors* **49** 1357–60
[5] Lashkova N A, Maximov A I, Ryabko A A, Bobkov A A, Moshnikov V A and Terukov E I 2016 *Semiconductors* **50** 1254–60
[6] Bobkov A A, Lashkova N A, Maximov A I, Moshnikov V A and Nalimova S S 2017 *Semiconductors* **51** 60–5
[7] Chuang Ch-H M, Brown P R, Bulović V and Bawendi M G 2014 *Nature Materials* **13** 796-801
[8] Wang Z L and Song J 2006 *J.Science* **312** 242–6

IOP Conf. Series: Journal of Physics: Conf. Series **993** (2018) 012024 doi:10.1088/1742-6596/993/1/012024

[9] Il'in A S, Fantina N P, Martyshov M N, Forsh P A, Chizhov A S, Rumyantseva M N, Gaskov A M and Kashkarov P K 2016 *Semiconductors* **50** 607–11
[10] Moreels I, Lambert K, Smeets D, Muynck D, Nollet T, Martins J S, Vanhaecke F, Vantomme A, Delerue Ch, Allan G and Hens Z 2009 *ACS Nano* **3** 3023–30
[11] Helander M G, Greiner Z, Wang Z B and Tang W M 2011 *J. Vac. Sci. Technol. A* **29** 011019
[12] Chevtchenko S A, Moore J C, Özgür Ü, Gu X, Baski A A and Morkoç H 2006 *Appl. Phys. Lett.* **89** 182111
[13] Lange I, Reiter S, Pätzel M, Zykov A, Nefedov A, Hildebrand J, Hecht S, Kowarik S, Wöll C, Heimel G and Neher D 2014 *Adv. Funct. Mater.* **24** 7014–24

IOP Conf. Series: Journal of Physics: Conf. Series **993** (2018) 012025 doi:10.1088/1742-6596/993/1/012025

Investigation of the influence of geometric parameters of carbon nanotube arrays on their adhesion properties

M V Il'ina, A A Konshin, O I Il'in, N N Rudyk, A A Fedotov and O A Ageev

Southern Federal University, "Nanotechnologies" Scientific and Educational Center, Shevchenko 2, Taganrog 347922, Russia

E-mail: mailina@sfedu.ru

Abstract. The results of experimental studies of adhesion of carbon nanotube (CNT) arrays with different geometric parameters and orientations using atomic-force microscopy are presented. The adhesion values of CNT arrays were determined, which were from 82 to 1315 nN depending on the parameters of the array. As a result, it was established that the adhesion of a CNT array increases with an increase in branching and disorientation of the array, as well as with the growth of the aspect ratio of CNTs in the array.

1. Introduction

The rapid development of nanotechnologies and the new opportunities for creating nanoscale structures have generated a great interest in artificial adhesion structures that mimic the parameters of biological objects (microfobs of gecko). The creation of artificial adhesion structures will solve a number of problems for micro- and nanoelectronics, as well as for the space industry.

Initially, polymeric columnar nanoscale structures were used as artificial adhesive structures. However, further studies have shown that the most promising material for creating such structures is aligned carbon nanotubes (CNTs) [1–8]. The adhesion of CNT arrays (10–500 N/cm^2) is many times greater than the adhesion of artificial polymer columnar nanostructures (3 N/cm^2) [1–8] due to the unique mechanical and geometric properties of nanotubes [9, 10].

The aim of this work is to study the influence of geometric parameters of CNT arrays (length and diameter of nanotubes, density and branching) on their adhesion properties.

2. Experimental details

The samples we studied were CNT arrays grown by plasma-enhanced chemical vapor deposition (PECVD) on a silicon wafer with a two-layer structure (Ni / Ti) [11]. For the study, 14 samples of four types were chosen: vertically aligned CNT bundles, individual vertically aligned CNTs, branched CNTs and disoriented CNTs, shown in figure 1 (a–d), respectively. The geometric parameters of the arrays were determined on the basis of statistical processing of scanning electron microscopy (SEM) images obtained using a Nova NanoLab 600 (FEI, Netherlands) and are presented in table 1.

The adhesion properties of carbon nanotube arrays were studied by atomic-force microscopy (AFM) using the NTEGRA Probe NanoLaboratory (NT-MDT). The AFM probe was a polysilicon probe with a conductive coating of tungsten carbide from the HA_HR/W_2C series with a radius of curvature of 35 nm and a stiffness of k = 34 N/m.

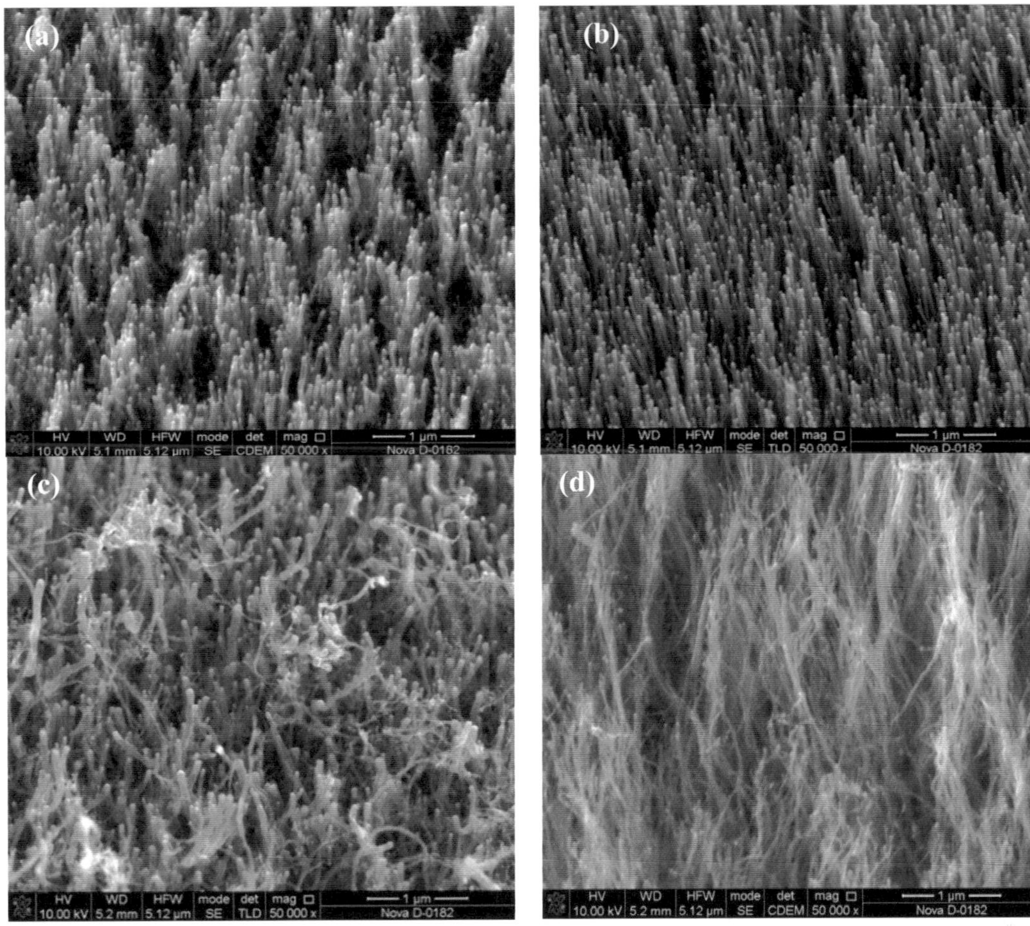

Figure 1 (a-d). SEM images of the four types of CNT arrays.

During the preliminary scanning of arrays in the semi-contact AFM mode, the surface of CNT arrays was determined and the measurement points were set. Then, the AFM force spectroscopy mode was used to measure two dependences: the dependence of the cantilever bending value on the degree of extension of the scanner's z-piezoelectric element (signal h) on the direct cantilever course (the red dependence in figure 2) and the same for the inverse cantilever course (the black dependence in figure 2). The characteristic experimental dependences of $J_{DFL}(h)$ obtained for CNT samples of the first and fourth types are shown in figure 2.

The dependences obtained using a force spectroscopy mode allow us to evaluate the adhesion of an CNT array F_a to the AFM probe surface [7]:

$$F_a = \Delta J_{DFL}(\mathrm{d}h / \mathrm{d}J_{DFL})k, \quad (1)$$

where ΔJ_{DFL} is the value equal to the difference of the J_{DFL} signal at the point of complete detachment of the probe from the CNT array and at the point of the maximum negative value of the J_{DFL} signal.

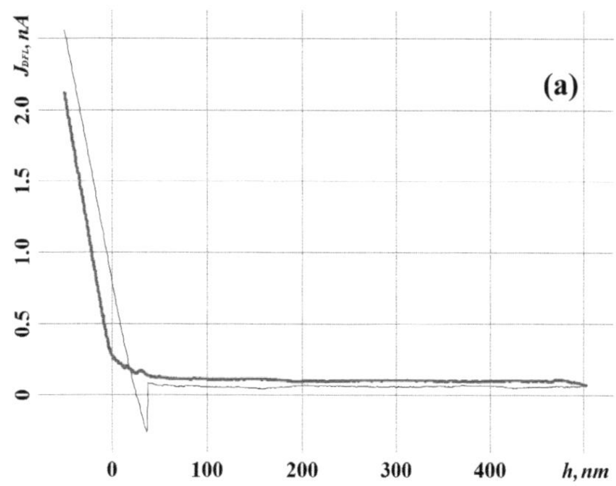

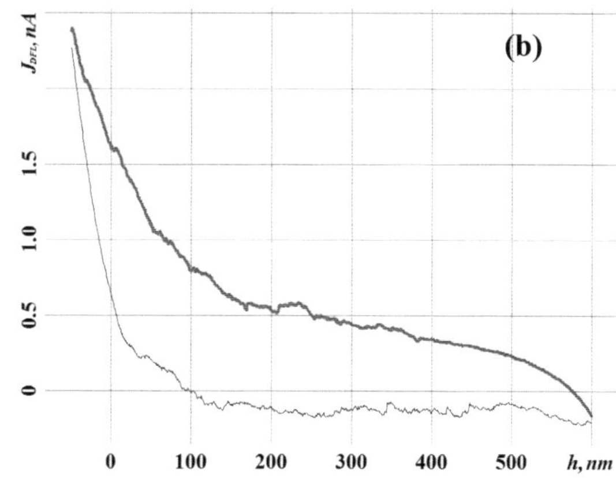

Figure 2. The experimental dependences $J_{DFL}(h)$ for the bundles of vertically aligned (a) and disoriented (b) CNTs.

The adhesion values calculated using formula (1) for the experimental samples are presented in table 1.

Table 1. The calculated adhesion values for experimental samples and their geometric parameters

Number of sample	Diameter (nm)	Height (μm)	Density ($1/\mu m^2$)	Adhesion (nN)	Average value of adhesion (nN)
			CNT bundles		
1	43.8	0.65	82	195.28	
2	35.6	1.21	72	255.71	294.77
3	51	0.69	69	433.31	
			Individual CNTs		
4	62	0.28	54	82.26	
5	56	0.55	59	179.58	
6	44.1	0.35	95	193.18	225.37
7	63.2	0.65	38	260.54	
8	44.9	0.71	60	411.29	
			Branched CNTs		
9	42.1	0.42	71	166.67	
10	50.1	15.97	44	1314.71	585.44
11	44.7	12.45	26	274.93	
			Disoriented CNTs		
12	34.6	2.11	89	349.71	
13	28.8	3.07	45	647.61	633.90
14	28.7	2.87	52	904.37	

3. Results and discussion

The analysis of the obtained results showed that the minimum value of adhesion (82.26 nN) was achieved for CNT sample 4, which is an array of individual vertically aligned CNTs with a minimum aspect ratio (the ratio of a CNT's length to its diameter) of ~ 4.5. The maximum value of adhesion (1314.71 nN) was demonstrated by CNT sample 10, which has having a branched structure, a high density and a maximum value of the aspect ratio of ~ 320. The analysis of the measurement results has also shown that the average adhesion values of experimental samples of disoriented CNT have the greatest value of adhesion. This is probably due to the fact that the disorientation of nanotubes contributes to an increase in the contact area between the AFM probe and the CNT array.

In addition, the adhesion measurements of individual CNTs (experimental samples 4 – 8) showed that the adhesion value increases from 82.26 to 411.29 nN with an increase in the aspect ratio from 4.5 to 15.8, respectively. At the same time, sample 6 was somewhat of an outlier for this dependence, which may be due to a significant increase in the density of CNTs in the array (by more than 1.5 times) that results in an increase in the number of CNTs interacting with the AFM probe as compared to samples 4, 5, 7 and 8.

The dependence of the adhesion value on the array density is confirmed by the results of force spectroscopy of sample 1 of vertically aligned CNT bundles, shown in figure 3.

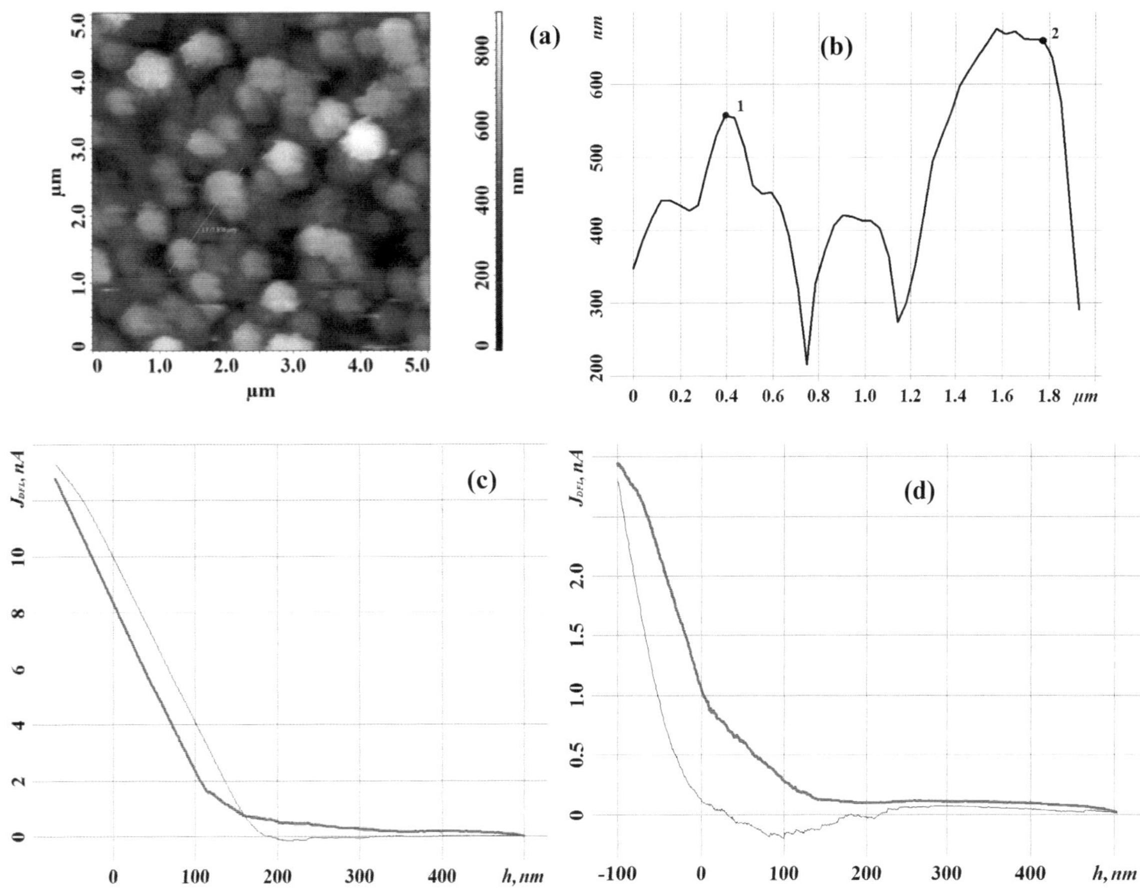

Figure 3. Results of the study of CNT sample 1: AFM image (a), a profilogram along the line (b) and force spectroscopy data at point 1 (c) and point 2 (d).

The obtained $J_{DFL}(h)$ dependences for bundles with diameters of 174 nm (point 1) and 452 nm (point 2) show that the adhesion of the AFM probe to a bundle of a larger diameter significantly exceeds the adhesion to a bundle of a smaller diameter (see figure 3(c-d)). In this case, the $J_{DFL}(h)$ dependence for a large-diameter CNT bundle had several peaks during both approach and retraction of the AFM probe due to the serial interaction of individual nanotubes with the AFM probe. Thus, it is shown that an increase in the number of CNTs interacting with an AFM probe leads to an increase in the adhesion of CNT arrays measured by a local method of AFM force spectroscopy.

The obtained results suggest that the adhesion of CNT arrays depends not only on the value of the aspect ratio and the structure of a CNT array, but also on the density of the array. Thus, it can be concluded that the greatest adhesion of CNT arrays to the AFM probe is observed for high density arrays of branched CNTs with high values of the aspect ratio.

4. Conclusion

This work experimentally studied the adhesion of carbon nanotube arrays of various types by using a force spectroscopy mode of AFM with a polysilicon probe with a radius of curvature of 35 nm. The value of the adhesion of CNT arrays varied from 82 to 1315 nN depending on their geometric parameters, which agrees well with the available literature data [1, 7, 8]. It is established that the maximum value of adhesion of CNT arrays to the AFM probe was achieved for the arrays that combine the branched or disoriented structure, the high density and the high aspect ratio of CNTs.

The results obtained can be used in the development of adhesion structures and elements of nanoelectronics based on carbon nanotubes.

Acknowledgment
The results were obtained using the equipment of the Scientific and Educational Center and the "Nanotechnology" Center for Collective Use at the Southern Federal University.

This work was financially supported by the Russian Foundation for Basic Research (project 16-29-14023) and an internal grant of the Southern Federal University (project VnGr-07/2017-26).

References

[1] Qu L and Dai L 2007 *Adv. Mater.* **19** 3844
[2] Rouhrig M, Thiel M, Worgull M and Houlscher H 2012 *Small* **8** 3009
[3] Autumn K, Liang Y A, Hsieh S T, Zesch W, Chan W P, Kenny Th W, Fearing R and Full R J 2000 *Nature* **405** 681
[4] Jeong H E, Lee S H, Kim P and Suh K Y 2008 *Colloid Surface A* **313-314** 359
[5] Majidi C, Groff R E, Autumn K, Baek S, Bush B, Gravish N, Maboudian R, Maeno Y, Schubert B, Wilkinson M and Fearing R S 2006 *Phys.Rev.Lett.* **97** 076103
[6] Qu L, Dai L, Stone M, Xia Z and Wang Z L 2008 *Science* **322** 238
[7] Yurdumakan B, Raravikar N R, Ajayan P M and Dhinojwala A 2005 *Chem. Comm.* **2005** 3799
[8] Tsai P-C, Jeng Y-R, Maoa C-P, Franklin K-TW and Hong C-N 2013 *Thin Solid Films* **545** 401
[9] Ageev O A, Il'in O I, Kolomiitsev A S, Konoplev B G, Rubashkina M V, Smirnov V A and Fedotov A A 2012 *Nanotechnol. Russ.* **7** 47
[10] Ageev O A, Blinov Yu F, Il'ina M V, Il'in O I, Smirnov V A and Tsukanova O G 2016 *Phys. Solid State* **58** 309
[11] Ageev O A *et al* 2016 *Advanced Materials: Manufacturing, Physics, Mechanics and Applications* (*Springer Proceedings in Physics* vol 175) ed I A Parinov *et al* (Heidelberg, New York, Dordrecht, London: Springer) chapter 40 pp 563–580

Conductive atomic force microscopy study of the photoexcitation effect on resistive switching in ZrO_2(Y) films with Au nanoparticles

A S Novikov[1], D O Filatov[2], D A Antonov[2], I N Antonov[3], M E Shenina[2] and O N Gorshkov[2,3]

[1] Department of Physics, Lobachevsky State University of Nizhny Novgorod, Nizhny Novgorod 603950, Russia
[2] Research and Educational Center for Physics of Solid State Nanostructures, Lobachevsky State University of Nizhny Novgorod, Nizhny Novgorod 603950, Russia
[3] Research Institute for Physics and Technology, Lobachevsky State University of Nizhny Novgorod, Nizhny Novgorod 603950, Russia

E-mail: nsv3333@yandex.ru

Abstract. We report on the experimental observation of the effect of optical excitation on resistive switching in ultrathin ZrO_2(Y) films with single-layered arrays of Au nanoparticles. The samples were prepared by depositing nanometer-thick Au films sandwiched between two ZrO_2(Y) layers by magnetron sputtering followed by annealing. Resistive switching was studied by conductive atomic force microscopy by measuring cyclic current-voltage curves of a probe-to-sample contact. The contact area was illuminated by radiation of a semiconductor laser diode with the wavelength corresponding to the plasmon resonance in an Au nanoparticle array. The enhancement of the hysteresis in cyclic current–voltage curves due to bipolar resistive switching under illumination was observed. The effect was attributed to heating of Au nanoparticles due to plasmonic optical absorption and a plasmon resonance, which enhances internal photoemission of electrons from the Fermi level in Au nanoparticles into the conduction band of ZrO_2(Y). Both factors promote resistive switching in a ZrO_2(Y) matrix.

1. Introduction

Resistive switching (RS) has been extensively studied in the last decade due to the prospects of its application in novel non-volatile memory devices [1]. The effect of RS consists in a reversible change of the resistance of a thin dielectric film sandwiched between two conductive electrodes (such a device has been called a *memristor* [2]) under the electric voltage applied between the electrodes. In recent years, the idea of using electromagnetic radiation as one of the methods to control the RS effect has attracted considerable attention [3, 4]. Optically controlled memristors are promising for various applications in novel integrated photonic devices and, even more, as the basis of a new field of science and technology – *memristive optoelectronics*.

In the present work, we have investigated the effect of optical illumination on local RS in ultrathin ZrO_2(Y) films (about 40 nm in thickness) with single sheet arrays of Au nanoparticles (NPs) using conductive atomic force microscopy (CAFM) [5]. Earlier, the observation of the effect of optical excitation on RS in Au/ZrO_2(Y)/Si metal-oxide-semiconductor stacks was reported [6]. The effect was attributed to a photovoltage at the ZrO_2(Y)/Si interface, which was added to the external voltage applied between an Au gate electrode and an *n*-Si substrate, thus promoting RS in a ZrO_2(Y) film. The goal of the present work was to study the effect of optical excitation on RS while excluding the substrate effect. For this purpose, we have studied ZrO_2(Y):NP-Au films deposited onto glass substrates covered by conductive transparent indium-tin-oxide (ITO) films. We attributed the observed effect of optical illumination to the excitation of plasmonic oscillations in Au NPs.

2. Experimental details

ZrO_2(Y) films with embedded Au NPs were deposited onto glass substrates covered by ~ 1 μm thick ITO films. ZrO_2(Y) layers were deposited by high-frequency magnetron sputtering in an Ar–O_2 gas mixture (50:50 % mol.) at a pressure of about 10^{-2} Torr from powder oxide targets using a Torr International® MSS-3GS vacuum system for thin film deposition. The molar fraction of the stabilizing oxide Y_2O_3 in the target material was about 0.12. The substrate temperature T_g was about 300 °C. Nanocomposite ZrO_2(Y):NP-Au films were prepared by the alternating deposition of ZrO_2(Y) and Au films followed by annealing [7]. First, underlying ZrO_2(Y) layers of about 20 nm thick were deposited onto ITO sublayer surfaces. Then, islanded Au films with a nominal thickness $d_{Au} \approx 1$ nm were deposited by direct current magnetron sputtering at $T_g = 200$ °C in Ar ambient. Finally, the islanded Au films were capped with ~20-nm-thick cladding ZrO_2(Y) layers deposited in the same conditions as the underlying ones. The samples were annealed in Ar ambient at 450 °C during 1 hour. The resulting structure of the nanocomposite film is shown schematically in figure 1. Earlier, the high-resolution cross-sectional transmission electron microscopy studies have shown the islanded Au films in ZrO_2(Y)/Au/ZrO_2(Y) stacks prepared as described above to coagulate into Au NPs during annealing in specified conditions [8]. Nearly spherical Au NPs with a diameter $D \approx 2$ nm were arranged almost in a single sheet inside ZrO_2(Y) films. The average spacing (l) between the centers of Au NPs in two-dimensional arrays was found to be about 4 nm. So far, the density parameter of an Au NP array $a = D/l$ was about 0.5. Also, the ZrO_2(Y) films with a thickness of about 40 nm were deposited onto ITO/glass substrates to serve as reference samples.

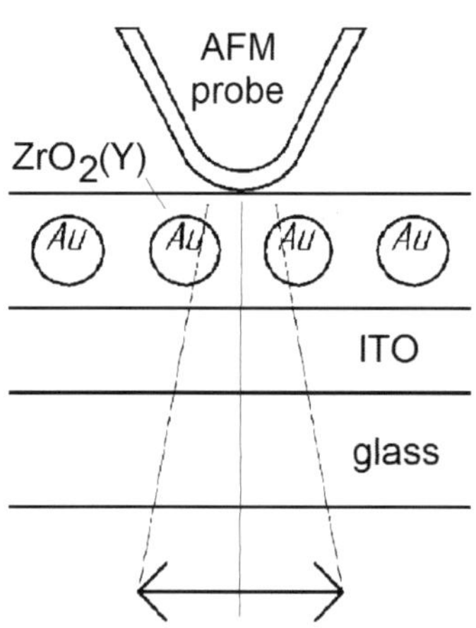

Figure 1. Schematic of the experiment to study the optical illumination effect on resistive switching in ZrO_2(Y):NP-Au films.

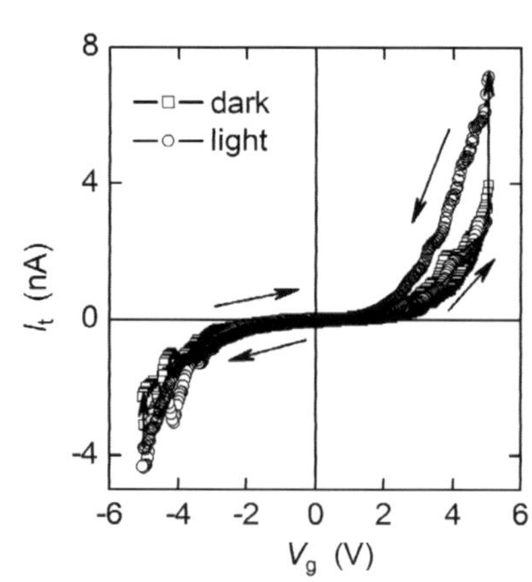

Figure 2. Cyclic *I–V* curves of the contact between the CAFM probe and ZrO_2(Y):NP-Au/ITO/glass film, measured in the dark and under photoexcitation.

The optical properties of ZrO_2(Y):NP-Au films were examined by optical transmission spectroscopy at 300 K using a Varian® Cary™ 6000i spectrophotometer. The RS investigations were carried out in an ultra-high vacuum (UHV) environment using an Omicron® UHV AFM/STM LF1™

atomic force/scanning tunneling microscope installed in an Omicron® MultiProbe™ RM UHV system. The base residual gas pressure inside the UHV chamber was about 10^{-10} Torr. The schematic layout of the experiment is shown in figure 1. The surface of a ZrO_2(Y):NP-Au film was scanned in contact mode by an NT MDT® NSG-11 DCP™ probe coated with a conductive diamond-like film. The gap voltage V_g has been applied to a conductive ITO sublayer with respect to a CAFM probe (virtually grounded), and the electric current flowing through the CAFM probe I_t was measured. The cyclic I–V curves of the probe-to-sample contact $I_t(V_g)$ were measured in every point of the AFM scan. The limits of the ramp voltage sweep in the course of recording the cyclic I–V curves V_{min} and V_{max} were set to +5 V and −5 V, respectively. These values are well above the threshold voltages of RS from the "OFF" state to the "ON" state (V_{set}) and back from the "ON" state to the "OFF" state (V_{reset}) in order to ensure RS of a nanocomposite ZrO_2(Y):NP-Au film during the bias voltage sweep cycle. The gap voltage V_g was swept from V_{min} up to V_{max} and back to V_{min} in a cyclic manner.

In order to study the effect of optical excitation on RS, the contact area between the CAFM probe and the ZrO_2(Y):NP-Au film was illuminated through a transparent glass substrate and a conductive ITO sublayer by a focused beam of a semiconductor laser diode (LD) attached to the AFM/STM chamber of an Omicron® UHV system (see figure 1). The LD output power was about 1 W, and the beam intensity was adjusted by a series of neutral light filters. The LD emission wavelength was about 660 nm, which matched the wavelength of the plasmon resonance (PR) in a dense Au NP array.

It is worth noting that one could expect the PR peak wavelength to be about 600 nm for separated spherical Au NPs with $D = 2$ nm embedded into a ZrO_2(Y) matrix, as follows from the Mie scattering theory [9], as well as from the results of earlier experiments [10]. The red shift of the PR peak to ≈ 660 nm in the optical transmission spectra of the sample studied in the present work was attributed to the collective plasmon excitations in a dense NP array [11, 12].

In order to protect the beam deflection sensor of an Omicron® AFM/STM LF1™ from damage by the powerful emission of a LD, the Si-based quadrant photodiode of the AFM sensor was protected by a colored glass light filter, which is transparent for the emission of an adjusting LD of the AFM sensor (with an emission wavelength of about 810 nm).

3. Results and discussion

Figure 2 shows the cyclic I–V curves for the contact between the CAFM probe and the ZrO_2(Y):NP-Au/ITO/glass film, measured under photoexcitation and in the dark. A butterfly-type hysteresis has been observed. The hysteresis was attributed to bipolar RS in the ZrO_2(Y):NP-Au film. The area of the hysteresis loop was essentially enhanced under illumination. This effect has been attributed to the excitation of collective plasmon oscillations in a dense Au NP array.

The excitation of plasmonic oscillations in Au NPs can affect RS in a ZrO_2(Y):NP-Au film in two ways.

The first one is heating of the NP material (and, correspondingly, of the ZrO_2(Y) matrix around NPs) due to plasmonic optical absorption [13]. Today's understanding of the RS mechanism in transition metal oxides is based on a concept of the migration of oxygen vacancies in the electric field applied to a dielectric film [14]. As a result, oxygen vacancies are arranged into conductive filaments growing through almost the entire thickness of the dielectric layer (the so-called forming process). In the particular case of Au NPs embedded into ZrO_2(Y) films, the arrangement of oxygen vacancies into filaments is promoted by the concentrated electric field near the surfaces, which is induced by the metal NPs [15, 16]. In view of all the above, local heating causes the migration of oxygen vacancies and hence the initiation of the filament growth near the NP surface.

The second possible mechanism of the illumination effect on RS could be related to the PR-enhanced internal photoemission of electrons from the Fermi level in Au NPs into the oxygen-vacancy-related α-band in ZrO_2(Y) barriers [13] or directly into the conduction band of the ZrO_2(Y) matrix. Free electrons in the conduction band can be accelerated in the strong electric field between the CAFM probe and the ITO sublayer, interact with the ZrO_2(Y) lattice, and release the accumulated energy to local heating (generation of phonons). Also, the generation of additional free electrons is possible via impact ionization (the avalanche breakdown of a ZrO_2(Y) film). The above phenomena can also initiate filament formation.

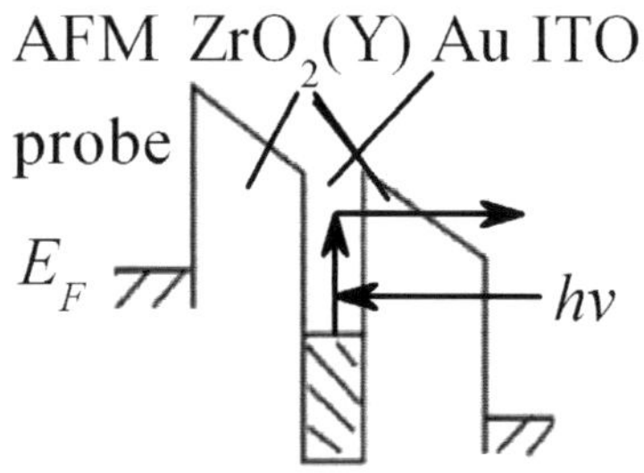

Figure 3. Qualitative band picture (300 K) of an Au NP in a ZrO_2(Y) film in the external electric field applied between the CAFM probe and the conductive ITO layer.

It is worth noting that the height of the potential barrier for electrons at the Au/ZrO_2(Y) interface (between the Fermi level in Au NPs and the conduction band edge in ZrO_2(Y)) is about 2.5 eV at 300 K [17]. At the same time, the photon energy $h\nu$ corresponding to an emission wavelength of 660 nm is about 1.9 eV, which is not enough for the direct internal photoemission of electrons into the conduction band of ZrO_2(Y) (see the band picture of an Au NP in a ZrO_2(Y) film in figure 3). However, the electrons could be potentially photoexcited into the conduction band of ZrO_2(Y) via photon-assisted tunneling through a triangle barrier at the interface of Au NPs with a ZrO_2(Y) matrix, as shown in figure 3. Note also that the triangle barrier height could be essentially decreased by the electric field between the CAFM probe and the ITO sublayer.

4. Conclusions

In the present study, we have observed experimentally that resistive switching is enhanced in a ZrO_2(Y) film with embedded Au nanoparticles under optical excitation at the wavelength corresponding to the plasmon optical absorption resonance in Au nanoparticles. The effect was attributed to the collective plasmon excitation in a dense Au nanoparticle array, which results in heating of Au nanoparticles and the surrounding ZrO_2(Y) matrix and in the plasmon-assisted internal photoemission of electrons into the conduction band of the ZrO_2(Y) matrix. Both factors promote the forming and resistive switching of ZrO_2(Y) films. To clarify the details of how optical excitation affects resistive switching in ZrO_2(Y) films with Au nanoparticles, further theoretical and experimental investigations are needed, including computer simulations of the effect of plasmonic optical absorption in Au nanoparticles on the migration of oxygen vacancies in a surrounding ZrO_2(Y) matrix.

Acknowledgements

The authors gratefully acknowledge the financial support of the Ministry of Education and Science of the Russian Federation (Project # 16.7864.2017). The CAFM measurements were performed using the shared research facilities of the Research and Educational Center for Physics of Solid State Nanostructures at Lobachevskii State University of Nizhnii Novgorod.

References

[1] Ielmini D and Waser R 2016 *Resistive Switching: From Fundamentals of Nanoionic Redox Processes to Memristive Device Applications* (Mannheim: Wiley-VCH)
[2] Strukov D B, Snider G S, Stewart D R and Williams R S 2008 *Nature* **453** 80
[3] Wang W, Panin G N, Fu X, Zhang L, Ilanchezhiyan P, Pelenovich V O, Fu D and Kang T W 2016 *Nature Sci. Rep.* **6** 31224
[4] Sun B, Wu J, Jia X, Lou F and Chen P 2017 *J. Sol-Gel Sci. Technol.* **75** 664
[5] Lee M H and Hwang C S 2011 *Nanoscale* **3** 490
[6] Tikhov S V, Gorshkov O N, Koryazhkina M N, Antonov I N and Kasatkin A P 2016 *Tech. Phys. Lett.* **42** 536

[7] Cho S, Lee H, Kub D Y, Leec T. S, Cheongc B, Kimc W M and Leec K S 2004 *Thin Solid Films* **447–448** 68
[8] Gorshkov O N, Antonov I N, Filatov D O, Shenina M E, Kasatkin A P, Pavlov D A and Bobrov A I 2016 *Tech. Phys. Lett.* **42** 36
[9] Mie G 1908 *Ann. Phys.* **25** 377
[10] Gorshkov O N, Pavlov D A, Trushin V N, Antonov I N, Shenina M E, Bobrov A I, Markelov A S, Dudin A Yu and Kasatkin A P 2012 *Tech. Phys. Lett.* **38** 185
[11] Saiki T 2017 *Appl. Phys. A* **123** 577
[12] Auguié B and Barnes W L 2008 *Phys. Rev. Lett.* **101** 143902
[13] Liskin D A, Filatov D O, Gorshkov O N, Gorshkov A P, Antonov I N, Shenina M E, Zubkov S Y and Sinutkin D Yu 2017 *J. Phys.Conference Series* **816** 012010
[14] Lee J S, Lee S and Noh T W 2016 *Appl. Phys. Rev.* **2** 031303
[15] Guan W, Long S, Jia R and Liu M 2007 *Appl. Phys. Lett.* **91** 062111
[16] Chang W-Y, Cheng K-J, Tsai J-M, Chen H-J, Chen F, Tsai M-J and Wu T B 2009 *Appl. Phys. Lett.* **95** 042104
[17] Filatov D O, Guseinov D V, Antonov I N, Kasatkin A P and Gorshkov O N 2014 *RSC Adv.* **4** 57337

Photoelectron scattering in a *p*-GaN(Cs,O) photocathode

S A Rozhkov[1,2], V V Bakin[1], S N Kosolobov[1], H E Scheibler[1,2] and A S Terekhov[1]

[1]Rzhanov Institute of Semiconductor Physics of the Siberian Branch of the Russian Academy of Sciences, 13 Lavrentieva ave., Novosibirsk 630090, Russia
[2]Novosibirsk State University, 2 Pirogova str., Novosibirsk 630090, Russia

E-mail: rozhkovs@isp.nsc.ru

Abstract. Photoelectron scattering in a *p*-GaN(Cs,O) photocathode was studied by photoelectron emission spectroscopy at low temperatures. It has been experimentally established that the transfer of photoelectrons from the bulk of a *p*-GaN(Cs,O) photocathode to vacuum is accompanied by the emission of a cascade of optical phonons in the bulk and at the GaN–vacuum interface. Based on how the high-energy features of the photoelectron energy distribution shift with the photon energy, we find $m_{hh} / m_0 = 2.3 \pm 0.3$.

1. Introduction

The GaN photocathodes with effective negative electron affinity (NEA) are one of the key elements of the ultraviolet position-sensitive single-photon detectors [1] and are promising candidates for electron sources in photoinjectors [2]. Despite the practical usefulness, the influence of certain physical processes on the low-energy photoelectron escape from a GaN NEA photocathode into vacuum is still unknown. One of these processes is inelastic photoelectron scattering in the bulk of GaN and near its emitting surface.

Nonequilibrium electron dynamics in GaN was studied earlier in many theoretical and experimental works (for example, [3–5] and [6–9], respectively), but their results are partly controversial. For instance, the authors of [8, 9] conclude that the electron scattering into the side valleys in GaN is stronger than the LO-phonon scattering for electrons with energies above 1 eV, while in [10] authors state that this conclusion contradicts the calculations of the band structure and hot electron transport in GaN based on a variety of experimental findings. In [10], the authors conclude that scattering into the side valleys should start at electron energies well above 2 eV. In addition to the above-mentioned discrepancies, the calculated [3, 4] and the measured [6] phonon scattering rates also substantially differ from each other. The above examples of inconsistencies do not allow us to rely entirely on the literature data to determine the role of inelastic scattering in the process of escape of photoelectrons into vacuum from a GaN NEA photocathode. We believe that the most reliable information about this role can be obtained directly from the energy distributions of photoelectrons ($n_e(\varepsilon)$) emitted from a GaN NEA photocathode to vacuum since they should contain information about the photoelectron distribution function in GaN. Earlier, $n_e(\varepsilon)$ was measured in [9, 11, 12], but in these works, measurements were performed with sufficiently low energy resolution and at room temperature, which could significantly broaden the important features of energy distributions. Recently, we measured the low-temperature photoemission spectra of a *p*-GaN(Cs,O) photocathode with high energy resolution [13]. The measurements of the quantum efficiency ($QE(\hbar\omega)$) and the longitudinal energy distribution of photoelectrons ($n_e(\varepsilon_{lon})$) were performed at 80 K,

IOP Conf. Series: Journal of Physics: Conf. Series **993** (2018) 012027 doi:10.1088/1742-6596/993/1/012027

which significantly reduced thermal broadening of the energy distributions and made it possible to detect the emission of a cascade of optical phonons by low-energy photoelectrons at the emitting surface of the *p*-GaN(Cs,O) photocathode.

In this work, the emission of optical phonons by hot photoelectrons in the *p*-GaN bulk and at the surface of a *p*-GaN(Cs,O) photocathode was observed and studied using photoelectron spectroscopy in a wide range of photoelectron energy.

2. Experimental details

The measurements of $QE(\hbar\omega)$ and $n_e(\varepsilon_{lon})$ were performed with a homemade planar photodiode where a transmission-mode *p*-GaN(Cs,O) photocathode and a metallic anode were mounted parallel to each other in a hermetically sealed titanium-aluminum case. The diameters of the photocathode and anode were 18 mm, which was larger than the distance between them by a factor of about 20. The electric field near the axis of the photodiode was sufficiently uniform, and we used the photodiode as an electron spectrometer with a uniform retarding field (URF spectrometer) [14]. The photocathode was fabricated from a *p*-GaN/AlN heterostructure grown on a (0001) sapphire substrate by MOCVD. The thickness of the wurtzite p-GaN layer of the photocathode was 85 nm, and the concentrations of magnesium and free holes (at 300 K) in this layer were 1×10^{19} and $\sim 2 \times 10^{17}$ cm^{-3}, respectively. The methods of cleaning the *p*-GaN surface and the deposition of the (Cs,O) layers were described in [15]. The measurements were performed using a setup consisting of a stabilized xenon lamp, a modulator of light, a monochromator with diaphragms on the input and output slits, quartz objectives, a liquid-nitrogen optical cryostat, a digital lock-in amplifier, several DAC–ADC units, and a computer. The photoemission was studied in the temperature range of 80–300 K. The spectral resolutions of the monochromator at the measurements of $QE(\hbar\omega)$ and $n_e(\varepsilon_{lon})$ were 2 and 15 meV, respectively. When measuring $QE(\hbar\omega)$, an accelerating voltage of 10 V was applied to the anode. To measure $n_e(\varepsilon_{lon})$ distributions, a retarding voltage with a controlled amplitude U_{ret} and an AC voltage with a constant amplitude of 30 meV were simultaneously applied to the anode. The amplitude of the AC component of the photocurrent as a function of U_{ret} was measured by a digital lock-in amplifier and recalculated to the $n_e(\varepsilon_{lon})$ distribution [14]. To obtain the derivative of the $n_e(\varepsilon_{lon})$ distribution, the second harmonic of an AC component of the photocurrent was simultaneously measured by the same digital lock-in amplifier. The energy resolution of the URF spectrometer used in this work was 30 meV. The sheet resistance of a *p*-GaN film at 80 K was ~ 1 MOhm/sq, so in order to avoid the surface charging and the corresponding $n_e(\varepsilon_{lon})$ distortion, the photoemission currents were kept below 20 nA.

3. Results

Figure 1 shows the results of $QE(\hbar\omega)$ measurements at 90 K. The shape of the *QE* spectrum is mainly determined by the shape of the absorption spectrum of the p-GaN layer of the photocathode. One can clearly see the threshold of band-to-band transitions, below which *QE* decreases exponentially. The position of the band gap energy (ε_g) for this photocathode is indicated by the arrow [13]. At $\hbar\omega > \varepsilon_g$, oscillations due to the interference of light in the *p*-GaN/AlN heterostructure are present in the spectrum. In optical transitions in the region $\hbar\omega < \varepsilon_g$, photoelectrons are excited both to the tails of the density of states of the conduction band of the *p*-GaN layer and to the band bending region (BBR) near the emitting surface due to the Franz-Keldysh effect.

We measured $n_e(\varepsilon_{lon})$ distributions at several $\hbar\omega_i$, which are marked by enumerated triangles in figure 1. Some of the measured $n_e(\varepsilon_{lon})$ distributions are shown in figure 2. All distributions were measured at 90 K. For all $\hbar\omega_i$, the energy distributions are broad functions in the range from the vacuum level (ε_{vac}) to the energy of the conduction band minima in the bulk of p-GaN (ε_{cb}). The position of ε_{cb} was determined according to [13]. It is seen that with an increase in $\hbar\omega_i$ the high-energy edge of the distributions shifts to higher energies and the number of hot photoelectrons increases substantially. Some $n_e(\varepsilon_{lon})$ distributions contain weak oscillations in the region $0.6 < \varepsilon_{lon} < 1.0$ eV (curves 2–4). To identify the nature of these weak features in the energy distributions, the derivatives of $n_e(\varepsilon_{lon})$ distributions were analyzed.

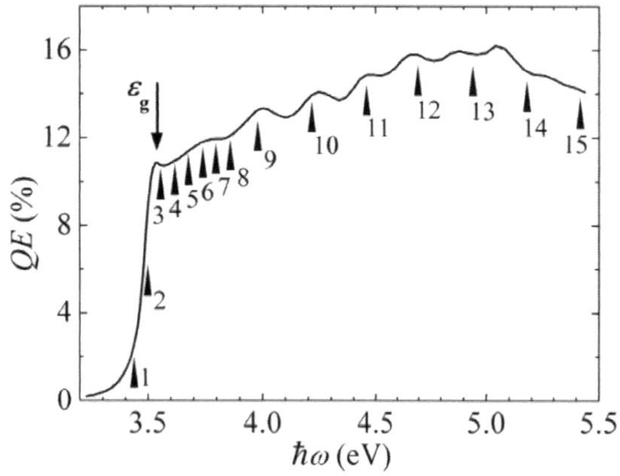

Figure 1. Quantum efficiency spectrum of a *p*-GaN(Cs,O) photocathode.

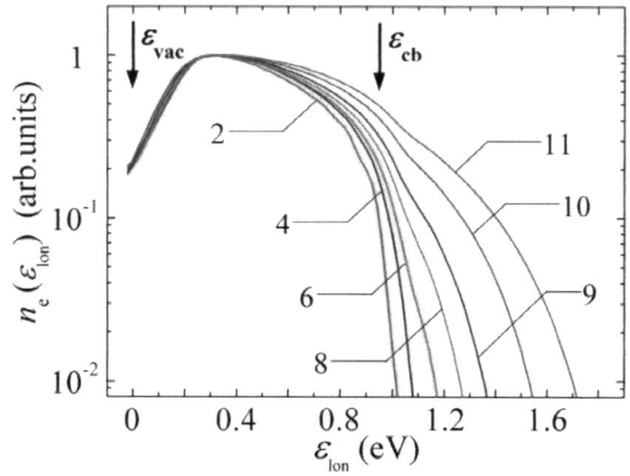

Figure 2. Longitudinal energy distributions of photoelectrons emitted by a *p*-GaN(Cs,O) photocathode.

Figure 3 shows the derivatives of $n_e(\varepsilon_{lon})$ multiplied by –1 for the $\hbar\omega_i$ values close to the threshold of the band-to-band transitions. For convenience, the curves are shifted vertically. The derivatives of $n_e(\varepsilon_{lon})$ at $\hbar\omega > \hbar\omega_1$ (curves 2–8) have an oscillating character. With increasing $\hbar\omega$, all features of the distributions shift to higher energies. It can be seen that the curves change qualitatively in the interval $\hbar\omega_4 - \hbar\omega_5$.

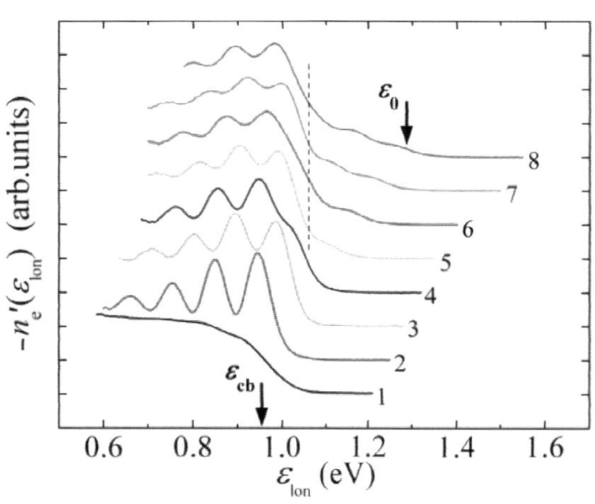

Figure 3. The derivative of longitudinal energy distributions of photoelectrons emitted by a *p*-GaN(Cs,O) photocathode, measured at low photon energies ($\hbar\omega_i = \hbar\omega_1 - \hbar\omega_8$).

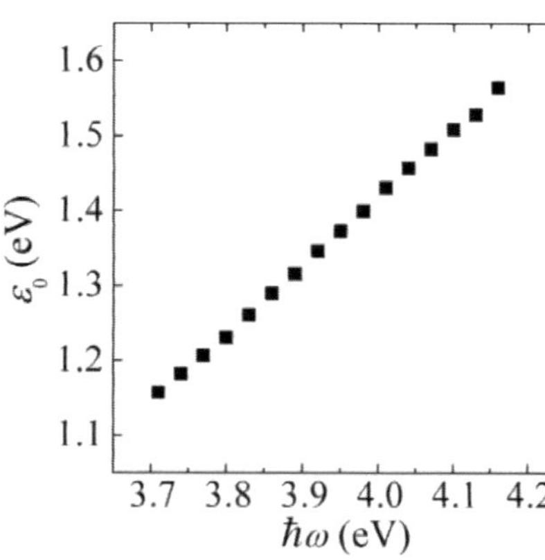

Figure 4. The position of a high-energy "step" vs the photon energy.

In the $\hbar\omega_1 < \hbar\omega < \hbar\omega_4$ range (curves 2–3), the derivatives consist of a series of equidistant peaks, the distance between which is 95 ± 3 meV. For $\hbar\omega_1 < \hbar\omega < \hbar\omega_2$, the amplitude of the peaks increases with an increase in $\hbar\omega$. At $\hbar\omega > \hbar\omega_2$ (curves 2–4), the peaks begin to broaden, and the amplitude of the high-energy peak begins to decrease so that for $\hbar\omega > \hbar\omega_4$ (curves 5–8) a high-energy peak transforms into a high-energy "step".

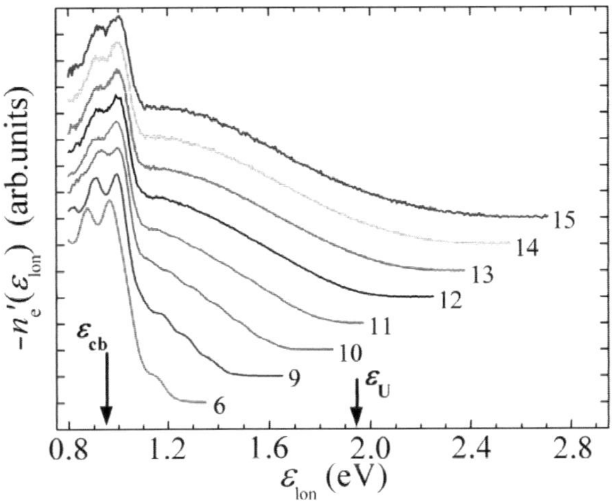

Figure 5. The derivative of longitudinal energy distributions of photoelectrons emitted by a *p*-GaN(Cs,O) photocathode, measured at high photon energies ($\hbar\omega_i = \hbar\omega_6, \hbar\omega_9 - \hbar\omega_{15}$).

For $\hbar\omega > \hbar\omega_4$ (curves 5–8), the derivatives can be divided into two regions. At $\varepsilon_{lon} \geq \varepsilon_{cb} + 0.1$ eV (marked vertically by the dashed line), the distributions have the form of weak equidistant "steps", the distance between which is 94 ± 2 meV. At $\varepsilon_{lon} \leq \varepsilon_{cb} + 0.1$ eV, the value of the derivative of $n_e(\varepsilon_{lon})$ increases sharply, and the curves consist of a series of strongly broadened peaks, similarly to the $\hbar\omega_1 < \hbar\omega < \hbar\omega_4$ region. At $\hbar\omega \geq \hbar\omega_5$, the position of the high-energy "step" ε_0, indicated by the arrow in figure 3 for curve 8, shifts linearly with an increase in $\hbar\omega$ (see figure 4), with a coefficient of 0.92 ± 0.01.

The derivatives of $n_e(\varepsilon_{lon})$ for higher $\hbar\omega$ are shown in figure 5. One can see that with a further increase in $\hbar\omega$, the form of the spectra does not change qualitatively. For all $\hbar\omega_i$ values in the energy range $\varepsilon_{lon} \geq \varepsilon_{cb} + 0.1$ eV, the high-energy "step" shifts to higher energies and monotonically broadens with increasing $\hbar\omega$. To describe this broadening, we subtracted the smooth component from the derivatives and took the Fourier transform in the energy region $\varepsilon_{lon} \geq \varepsilon_{cb} + 0.1$ eV. Figure 6 shows the dependence of the Fourier harmonic amplitude ($A_{Fourier}$) on $\hbar\omega$. Similarly, the dependence of $A_{Fourier}$ on the temperature was measured (figure 7) for a constant value of the initial photoelectron energy in the conduction band. Also, we can note that in the energy range $1.8 < \varepsilon_{lon} < 2.1$ eV, where the minimum of the side U-valleys of GaN is presumably located [8, 9] (marked by "ε_U" in figure 5), no specific features in the energy distributions (curves 12–15) were found.

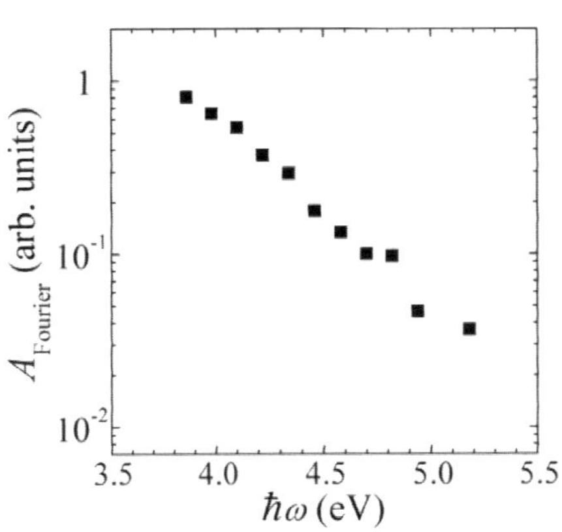

Figure 6. The dependence of the Fourier amplitude of high-energy "steps" on the photon energy. $T = 90$ K.

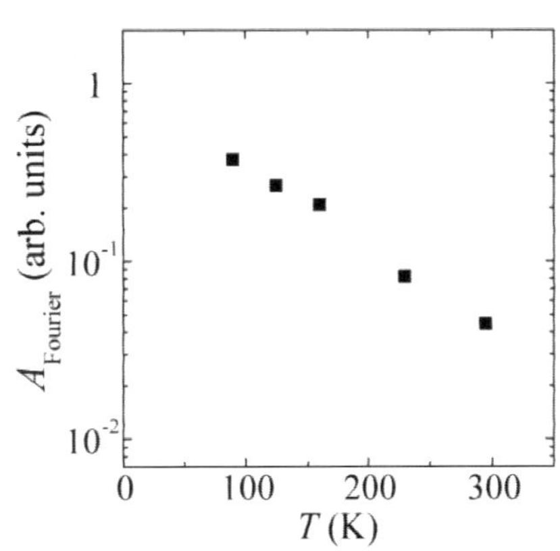

Figure 7. The temperature dependence of the Fourier amplitude of high-energy "steps". $\hbar\omega(T) = \varepsilon_g(T) + 0.68$ eV.

4. Discussion

We believe that the observed features of the distributions and their behavior have a clear physical interpretation. The distance between the features in the distributions is close to the energy of the LO phonons ($\hbar\Omega_{LO} = 92$ meV) in GaN, so we associate the features with the emission of optical phonons by photoelectrons. The high-energy peaks in the derivative of the $n_e(\varepsilon_{lon})$ distributions in the $\hbar\omega_1 < \hbar\omega < \hbar\omega_4$ range correspond to photoelectrons that escape to vacuum without emission of optical phonons. The lower peaks correspond to the photoelectrons that emit one or more optical phonons. We should point out that oscillations in the distributions are observed at $\hbar\omega \approx \varepsilon_g$ ($\hbar\omega_1 < \hbar\omega < \hbar\omega_3$), when the energy of photoelectrons in the quasineutral part of p-GaN is less than the energy of LO phonons, therefore in this case the emission of optical phonons occurs only in the BBR and on the vacuum side of the interface [13]. The amplitude of the high-energy peak at $\hbar\omega_1 < \hbar\omega < \hbar\omega_4$ depends on the mean free path of the photoelectrons in GaN. With an increase in $\hbar\omega$, electrons gain sufficient energy to emit an LO phonon in the p-GaN bulk. In this case, the energy mean free path decreases sharply and, consequently, the amplitude of the high-energy peak decreases (figure 3, curves 3–5).

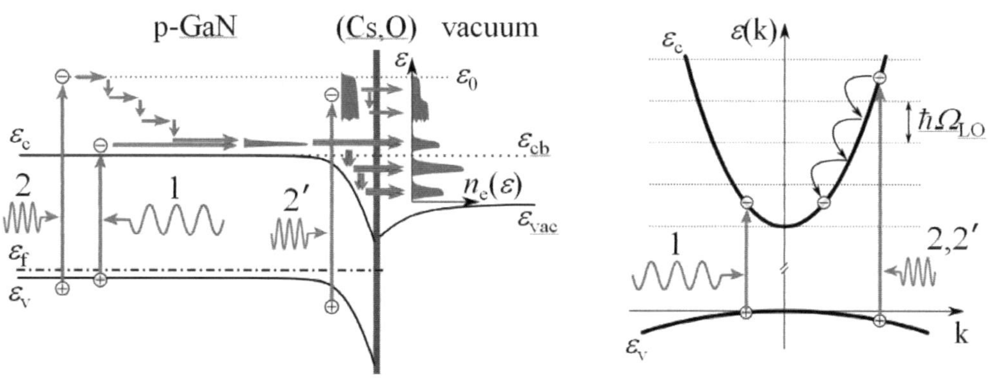

Figure 8. Energy diagram of a *p*-GaN(Cs,O) photocathode.

Figure 8 illustrates the process of photoemission in a *p*-GaN(Cs,O) photocathode. At low photon energies ($\hbar\omega_1 < \hbar\omega < \hbar\omega_3$), when the photoelectron energy in the quasineutral part of *p*-GaN is less than the energy of the LO phonon (process "1" in figure 8), electrons reach the BBR, gain kinetic energy in the near-surface electric field and escape to vacuum while emitting a cascade of optical phonons. At high photon energies ($\hbar\omega > \hbar\omega_4$), the majority of photoelectrons produced in the quasineutral part of p-GaN first emits a cascade of LO phonons (process "2" in figure 8), and then reaches the BBR with low kinetic energy. Subsequently, these electrons emit a cascade of phonons and escape to vacuum. Along with this, a small fraction of photoelectrons is generated near the surface in the BBR (process "2′" in figure 8) at the distances of several mean free paths from the surface and escape to vacuum while emitting several phonons. In contrast to the processes "1" and "2", these photoelectrons do not have a narrow initial energy distribution because it is smeared by the near-surface electric field. Photoelectrons excited at different positions in the BBR have different energies, therefore at high $\hbar\omega$ we observe series of "steps". The position of the high-energy "step" in the derivatives of $n_e(\varepsilon_{lon})$ (ε_0 in figure 3) coincides with the initial energy of photoelectrons in the conduction band. In the model of parabolic isotropic bands, the electron energy (ε) increases with increasing $\hbar\omega$ according to the following law:

$$\varepsilon(\hbar\omega) = (\hbar\omega - \varepsilon_g)/(1 + m_e / m_h), \quad (1)$$

where m_e is the electron effective mass and m_h is the hole effective mass. Using the value of the electron mass in GaN ($m_e / m_0 = 0.20$) [16] and the slope coefficient of the $\varepsilon_0(\hbar\omega)$ dependence (figure 4), we can calculate the effective mass of heavy holes in GaN. For energies of heavy holes in the valence band $\varepsilon_{hh} > 15$ meV, we find $m_{hh} / m_0 = (2.3 \pm 0.3)$. It is known that the valence band structure in wurtzite GaN is non-parabolic and anisotropic near the valence band edge, and that the hole effective mass increases with increasing energy [17]. In [18], an increase in the hole effective mass from 1.2 m_0 to 1.8 m_0 was measured in the interval of hole energies 5–11 meV by magnetoreflectance spectroscopy. We measured the value of the heavy hole mass at slightly higher hole energies and it agrees well with the above data.

With an increase in $\hbar\omega$ and temperature, the "contrast" of all high-energy features in the derivatives of $n_e(\varepsilon_{lon})$ decreases. Apparently, this is due to the broadening of the energy distribution of photoelectrons in *p*-GaN. Possible physical reasons for this broadening may be related to the anisotropy of the GaN band structure, the lifetime broadening of electronic states, and acoustic phonon scattering of photoelectrons during their motion in the BBR. In order to determine the main contribution, we plan to model the generation and transport of photoelectrons in *p*-GaN at different electron energies and temperatures.

5. Conclusions

The photoemission spectra of a *p*-GaN(Cs, O) photocathode at low temperatures were measured. In energy distributions of emitted electrons, the features associated with the optical phonon cascade emission by hot photoelectrons both in the bulk and at the surface of a *p*-GaN(Cs,O) photocathode were observed for the first time. It is established that the emission of optical phonons is the dominant inelastic scattering channel in *p*-GaN(Cs,O)-photocathodes, and it significantly affects their photoemission characteristics. The value of the heavy hole mass in GaN is determined from the dependence of the position of the high-energy features of the energy distributions on $\hbar\omega$. No features in the energy distributions related to photoelectron scattering to side valleys of GaN were observed.

Acknowledgments

We are grateful to V.V. Lundin (Ioffe Institute, Russian Academy of Sciences) for providing heterostructures for fabricating photocathodes.

This work was supported by the Russian Foundation for Basic Research (project no. 16–02–00534).

References

[1] Siegmund O, Vallerga J, McPhate J, Malloy J, Tremsin A, Martin A, Ulmer M and Wesswls B 2006 *Nucl. Instrum. Methods Phys. Res.* A **567** 89

[2] Bazarov I V, Dunham B M, Liu X, Virgo M, Dabiran A M, Hannon F and Sayed H 2009 *J. Appl. Phys.* **105** 083715

[3] Yamakawa S, Akis R, Faralli N, Saraniti M and Goodnick S M 2009 *J. Phys.: Condens. Matter* **21** 174206

[4] Bertazzi F, Moresco M and Bellotti E 2009 *J. Appl. Phys.* **106** 063718

[5] Zhukov V P, Tyuterev V G, Chulkov E V and Echenique P M 2016 *J. Appl. Phys.* **120** 085708

[6] Tsen K T, Ferry D K, Botchkarev A, Sverdlov B, Salvador A and Morkoç H 1997 *Appl. Phys. Lett.* **71** 1852

[7] Shishehchi S, Rudin S, Garrett G A, Wraback M and Bellotti E 2013 *J. Appl. Phys.* **114** 233106

[8] Marcinkevičius S, Uždavinys T K, Foronda H M, Cohen D A, Weisbuch C and Speck J S 2016 *Phys. Rev.* B **94** 235205

[9] Piccardo M, Martinelli L, Iveland J, Young N, DenBaars S P, Nakamura S, Speck J S, Weisbuch C and Peretti J 2014 *Phys. Rev.* B **89** 235124

[10] Bertazzi F, Goano M, Zhou X, Calciati M, Ghione G, Matsubara M and Bellotti E 2015 *Appl. Phys. Lett.* **106** 061112

[11] Liu Z, Machuca F, Pianetta P, Spicer W E and Pease R F W 2004 *Appl. Phys. Lett.* **85** 1541

[12] Pakhnevich A A, Bakin V V, Yaz'kov A V, Scheibler H E, Shevelev S V, Tereshchenko O E, Yaroshevich A S and Terekhov A S 2004 *JETP Letters* **79** 479

[13] Rozhkov S A, Bakin V V, Gorshkov D V, Kosolobov S N, Scheibler H E and Terekhov A S 2016 *JETP Letters* **104** 135
[14] Orlov D A, Andreev V E and Terekhov A S 2000 *JETP Letters* **71** 151
[15] Tereshchenko O E, Scheibler H E, Yaroshevich A S, Shevelev S V, Terekhov A S, Lundin V V, Zavarin E E and Besyul'kin A I 2004 *Phys. Solid State* **46** 1949
[16] Drechsler M, Hoffmann D M, Meyer B K, Detchprohm T, Amano H and Akasaki I 1995 *Jpn. J. Appl. Phys.* **34** L1178
[17] Kim K, Lambrecht W R L, Segall B and van Schilfgaarde M 1997 *Phys. Rev.* B **56** 7363
[18] Shields P A, Nicholas R J, Peeters F M, Beaumont B and Gibart P 2001 *Phys. Rev.* B **64** 081203

Bipolar resistive switching in metal-insulator-semiconductor nanostructures based on silicon nitride and silicon oxide

M N Koryazhkina[1], S V Tikhov[1], A N Mikhaylov[1], A I Belov[1], D S Korolev[1], I N Antonov[1], V V Karzanov[1], O N Gorshkov[1], D I Tetelbaum[1], P Karakolis[2,3] and P Dimitrakis[2]

[1] Lobachevsky University, 23/3 Gagarin prospect, 603950 Nizhny Novgorod, Russia
[2] Institute of Nanoscience and Nanotechnology, NCSR "Demokritos", Agia Paraskevi 15341, Greece
[3] Department of Physics, University of Patras, GR 26500 Patras, Greece

E-mail: mahavenok@mail.ru

Abstract. Bipolar resistive switching in metal-insulator-semiconductor (MIS) capacitor-like structures with an inert Au top electrode and a Si_3N_4 insulator nanolayer (6 nm thick) has been observed. The effect of a highly doped n^+-Si substrate and a SiO_2 interlayer (2 nm) is revealed in the changes in the semiconductor space charge region and small-signal parameters of parallel and serial equivalent circuit models measured in the high- and low-resistive capacitor states, as well as under laser illumination. The increase in conductivity of the semiconductor capacitor plate significantly reduces the charging and discharging times of capacitor-like structures.

1. Introduction

Memristors [1] or memristive nanomaterials (devices) are the basis for new-generation non-volatile memory devices (Resistive Random Access Memory), the operation of which relies on switching between at least two stable states of a material at different applied voltages: High Resistance State (HRS) and Low Resistance State (LRS) [2]. Memristive devices are usually implemented on the basis of a metal-insulator-metal (MIM) and (rarely) a metal-insulator-semiconductor (MIS) capacitor-like structure. The latter configuration is the most compatible with the traditional CMOS process, especially if the "native" Si-based insulators (SiO_2 and Si_3N_4) are used. In order to make resistive switching (RS) possible in a metal-insulator structure, the electroforming process is usually required, which consists of applying a voltage greater than a certain value (the forming voltage) and results in irreversible changes in electrical properties of a capacitor due to the local rearrangement of the insulator atomic structure. The MIS configuration with a relatively high-resistance semiconductor substrate (capacitor plate) makes it possible to efficiently investigate the mechanisms of ionic and electronic phenomena occurring in the insulator during electroforming and RS [3], as well as to get the resistive memory effect manipulated by light due to a possible change in resistance of the semiconductor substrate [4]. Usually, heavily doped silicon substrates with negligible resistance, which are similar to metal capacitor plates, are used for the fabrication of memristive MIS capacitors, including capacitors with silicon nitride layers [5-9]. The memristive devices based on Si_3N_4 are characterized by both unipolar and bipolar RS, and active metal electrodes like Ni or Cu are most commonly used, as their material is responsible for the electrochemical RS mechanism.

In the present work, the bipolar RS in the MIS capacitors on the basis of Si_3N_4 with a chemically inert top electrode (Au) has been investigated in relation to the intrinsic phenomena in a "metal-free" insulator-semiconductor nanostructure. The measurements of small-signal parameters of equivalent circuit models showed [10-12] that the resistance of a semiconductor capacitor plate has a great impact on the memristor characteristics when the thickness of an insulator nanolayer is comparable with the thickness of the space charge region in a Si substrate.

2. Material and methods

Memristive capacitor-like devices were deposited on a n^+-Si substrate with an equilibrium density of electrons $n_0 \approx 3{\cdot}10^{19}$ cm^{-3} and an area of 15×15 mm^2. Figure 1 shows the schematic cross-section of these devices. A Si_3N_4 layer with a thickness of 6 nm was deposited on a substrate by the LPCVD method at a temperature of 800 °C. In some of the devices, a 2-nm-thick SiO_2 layer was preliminarily formed on a Si substrate by thermal oxidation of Si in dry oxygen at 900 °C. The top Au electrodes (40 nm) with an area of 10^{-2} or 10^{-3} cm^2 were deposited through a shadow mask by the method of DC-magnetron sputtering at a substrate temperature of 200 °C. On the back-side of a Si substrate, the Al contact was deposited using an e-gun evaporator.

Current-voltage (*I-V*) characteristics were measured by using an Agilent B1500A semiconductor device analyzer at room temperature with voltage sweep amplitudes up to 10 V and a sweep rate of 0.3 V/s. The sign of bias on the device corresponds to the potential of the top electrode (Au). Different current compliance (CC) was used for negative voltage sweeps. The temperature dependencies of current (*I-T*) were measured in a range of 300 ÷ 500 K for the devices biased at $V = 0.5$ V. Small-signal characteristics of the capacitors were measured at a test voltage $V = 20$ mV in a frequency range of $10^3 \div 10^7$ Hz in parallel and serial equivalent circuit models: parallel capacitance (C_p), parallel conductivity reduced to a circular frequency (G_p/ω), parallel resistance (R_p), serial resistance (R_s) and insulator losses ($tan\ \delta$), which do not depend on the type of equivalent circuit. Capacitors with an area of ~ 10^{-2} cm^2 were used in measurements. Before electroforming, some capacitors were also measured under illumination with a blue MASTER-PRO LSB462 laser at a wavelength of 462 nm and a power density of ~ 1 W/mm^2.

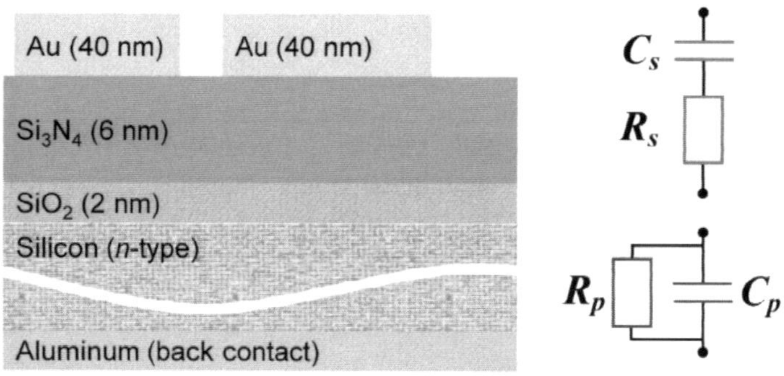

Figure 1. Schematic cross-section of the $Au/Si_3N_4/SiO_2/n^+$-Si capacitor and its equivalent RC circuits used for the determination of small-signal parameters (serial capacitance C_s and resistance R_s, parallel capacitance C_p and resistance R_p).

3. Result and discussion

Let us first consider the experimental results obtained for the $Au/Si_3N_4/n^+$-Si capacitors without a SiO_2 interlayer. Before electroforming, the Si_3N_4 layers were characterized by relatively good insulator properties: an electrical resistivity of ~ $3{\cdot}10^{15}$ Ω·cm and a breakdown field density of ~ $7{\cdot}10^6$ V/cm. The value of the relative permittivity at 1 kHz was ~ 5, and the values of $tan\ \delta$ were ~ 0.01 at the same frequency. The current transport mechanism at room temperature in the initial state (IS) can be explained by the Pool-Frenkel effect [13], as the *I* - *V* characteristics are quite symmetric with respect to the voltage polarity (figure 2) and are linearized in the ln $I - V^{1/2}$ plot with a slope corresponding to the Pool-Frenkel emission (figure 3). The depth of deep traps in Si_3N_4 responsible for the observed Pool-Frenkel effect was calculated from *I-V* characteristics and amounted to ~ 1 eV from the bottom of the conduction band.

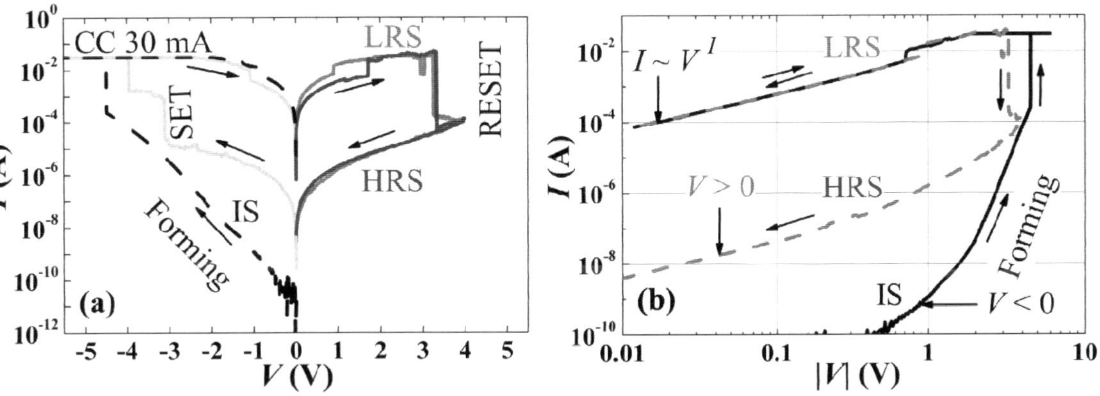

Figure 2. Typical *I-V* characteristics of the Au/Si_3N_4/n^+-Si capacitor, measured in the processes of electroforming and RS and plotted in semi-logarithmic (a) or double-logarithmic (b) scales.

Typical *I-V* characteristics of a Au/Si_3N_4/n^+-Si capacitor-like device (figure 2) demonstrate the electroforming process, during which the capacitor state is changed from the IS to LRS, and the bipolar RS is changed from LRS to HRS (RESET transition) at a positive bias and back from HRS to LRS (SET transition) at a negative bias. The electroforming manifests itself in the form of a current jump at around –5 V during the first negative voltage sweep. The *I-V* curve in the LRS after a complete SET transition (with no CC or a low CC at 30 mA) is described by the Ohm's law (figure 2). The corresponding temperature dependence in the range of 300 ÷ 500 K (figure 4) is typical of electron scattering on impurity ions in semiconductors ($I \sim T^{3/2}$). When the SET transition is not complete (with a high CC at 10 mA), the linear *I-V* relation in LRS becomes quadratic with an increase in voltage. In this regard, the resistance of a silicon capacitor plate mainly provides the resistance in LRS after a complete SET transition, whereas after an incomplete SET transition the resistance of an insulator is included in the total resistance.

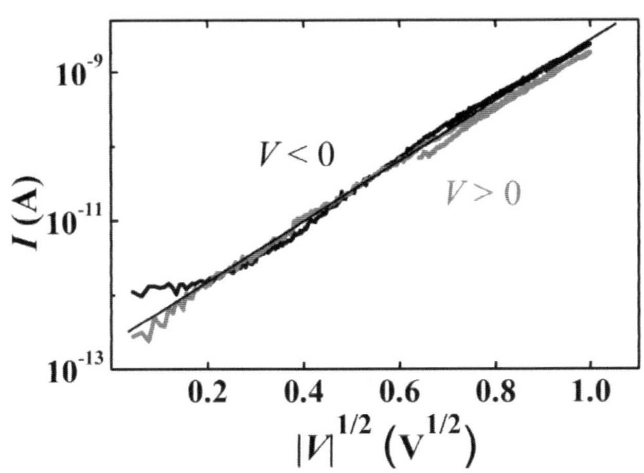

Figure 3. The ln $I - V^{1/2}$ plots for different polarities in IS.

IOP Conf. Series: Journal of Physics: Conf. Series **993** (2018) 012028 doi:10.1088/1742-6596/993/1/012028

After the RESET transition, the current in HRS is three orders of magnitude smaller than the current in LRS, but more than 2 orders of magnitude higher than the current in IS prior to electroforming. The latter result indicates the irreversible change of insulator properties after electroforming due to the occurrence of filaments (conductive channels shorting the capacitor [10]). In the low-conductive state, the I - V curve is also characterized by a linear law at low voltages, which transforms to a quadratic law at higher voltages. Such a quadratic I - V dependence is typical of the space charge limited current mechanism [13, 14] but it is hard to rigorously confirm this for memristive nanostructures because it is impossible to vary the thickness in the location of filament rupture.

The temperature dependence of a current in HRS (as well as in LRS obtained with CC at 10 mA) shown in figure 4 reveals the power growth with a rate close to 1, which probably corresponds to tunnel electron jumps through barriers arising in completely or partially broken filaments (the tunneling process in MIM structures is usually described by the quadratic I - T dependence [15]).

The observed bipolar nature of RS is typical of the valence-change oxide materials, in which the growth and rupture of filaments are related to the oxidation-reduction (redox) phenomena with the participation of oxygen vacancies (ions) [16]. At the same time, it is known that Si_3N_4 films are used to prevent the ion drift of impurities in insulator layers of MIS transistors [17], so the question of the relationship between the filament growth in Si_3N_4 and ion migration induced by the electric field requires special consideration. In particular, it is necessary to simulate the stochastic processes of breaking Si-N bonds and the formation of nitrogen vacancies or ions by the Monte-Carlo approach, as has already been done for SiO_2-based memristive devices [18]. In any case, the mechanism of the filament formation in Si_3N_4 can be associated with a high concentration of defects in these layers (up to 10^{19} cm^{-3} [19]). These defects can be ionized in the regions of electric field concentrators (e.g. represented by the roughness of an Au electrode) and form filaments between capacitor plates. When the high defect density in a filament is achieved, the electronic conductivity becomes possible through a filament due to the charge injection from electrodes and subsequent hopping between the defect sites.

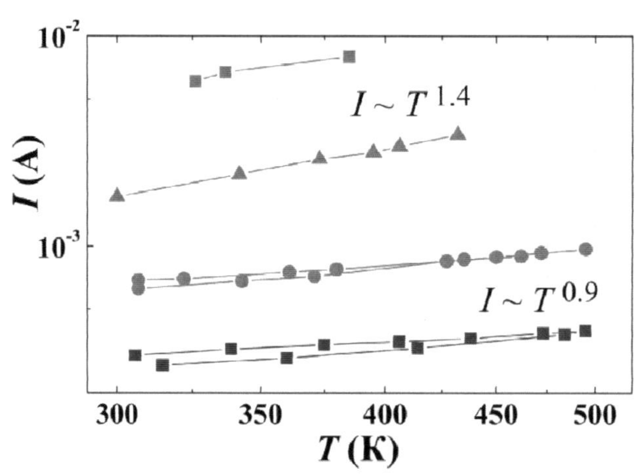

Figure 4. Typical I - T dependencies for the Au/Si_3N_4/n^+-Si capacitor, measured at $V = 0.5$ V in HRS (blue, multiplied by a factor of 100) and LRS (red) with different CC during the proceeding SET transition: no CC (■), CC = 10 mA (●), CC = 30 mA (▲).

Important information about the origin of resistive states can be obtained from the analysis of the results of small-signal measurements. In figure 5, the frequency dependencies of the parameters of equivalent circuits in IS, HRS and LRS are plotted. In the case of IS, the data are also obtained under illumination of the capacitor by laser emission at a wavelength of 462 nm.

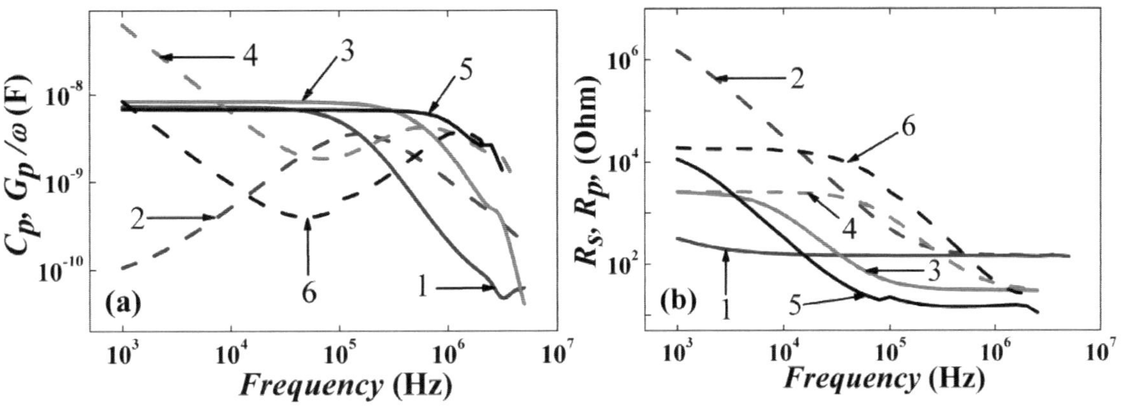

Figure 5. The frequency dependencies of C_p (solid line) and G_p/ω (dashed line) (a); R_s (solid line) and R_p (dashed line) (b) for the Au/Si_3N_4/n^+-Si capacitor (zero voltage bias; AC test signal amplitude is 20 mV). Curves 1, 2 (blue) are measured in HRS and IS; 3, 4 (red) are measured in LRS; 5, 6 (black) are measured in IS under laser illumination.

The frequency dispersion of capacitance and conductivity (figure 5(a)) is related to the finite charging time of the capacitor due to the presence of the serial resistance of capacitor plates. It is known that charging of a capacitor is described by the expression:

$$V = E\left(1 - e^{-\frac{t}{RC}}\right), \quad (1)$$

where V is the voltage on the capacitor plates during charging, E is the charging voltage of a capacitor, t is the charging time, and R is the serial resistance. In our case, $R = R_{s\infty}$ at a high frequency of $2\cdot10^6$ Hz (figure 5(b)). The dependencies of G_p/ω on C_p (Cole-Cole plots), not shown here, have demonstrated [11] that such curves are close to semicircles, thus the process of capacitance relaxation is characterized by a single time $\tau = R_{s\infty}\cdot C_{p0}$, where C_{p0} is the low-frequency capacitance in the saturation region ($C_p = C_s$ at low frequencies). This time value can be estimated from the capacitance frequency dependence at a level of $0.7\cdot C_{p0}$ or in the maximum of the G_p/ω frequency dependence as $1/2\pi f$ (figure 5(a), curves 1, 2), and the estimations give $\tau \approx 10^{-6}$ s for IS and HRS. The τ value actually determines the performance of a memristor when reading its state in IS and HRS. Of course, this performance can be increased proportionally with a decrease in capacitance, i.e. in the capacitor area, and a decrease in $R_{s\infty}$. After switching to LRS (curves 3 and 4 in figure 5), as well as under laser illumination (curves 5 and 6 in figure 5), the value of τ decreases proportionally with a decrease in the serial resistance $R_{s\infty}$ (4 and 10 times decrease, respectively), since the C_{p0} value practically does not change. Also, the ohmic losses in LRS and under illumination grow in the range of low frequencies. In the first case, these losses are related to partial shorting of a capacitor by filaments and, in the second case, with the emergence of constant capacitor leakage due to the external photoelectric effect that originated from the emission of electrons from the Au electrode. In both cases, the value of parallel resistance is decreased at low frequency (10^3 Hz) (figure 5(b), curves 4 and 6) that characterizes the insulator properties. The magnitude of this resistance change in HRS and LRS is equivalent to the difference of currents in HRS and LRS.

The change in $R_{s\infty}$ can be associated with the changes in a semiconductor capacitor plate. The initial state of the semiconductor surface before electroforming apparently corresponds to a depleted layer on a semiconductor surface, which is formed as a result of a contact potential difference equal to a difference of electron work functions of gold and silicon. The work function of silicon for the given doping level is almost equal to its electron affinity energy (~ 4 eV) [20]. The average work function of Au is ~ 5.2 eV [21]. Thus, in the space charge region of a semiconductor (SCR), which is serially connected to the insulator, the potential difference V_c of up to 1 V can drop in the absence of external voltage. Then, one can estimate the SCR thickness d in Si by using the following formula [22]:

$$d = \sqrt{2\varepsilon_0\varepsilon_s V_c / q n_0}, \quad (2)$$

where ε_0 is the absolute insulator permittivity of vacuum, ε_s is the relative permittivity of Si, and q is the charge of electron. The estimation gives $d \approx 10$ nm. Thus, the SCR thickness in a semiconductor substrate is comparable with the thickness of an insulator layer, and the SCR resistance can affect the characteristics of a memristive device. Light illumination increases the SCR conductivity due to a decrease in its thickness induced by photogenerated carriers. After the SET transition ($V < 0$), the positive charge can be accumulated near the insulator-semiconductor interface due to the capture of holes from a semiconductor, which increases the SCR conductivity and reduces its thickness, resulting in a reduced contribution of the serial resistance to the total resistance of capacitor plates. Moreover, the accumulation of positive charge favors the shorting of a capacitor by forming filaments. The back RESET transition ($V > 0$) leads to the accumulation of negative charge due to the capture of electrons in the defect states in an insulator, which increases the serial resistance due to the formation of a potential barrier for electrons.

Au/Si_3N_4/SiO_2/n^+-Si capacitors also demonstrate bipolar RS and similar regularities. However, the presence of a SiO_2 interlayer affects the total serial resistance and thus modifies the character of τ change. The estimation gives $\tau \approx 2.6 \cdot 10^{-5}$ s for IS and 10^{-7} s for HRS. These changes are also proportional to the changes in $R_{s\infty}$. After switching to HRS, the negative charge from Si is captured at the Si_3N_4/SiO_2 interface, leading to an increase in conductivity of a SiO_2 layer, reduction of the serial resistance of a semiconductor plate, and an increase in the capacitor performance. In addition, the capacitors with a double-layer insulator do not withstand prolonged cyclic switching, unlike Au/Si_3N_4/n^+-Si capacitors, for which the number of cycles in the continuous switching mode is at least 100.

The degradation of RS is revealed in the form of a nonreversible transition of the Si_3N_4-based capacitor to LRS, the origin of which should be a subject of further investigation. One of the most probable mechanisms can be related to hard thermal breakdown of a thin dielectric layer (permanent material change) due to high current and local heating.

4. Conclusions

Bipolar resistive switching has been observed in Au/Si_3N_4/n^+-Si and Au/Si_3N_4/SiO_2/n^+-Si capacitor-like MIS structures with insulator nanolayers represented by thin Si_3N_4 (6 nm) and SiO_2 (2 nm) films. The origin of the memristive behavior is related to the formation and rupture of conductive filaments in insulator layers. It is established that electronic processes are responsible for the effect of a highly doped n^+-Si ($3 \cdot 10^{19}$ cm^{-3}) semiconductor plate of a capacitor and a SiO_2 interlayer on the small-signal parameters of parallel and serial equivalent circuit models after electroforming and subsequent switching between the low-resistance and high-resistance states, as well as at illumination of a capacitor by blue laser emission. The corresponding change in the parameters of the semiconductor space charge region and the charge capture at the Si_3N_4/SiO_2 interface can affect the resistive state and should be taken into account in the interpretation of resistive switching. The observed increase in conductivity of a semiconductor plate and a SiO_2 interlayer significantly lowers the capacitor charging and discharging time, which determines the capacitor reading performance (response time of a memristive device).

Acknowledgments

The study is supported by the Ministry for Education and Science of the Russian Federation (RFMEFI58717X0042).

References

[1] Chua L O 1971 *IEEE Trans. Circuit Theory* **18** 507

[2] Waser R, Ielmini D, Akinaga H, Shima H, Wong H-S P, Yang J J, Yu S 2016 *Resistive Switching: From Fundamentals of Nanoionic Redox Processes to Memristive Device Applications* ed. Ielmini D and Waser R (Weinheim, Wiley-VCH) 1
[3] Tikhov S V, Gorshkov O N, Antonov I N, Kasatkin A P and Koryazhkina M N 2014 *Technical Physics Letters* **40** 837
[4] Tikhov S V, Gorshkov O N, Koryazhkina M N, Antonov I N and Kasatkin A P 2016 *Technical Physics Letters* **42** 536
[5] Jiang X et al 2014 *Journal of Applied Physics* **116** 123705
[6] Kim S, Kim M-H, Cho S and Park B-G 2016 *IEICE Trans. Electron.* **E 99-C** 547
[7] Kim S, Cho S, Ryoo K-C and Park B-G 2015 *J. Vac. Sci. Technol. B* **33** 062201
[8] Kim S, Jung S, Kim M-H, Cho S and Park B-G 2015 *Solid-State Electronics* **114** 94
[9] Berdnikov A E, Gusev V N, Mironenko A A, Popov A A, Perminov A V, Rudy A C and Chernomordick V D 2013 *Semiconductors* **47** 641
[10] Tikhov S V, Gorshkov O N, Antonov I N, Kasatkin A P, Korolev D S, Belov A I, Mikhaylov A N and Tetel'baum D I 2016 *Technical Physics* **61** 745
[11] Oreshkin P T 1977 *Tutorial on physics of semiconductors and dielectrics* (Moscow and Leningrad, Higher School) [In Russian]
[12] Epshtein S L 1965 *Measurement of capacitor characteristics* (Moscow and Leningrad, Energy) [In Russian]
[13] Nasyrov K A and Gritsenko V A, 2013 *Physics-Uspekhi (Advances in Physical Sciences)* **56** 999
[14] Lampert M A and Mark P 1970 *Current injection in solids* (New York and London, Academic Press)
[15] Elinson M I, Stepanov G V, Perov P I, Pokalyakin V I 1966 *Problems of Thin-Film Electronics* ed. Zernov D V, Elinson M I, Sandomirskiy V B (Moscow, Sovetskoe Radio) [in Russian] 5
[16] Mikhaylov A N et al 2016 *Physica Status Solidi C* **13** 870
[17] Morosanu C-E 1980 *Thin Solid Films* **65** 171
[18] Sadi T, Wang L, Gao D, Mehonic A, Montesi L, Buckwell M, Kenyon A, Shluger A and Asenov A 2016 *Proc. IEEE International Conference on Simulation of Semiconductor Processes and Devices* 149
[19] Shaposhnikov A V, Petrov I P, Gritsenko V A and Kim C W 2007 *Physics of the Solid State* **49** 1628
[20] Bechstedt F and Enderlein R 1988 *Semiconductor Surfaces and Interfaces* (Berlin, Akademie-Verlag)
[21] Rhoderick E *H* 1978 *Metal-semiconductor contacts* (Oxford, Clarendon Press)
[22] Ovsyuk V N 1984 *Electronic processes in semiconductors with space charge regions* (Novosibirsk, Science (Siberian branch)) [In Russian]

IOP Conf. Series: Journal of Physics: Conf. Series **993** (2018) 012029 doi:10.1088/1742-6596/993/1/012029

Study of PV characteristics of $Al_xGa_{1-x}As/GaAs$ photodiodes

V S Kalinovskiy[1], E V Kontrosh[1] , G A Gusev[2], A N Sumarokov[3], G V Klimko[1], S V Ivanov[1], V S Yuferev[1], T S Tabarov[1] and V M Andreev[1]

[1]Ioffe Institute, 26 Polytekhnicheskaya, St. Petersburg, Russian Federation, 194021
[2]Peter The Great St. Petersburg Polytechnic University, 29 Polytekhnicheskaya, St. Petersburg, Russian Federation, 195251
[3]The First Electrotechnical University, Pr. Popov 5, St. Petersburg, Russian Federation, 197376

E-mail: vitak.sopt@mail.ioffe.ru

Abstract. Simulation of the structure of multijunction photodiodes has been performed on the assumption that an equal of the number of photons is absorbed in them, and PV characteristics have been stadied $Al_xGa_{1-x}As/GaAs$ photodiodes fabricated by the molecular-beam epitaxy technique. According to the performed measurements and analysis of the results, it has been shown that the efficiency of single-junction photodiodes in converting monochromatic optical radiation at a wavelength of 810 nm reaches 50%. The relationship between the "saturation" currents calculated from the dark *I - V* characteristic for the diffusion current flow (Shockley) in the space charge region of a photodiode *p-n* junction and the obtained values for the efficiency of converting optical radiation in the wavelength range of 700 - 900 nm has been shown. When the "saturation" current flow for the diffusion mechanism rises by an order of magnitude, the efficiency value at excitation by monochromatic radiation at 810 nm and 780 nm drops by 25% and 11%, respectively.

1. Introduction

In recent years, application of multijunction nano-dimensional heterostructures of III-V photoactive *p-n* junctions with forbidden gaps increasing towards the frontal surface and connected by tunnel diodes was, mainly, considered for creating efficient sunlight converters – multijunction solar cells (MJ SCs) [1]. At present, MJ nano-dimensional heterostructures find their application also in creating high-efficient photodiodes (PDs) converting monochromatic radiation [2, 3]. Unlike an MJ SC, a monolithic PD consists of several *p-n* junctions with similar forbidden gaps of the approximate composition, but with different geometries of junctions connected by tunnel diodes. In the PV mode of PD operation, this allows conversion of monochromatic radiation with higher voltage of the output signal and improvement of conditions for correlation with the load. In the present paper, the structure of a monolithic triple-junction *p-i-n* $Al_xGa_{1-x}As/GaAs$ PD has been simulated. Dark and load *I-V* characteristics of single-junction *p-i-n* $Al_xGa_{1-x}As/GaAs$ PDs were investigated experimentally under excitation by continuous monochromatic radiation at a wavelength of 830 nm at a power of up to 400 mW and at 780 nm at a power of up to 1.3 W. Spectral characteristics of the external quantum yield of a *p-i-n* $Al_xGa_{1-x}As/GaAs$ PD with a small photosensitive surface (500 μm in diameter) were recorded.

2. Simulation of multijunction and single-junction $Al_xGa_{1-x}As/GaAs$ PD structures

A model was calculated for a monolithic triple-junction *p-i-n* PD based on the $Al_xGa_{1-x}As/GaAs$ nanoheterostructure. The schematic of the MJ PD is presented in figure 1. Each junction thickness was determined under the condition of equality of the number of absorbed radiation photons according to the expressions (1-5) presented below. The total amount of absorbed radiation in a PD was determined from the expression:

$$I_0 = I - I_{out}, \tag{1}$$

where I is the incident radiation intensity (number of photons per 1 cm^2), and I_{out} is the intensity of radiation that passes through. The amount of radiation absorbed in the ith junction of an MJ PD with allowing for absorption only in p, i and n layers is:

$$I_{0i} = I_i(1 - e^{(-\alpha_{pi} l_{pi} - \alpha_{ii} l_{ii} - \alpha_{ni} l_{ni})}), \tag{2}$$

where I_i is the intensity of radiation entering a junction; α is the absorption coefficient of a corresponding layer (p, i or n), and l is the thickness of a corresponding layer (p, i or n).
The amount of radiation absorbed in the first junction at $i = 1$ with allowance for optical losses in the wideband window is:

$$I_1 = I - I_{win}, \tag{3}$$

where I_{win} is the amount of radiation absorbed in the wideband window located in the first junction structure.

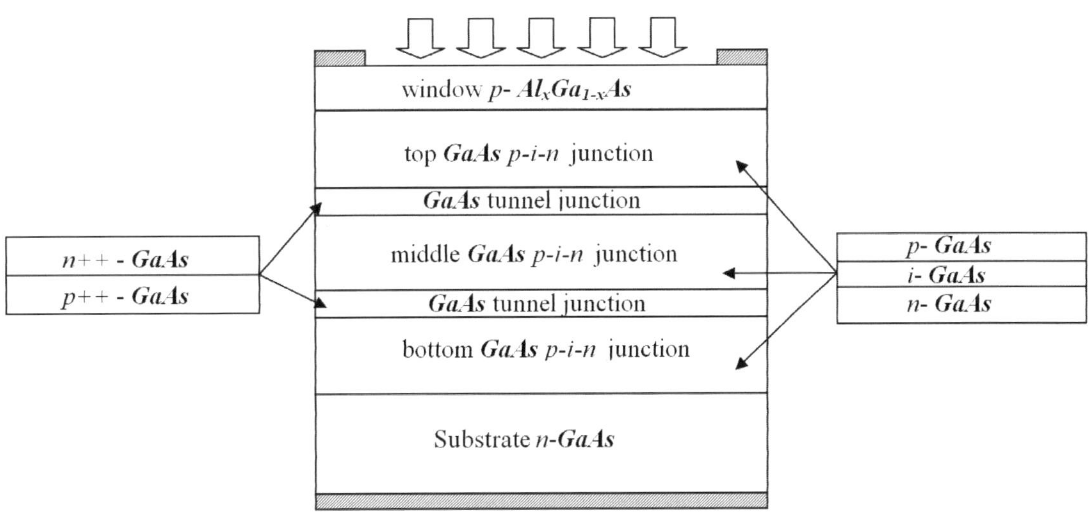

Figure 1. Schematic of a triple-junction $Al_xGa_{1-x}As/GaAs$ PD.

The amount of radiation absorbed in the below located junctions at $1 < i \le N$ with allowance for optical absorption in tunnel n^{++}-p^{++} junctions is:

$$I_i = I_{i-1} e^{(-\alpha_{p,(i-1)} l_{p,(i-1)} - \alpha_{i,(i-1)} l_{i,(i-1)} - \alpha_{n,(i-1)} l_{n,(i-1)})} - I_{TD_{i-1}}, \tag{4}$$

where I_{TD} is the amount of radiation absorbed in a tunnel diode.

Then, the portion of radiation absorbed in each junction with allowance for photocurrent equality in all MJ PD junctions is:

$$I_{0i} = \frac{I - I_{win} - \sum_{i=1}^{N-1} I_{TD,i} - I_{out}}{N} \quad (5)$$

The calculated thicknesses of *p, i, n* layers for each junction and doping levels are presented in figure 2. Numbers 1, 2 and 3 denote junctions, and number 4 represents connecting tunnel junctions in correspondence with the structural schematic presented in figure 1.

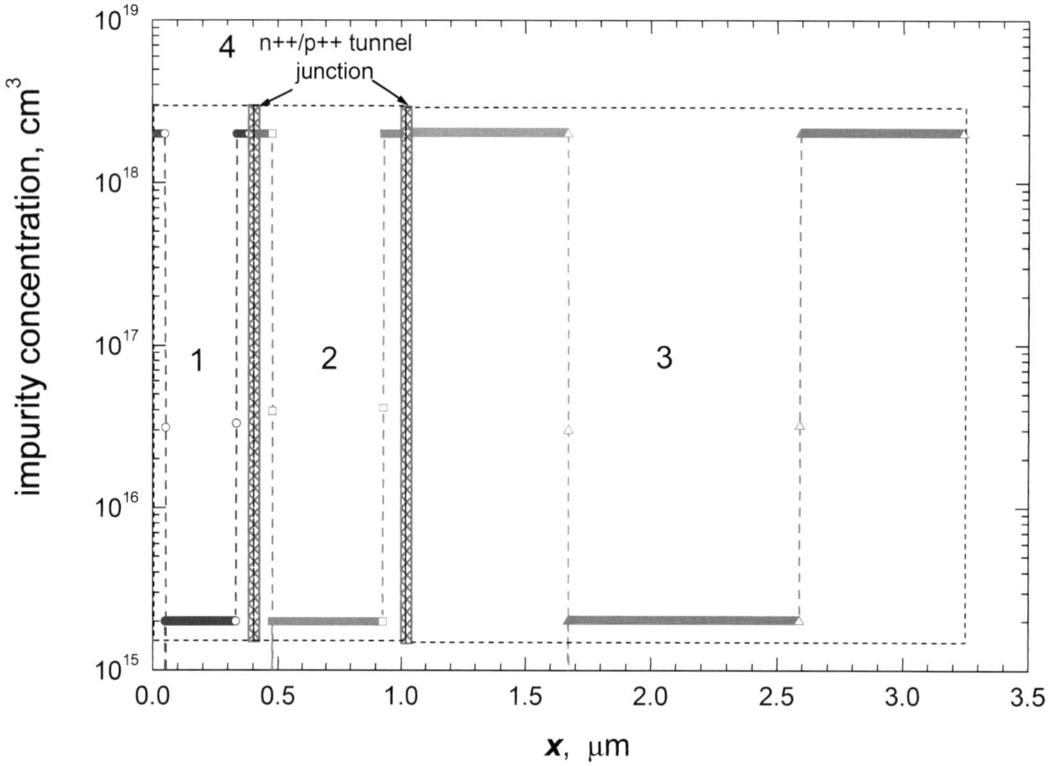

Figure 2. Schematic of a triple-junction monolithic *p-i-n* $Al_xGa_{1-x}As/GaAs$ PD with indicated calculated thicknesses of *p-i-n* junctions and n^{++}-p^{++} connecting diodes along the x-axis, and rated doping levels shown along the y-axis: (1) top, (2) middle, (3) bottom, (4) tunnel junctions.

3. Static and dynamic characteristics of single-junction *p-i-n* $GaAs/Al_xGa_{1-x}As$ PDs

The heterostructure of $Al_{0.12}Ga_{0.88}As/GaAs$ fabricated by molecular-beam epitaxy was used to investigate characteristics of single-junction *p-i-n* PDs with frontally entering monochromatic radiation (figure 3). PD chips sized 1 mm × 1 mm were formed on the grown epitaxial wafer.

Directly on the epitaxial wafer, dark *I-V* characteristics of single-junction PDs were recorded within the forward voltage bias range of 0 - 1.5 V. The forward dark *I-V* characteristics were analyzed according to the procedure described in [4, 5], using a three-component exponential model of a PD *I-V* characteristic:

$$J = \sum_{i=1}^{3} J_{0i}\left(\exp\left(\frac{V_\varphi}{A_i\varepsilon}\right) - 1\right) \tag{6}$$

Obtained were the values of the pre-exponential factors ("saturation" currents for the main mechanisms of minority charge carrier transport), J_{0i}, diode coefficients (ideality coefficients), A_i, and structure series resistance, R_s, for three minority charge carrier transport mechanisms in the space charge region of a *p-i-n* junction: tunnel-trap ("overabundant") with a diode coefficient A_t greater than 2, recombination (Sah-Noyce-Shockley) with $A_r = 2$ and diffusion (Shockley) with $A_d = 1$.

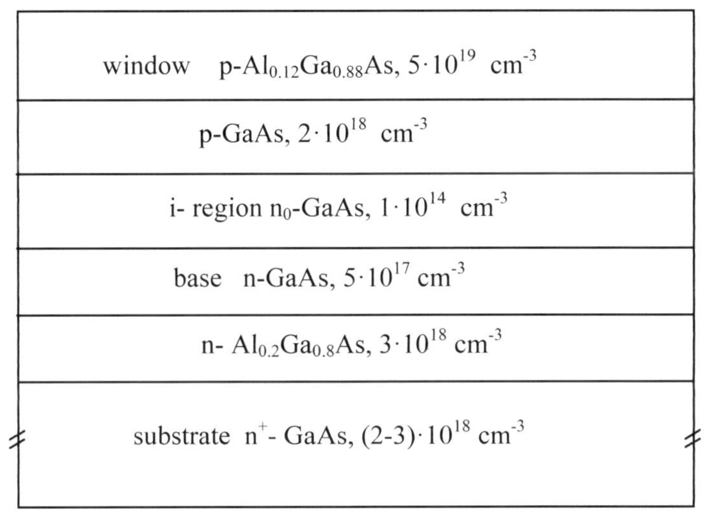

Figure 3. Heterostructure of $Al_xGa_{1-x}As$ /GaAs *p-i-n* PDs fabricated by the MBE technique.

From the above-mentioned PDs whose characteristics were recorded and analyzed, those with different values of pre-exponential factors corresponding to the diffusion component J_{0d} were selected. The selection was made for estimating the influence of the pre-exponential factors J_{0d} on the single-junction PD efficiency in the case of converting monochromatic radiation in the range of 700 – 900 nm. The dark *I-V* characteristics of the selected PDs are shown in figure 4 and marked as b2, d13, and b10, which corresponds to the indices on the epitaxial wafer. Table 1 presents the rated values of "saturation" currents J_{0t}, J_{0r}, J_{0d} (pre-exponential factors) for the tunnel-trap, recombination and diffusion current flow mechanisms and, accordingly, the specific series resistance R_s of a *p-i-n* junction structure, all obtained in analyzing experimental dark *I-V* characteristics (figure 4).

To determine the spectral range of the sensitivity of a tested PD epitaxial structure, external quantum efficiency characteristics were recorded in the wavelength range of 700 – 900 nm. The dependences are presented in figure 5.

For the selected PDs, load *I-V* characteristics were recorded upon exposure to monochromatic 0.1-0.4 W radiation at a wavelength of 830 nm. The efficiency values calculated from the load *I-V* characteristics in dependence on the incident radiation power (0.1 – 0.4 W) are presented in figure 6 (curves 1, 2 and 3) and table 1. Based on the data presented in figure 6 and table 1, it follows that the b2 PD with a minimal "saturation" current density value $J_{0d} = 1\cdot10^{-20}$ A/cm^2 has the maximum value of the efficiency in conversion of constant laser radiation (50%) at a wavelength of 830 nm and a power of 0.4 W. The b10 PD with the maximum value of $J_{0d} = 1\cdot10^{-19}$ A/cm^2 and $R_s > 10$ mOhm·cm^2 has the minimum efficiency of 37.8% at an incident laser radiation power of 0.4 W.

Figure 6 and table 1 present rated values of the efficiency for peak values of pulse amplitude of the photo-response for tested single-junction $Al_xGa_{1-x}As$/GaAs *p-i-n* PDs. According to the obtained characteristics (figure 6, curves 4, 5 and 6, table 2), the b2 PD having the minimum "saturation" current density value converts pulsed radiation at a wavelength of 780 nm with the maximum efficiency of 35%. The decrease in the PD efficiency under excitation at a wavelength of 780 nm, compared to that at continuous radiation impact at 830 nm, occurs, first of all, by lowering the PD spectral sensitivity at 780 nm (figure 5).

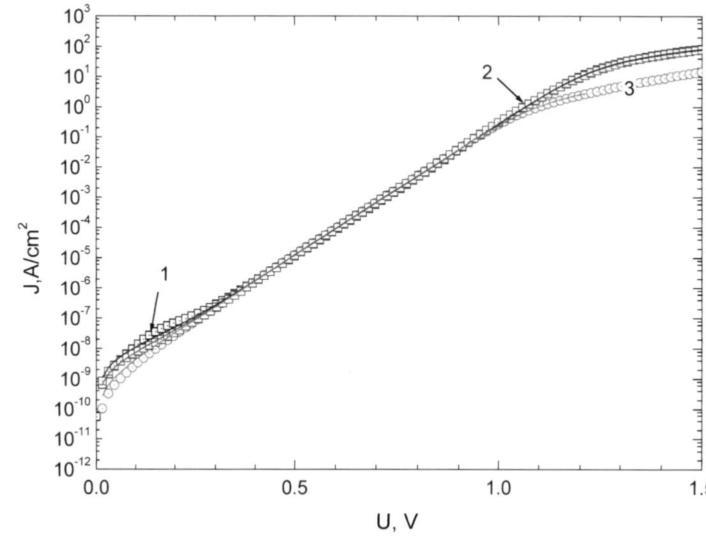

Figure 4. Dark *I-V* characteristics for the selected PDs b2 (curve 1), d13 (curve 2), b10 (curve 3) with different values of pre-exponential factors for three current flow mechanisms: tunnel-trap, J_{0t}, (A > 2); recombination, J_{0r}, (A = 2); diffusion, J_{0d}, (A = 1), and the value of the resistance R_S.

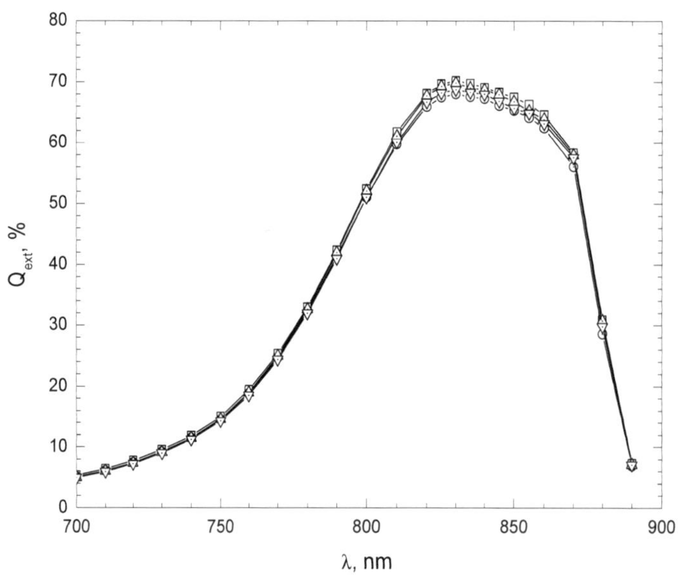

Figure 5. Spectral characteristics of the external quantum efficiency of the structure of an $Al_xGa_{1-x}As/GaAs$ *p-i-n* PD with a photoactive surface having a diameter of 500 μm.

Table 1.

No of PD	*Efficiency*, % (λ = 835 nm)			R_s, (mOhm·cm^2)	J_{0t}, (10^{-8} A/cm^2) (A > 2)	J_{0r}, (10^{-10} A/cm^2) A = 2)	J_{0d}, (10^{-20} A/cm^2) (A = 1)
	$P_{опт}$ = 0.1 (W)	0.18 (W)	0.4 (W)				
b2	47.0	49.0	50.0	< 3	1.0	5.7	1.0
d13	44.0	45.4	46.0	< 3	0.1	5.5	2.0
b10	39.4	39.0	37.8	>10	0.8	5.8	10

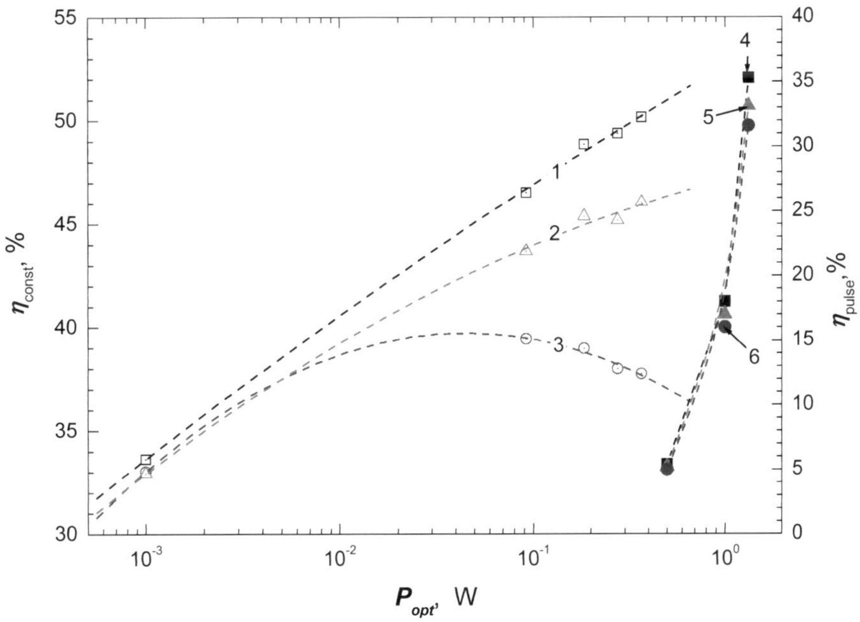

Figure 6. Dependence of the efficiency on incident power for PDs b2 (curves 1,4), d13 (curves 2,5) and b10 (curves 3,6) in a PV mode at excitation by monochromatic radiation: curves 1, 2, 3, at continuous excitation at a wavelength of 830 nm (25 ^{0}C); curves 4, 5, 6, at excitation at a wavelength of 780 nm.

Table 2.

No of PD	Efficiency, % (λ = 780 nm)		
	P_{opt} = 0.5 (W)	1.0 (W)	1.33 (W)
b2	5.4	18	35.3
d13	5.2	17	33.1
b10	5	16	31.6

The parameters calculated from dark and load I-V characteristics and fast operation characteristics of single-junction PDs were used to build the dependences of the efficiency on "saturation" current corresponding to the diffusion mechanism for the minority charge carrier transport in the continuous generation mode (figure 7, curve 1) and in the pulsed one (figure 7, curve 2).

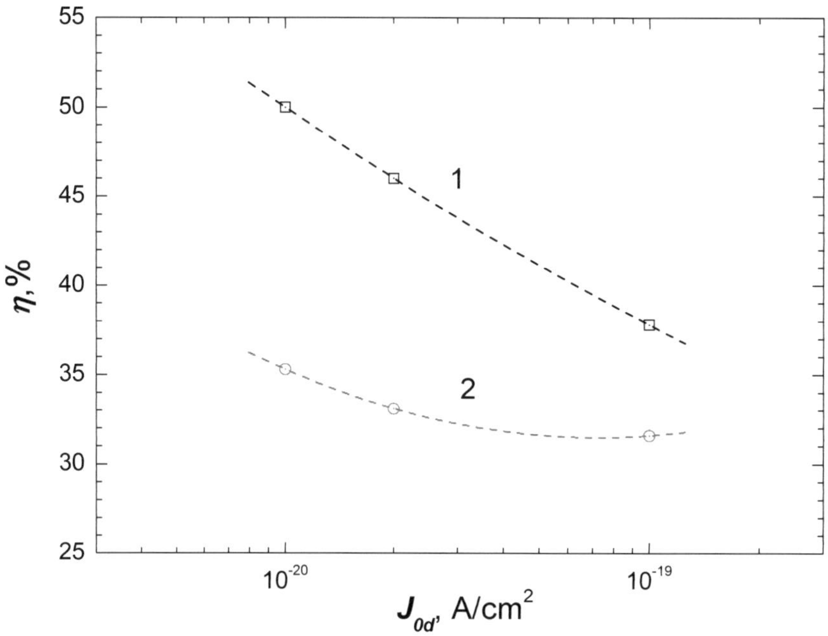

Figure 7. Dependences of the PD efficiency on "saturation" current for the diffusion mechanism, obtained from dark current-voltage characteristics of photodiode samples (b2, d13, b10) in the continuous generation mode (curve 1) and in the pulsed one (curve 2).

From the dependences in figures 6 and 7, the relationship between the "saturation" currents for the diffusion mechanism of minority charge carrier flow in the space charge region of photoactive *p-i-n* junctions and the efficiency of optical radiation conversion is well pronounced. The smaller the values of pre-exponential factors (J_{0d} for the diffusion current flow mechanism), the smaller the PD efficiency. The growth of J_{0d} by an order of magnitude decreases the efficiency by more than 10% in the laser excitation mode at a wavelength of 838 nm and by 4% for pulsed excitation at 780 nm.

4. Conclusion

A rated model of a triple-junction $Al_xGa_{1-x}As/GaAs$ photodiode for converting powerful monochromatic radiation in the wavelength range of 700 - 900 nm is proposed.

Investigation of static and dynamic characteristics of single-junction *p-i-n* $Al_xGa_{1-x}As/GaAs$ PDs has been carried out. According to this investigation, it has been shown that the greater the "saturation" current values J_{0d} corresponding to the diffusion current flow, the smaller the PD efficiency value at excitation by monochromatic radiation. For the tested *p-i-n* PDs, the growth of the "saturation" currents J_{0d} by an order of magnitude causes the efficiency to decrease by more than 10% in the mode of excitation by continuous monochromatic radiation at a wavelength of 830 nm and by 4% in the modes of excitation by monochromatic pulse radiation at 780 nm.

Acknowledgements

The work has been supported by the Russian Science Foundation (Agreement No 17-79-30035).

References

[1] Alferov Zh I, Andreev V M and Rumyantsev V D Solar photovoltaics: Trends and prospects 2004 *Semiconductors* **38**, 899.

[2] Fafard S, York M C A, Proulx F, Valdivia C E, Wilkins M M, Arès R, Aimez V, Hinzer K and Masson D P Ultrahigh efficiencies in vertical epitaxial heterostructure architectures 2016 *Applied Physics Letters* **108** 071101.

[3] Schubert J, Oliva E, Dimroth F, Guter W, Loeckenhoff R and Bett A. W. High-Voltage GaAs Photovoltaic Laser Power Converters 2009 *IEEE Transactions on Electron Device* **56** 170

[4] Andreev V M, Evstropov V V, Kalinovskiy V S, Lantratov V M and Khvostikov V P 2009 Current flow and potential efficiency of solar cells based on GaAs and GaSb *p-n* junctions *Semiconductors* **43** 644

[5] Andreev V M *et al* 2010 Current flow and efficiency of Ge p-n junctions in triple-junction GaInP/Ga(In)As/Ge solar cells for space applications *Proc. 25th European Photovoltaic Solar Energy Conference and 5th World Conference on Photovoltaic Energy Conversion* (Valencia) 979

Investigating the effect of silicon surface chemical treatment on Al/Si contact properties in GaP/Si solar cells

D Kudryashov[1] and A Gudovskikh[1,2]

[1] St. Petersburg national Academic University RAS, 194021 St. Petersburg, Russia
[2] St. Petersburg Electrotechnical University "LETI", 197376 St. Petersburg, Russia

E-mail: kudryashovda@spbau.ru

Abstract. In the present work, experimental studies have been carried out to reveal how chemical treatment of a silicon surface affects the properties of the Al/Si contact. It has been shown that for p-type monocrystalline silicon substrates with a resistivity of 10 ohm·cm, it is possible to form an ohmic Al/Si contact by magnetron sputtering of an aluminum thin film and its further annealing at temperatures of 400 - 450 °C. In the range of annealing temperatures of 250 - 400 °C, the Si substrate treatment in the HF solution leads to a significant increase in currents on the current-voltage curves of the Al/Si contact, while in the range of 450 - 700 °C, the effect of chemical treatment of the silicon is not detected.

1. Introduction

The rapid development of solar cell technology is motivated by increasing attention to the problems of environmental protection and hydrocarbon depletion. At the moment, the highest efficiency of 46% has been achieved for solar cells based on mechanically stacked GaInP/GaAs and GaInAsP/GaInAs systems [1]. Such a complicated structure is necessary because nowadays the lattice-matched solar cell technology is near its limit and researchers have to search for new ways to increase solar cell efficiency. One of these approaches focuses on the use of Si heterojunctions and Si/GaP superlattices. According to the theoretical estimations, a 5-nm silicon layer inside GaP leads to the formation of a quantum well with an optical bandgap of 1.7 eV. At the same time, a superlattice composed of Si/GaP layers becomes a direct bandgap [2]. Another approach involves the direct growth of diluted nitride layers on silicon substrates. It is found that a small amount of nitrogen in GaP narrows the bandgap with its transition from indirect to direct. Both approaches ensure the growth of materials with a certain bandgap and thus help to get rid of the thermalization effect and increase the solar cell efficiency. However, in both cases, the problem of the formation of a backside ohmic contact is still unresolved. The point is that an ohmic contact for crystalline solar cells is typically formed by aluminum annealing at temperatures of 700 - 1000 °C [3]. However, these conditions lead to degradation of GaPNAs and Si/GaP layers. The aim of this work is to investigate the effect of silicon surface chemical treatment on Al/Si contact properties in GaP/Si solar cells.

2. Experimental details

Aluminum contacts were deposited on p-Si (100) wafers with a resistivity of 10 Ohm·cm by a magnetron sputtering setup (BOC EDWARDS Auto 500RF) at room temperature. To eliminate the possible resistivity variations for different wafers in the box, the whole substrate was used for contact formation. The distance between an Al target and a substrate was 100 mm. The sputtering chamber was evacuated to less than $5 \cdot 10^{-6}$ mbar prior to deposition. Pure argon (99.999 %) was used as the working gas. Prior to its deposition, one part of each sample was treated with HF solution (1:3) for 1 minute and subsequently rinsed in deionized water followed by N_2 purge. After chemical treatment, the samples were immediately transferred into a vacuum chamber. Al pads were deposited via a metallic mask with 4×4 mm squares indented to 1.5 mm (figure 1). The mask was located at some distance from the wafer edge to reduce the effect of possible substrate doping variations.

Figure 1. Aluminum contact pads deposited on silicon wafers.

Aluminum films were deposited in two steps. They were sputtered initially into a layer of 80 nm at a power of 50 W and then into a layer of 100 nm at a power of 100 W. The above procedure made it possible to form a uniform Al layer and, at the same time, minimize the damage to the silicon surface layer due to bombardment by high energy particles. The aluminum layer thickness was verified using an AMBiOS XP-1 profilometer. Annealing of Al contacts was performed with a JIPelec Jetfirst 100 rapid temperature annealing setup in a nitrogen atmosphere. The current density-voltage (IV) curves were measured using a Keithley 2400 source meter in the dark at room temperature.

3. Results and discussion

Figure 2 demonstrates the difference between current-voltage curves measured for deposited Al/Si contacts for untreated and HF-treated samples. All samples show similar characteristics, indicating good homogeneity of the silicon wafers. It can be seen that the current values for Si substrates immersed in HF prior to Al deposition are 2.5 times higher than for the untreated samples.

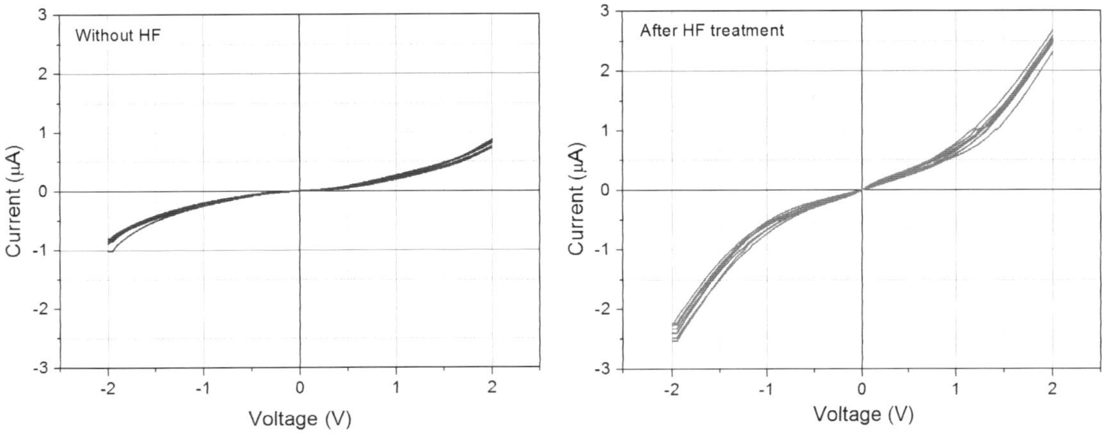

Figure 2. IV curves measured between Al/Si contacts for different samples: HF-treated (right) and untreated (left) Si wafers.

IOP Conf. Series: Journal of Physics: Conf. Series **993** (2018) 012030 doi:10.1088/1742-6596/993/1/012030

Then all the samples were annealed in a nitrogen environment at different temperatures. The annealing time was set inversely proportional to the annealing temperature in order to maintain an approximately constant depth of aluminum diffusion into the silicon substrate (figure 3).

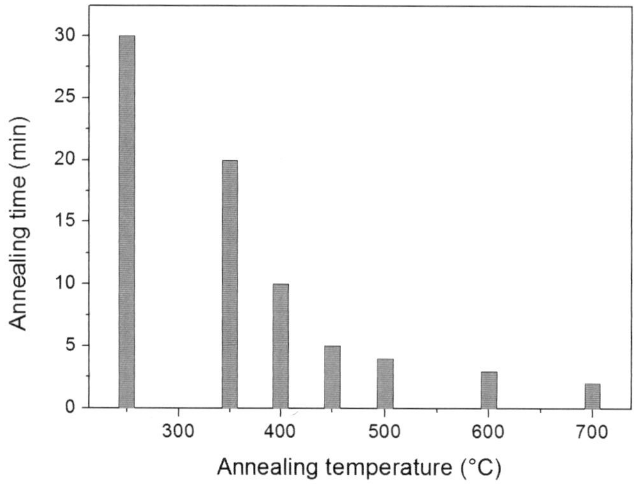

Figure 3. Relationship between annealing time and temperature.

The experimental data for annealed samples are shown in figure 4. It can be seen that a rapid increase of currents with increasing annealing temperature is observed in the temperature range of 250 - 450 °C. The samples annealed at 500 - 700 °C show almost identical IV characteristics.

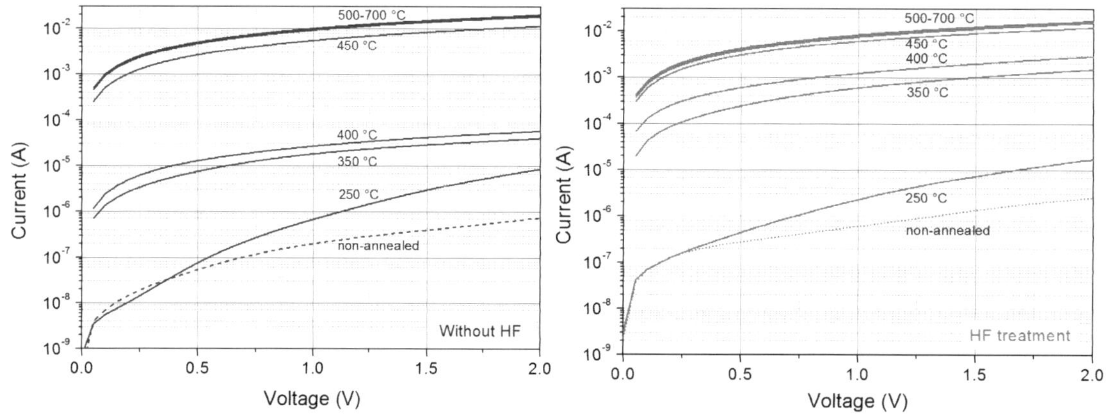

Figure 4. Current-voltage curves (the positive sweep) for contacts at different annealing temperatures for HF-treated (right) and untreated (left) samples.

The HF-treated samples annealed under the same annealing conditions show higher currents at the temperature range of 250 - 400 °C compared to untreated samples. The current values at much higher temperatures are almost the same (figure 5). The maximum difference in current values is observed in the range of 250 - 400° C. In this range, the diffusion processes have already started for HF-treated samples. On the oxidized (untreated) surface, some additional energy is still needed to overcome the diffusion barrier.

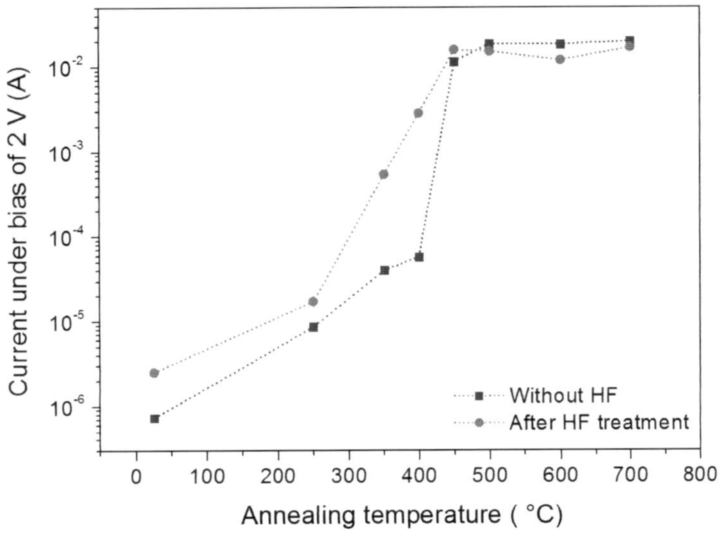

Figure 5. Influence of annealing temperature on the Al/Si contact current at a bias of 2 V.

It should be noted that the samples annealed in a nitrogen environment at temperatures above 600 °C show a visible contact degradation and a change in the shade of the metal film. This is probably due to the interaction between aluminum and nitrogen. All the samples annealed above 600 °C had a yellow shade. Below this temperature, the Al/Si contacts remained stable.

4. Conclusions

The experimental results obtained demonstrate the effect of chemical treatment of the silicon surface on the properties of the Al/Si contact. For p-type monocrystalline silicon wafers with a resistivity of 10 ohm·cm, it is possible to fabricate an ohmic Al/Si contact by magnetron sputtering of an Al thin film and its further annealing at temperatures of 400 - 450 °C. In the range of annealing temperatures of 250 - 400 ºC, the Si substrate treatment in HF solution leads to a significant increase in currents on the current-voltage curves of the Al/Si contact, while in the range of 450 - 700 ºC the influence of chemical treatment of the silicon has not been detected.

Acknowledgment
This work was supported by RFBR (project no. 17-08-00474 A)

References

[1] Green M A, Hishikawa Y, Warta W, Dunlop E D, Levi D H, Hohl-Ebinger J and Ho-Baillie A W H 2017 *Prog Photovolt Res Appl.* **25** 668
[2] Lin-Chung P J 1991 *MRS Proceedings* **220** 589
[3] Kaminski A, Vandelle B, Fave A, Boyeaux J P, Le Quart Nam, Monna R, Sarti D and Laugier A 2002 Solar Energy Materials and Solar Cells **72** 373

IOP Conf. Series: Journal of Physics: Conf. Series **993** (2018) 012031 doi:10.1088/1742-6596/993/1/012031

Quantum-cascade lasers in the 7-8 μm spectral range with full top metallization

A S Kurochkin[1,2], A V Babichev[1,2,3], D V Denisov[1], L Ya Karachinsky[1,2,3], I I Novikov[1,2,3], A N Sofronov[4], D A Firsov[4], L E Vorobjev[4], A Bousseksou[5] and A Yu Egorov[1,2]

[1] Connector Optics LLC, St. Petersburg 194292, Russia
[2] ITMO University, St. Petersburg 197101, Russia
[3] Ioffe Institute, St. Petersburg 194021, Russia
[4] Peter the Great St. Petersburg Polytechnic University, St. Petersburg 195251, Russia
[5] Center of Nanoscience and Nanotechnology (C2N), Univ. Paris Sud and Paris-Saclay, 91405 Orsay cedex, France

E-mail: alexander.kurochkin@connector-optics.com

Abstract. The paper demonstrates the generation of multistage quantum-cascade lasers (QCL) in the 7-8 μm spectral range in the pulse generation mode. The active region structure we used is based on a two-phonon resonance scheme. The QCL heterostructure based on a heteropair of $In_{0.53}Ga_{0.47}As/Al_{0.48}In_{0.52}As$ solid alloys was grown by molecular beam epitaxy and includes 50 identical stages. A waveguide geometry with top cladding with full top metallization (surface-plasmon quantum-cascade lasers) has been used. The developed QCLs have demonstrated multimodal generation in the 7-8 μm spectral range in the pulse mode in the 78-250 K temperature range. The threshold current density for a 1.6 mm long laser and a 20 μm ridge width amounted to ~ 2.8 kA/cm^2 at a temperature of 78 K. A temperature increase to 250 K causes a long-wave shift of the wavelength from 7.6 to 7.9 μm and a j_{th} increase to 5.0 kA/cm^2.

1.Introduction

Quantum-cascade lasers (QCL) that emit in the 7-8 μm spectral range can be effectively applied for remote gas analysis. In particular, the lasers with a 7.3-7.9 μm wavelength are used for detection of SO_x, CH_4, H_2S, C_2H_2, N_2O, and trinitrotoluene [1-3].

Methane detection (at a 7.8 μm lasing wavelength) based on QCL is used for remote leak control at gas pipelines and chemical plants. Detection of various liquids at precipitation on the QCL surface is also of interest [4].

A number of structures of a QCL active region emitting in the 7-8 μm spectral range have been described in literature. Most frequently mentioned structures are based on a two-phonon [5] and a three-phonon resonance [6].

The main technologies for growing QCL heterostructures are molecular beam epitaxy (MBE) and metal-organic chemical vapor deposition (MOCVD). MBE makes it possible to implement lasers with a higher efficiency [7]. The use of industrial MBE production systems has a positive effect on the structural quality of the developed heterostructures [8-10] and enables implementation of multistage

QCL heterostructures with a high identity of the chemical composition and thicknesses of the cascade-forming layers.

This paper presents the results on generation and optical characteristics of multistage QCLs that emit in the pulse mode in the 7-8 μm spectral range in the 78-250 K temperature range.

2. Experiment and results

The QCL heterostructure was fabricated by Connector Optics LLC using a Riber 49 MBE production system with a valved As cracker source and ABI 1000 brand sources for generating gallium and indium flows. The InP substrate with a (100) orientation was doped with sulfur to a level of $3\cdot10^{18}$ cm^{-3}. The waveguide bottom cladding was developed based on InP (2 μm thick) and $In_{0.53}Ga_{0.47}As$ (0.5 μm thick) layers, which were grown on a substrate in succession. The active region with 50 quantum stages was developed based on lattice-matched $In_{0.53}Ga_{0.47}As$ quantum wells with $Al_{0.48}In_{0.52}As$ potential barriers. The cascade was developed using the two-phonon resonance charge-carrier extraction. After the development of the cascades, $In_{0.53}Ga_{0.47}As$ layers with $1\cdot10^{17}$ cm^{-3} and $1\cdot10^{19}$ cm^{-3} doping levels and with 0.1 μm and 0.02 μm thicknesses, respectively, were formed in succession. A detailed description of the QCL heterostructure design is shown in [10].

A waveguide geometry with top cladding with full top metallization (surface-plasmon quantum-cascade lasers) has been used for QCL generation [11, 12]. Such a waveguide design is promising for photonic crystal laser generation.

The laser ridge generation started with wet etching of a deep mesa (based on $HBr:HNO_3:H_2O$ solution). The etching mask was a silicon oxide layer etched locally in a fluorine-based plasma under an AZ 5214 photoresist mask. After the generation of the mesa profile and removal of the silicon oxide mask, the passivation of a mesa side wall was effected by depositing a silicon nitride (Si_3N_4) layer by plasma-chemical deposition. The windows in the Si_3N_4 layer were opened for metallization by reactive ion etching in a CHF_3/O_2 gas mix under a photoresist mask. The top metallization was performed through a photoresist mask by electron beam deposition of consecutive layers of titan and gold with 20 and 200 nm thicknesses, respectively. Before the bottom contact metallization, the substrate was lapped down to a thickness of 150 μm. For bottom metallization, a sequence of titan/gold layers was used with 20 and 200 nm thicknesses, respectively. The ridges were obtained by mechanical cleavage. The ridge laser width (near the heterostructure surface) amounted to 20 μm. The ridge laser length amounted to 1.6 mm. No high-reflection and anti-reflection coating was applied to the laser facets. Mounting was performed "heterostructure up" onto a silver-plated copper heatsink.

Studies of the experimental QCL samples mounted on a heatsink were performed in the 78-250 K temperature range. The current-voltage characteristics were measured at current pulses flashing with a 100 ns duration and an 84 kHz frequency (the duty cycle of the pulses was 0.84 %). The QCL lasing spectra were recorded using a Fourier spectrometer and an HgCdTe photodetector.

Figure 1 shows the current-voltage and power-voltage characteristics measured at various temperatures. The threshold current density, j_{th}, for the examined QCL amounted to 2.8 kA/cm^2 at a temperature of 78 K. A temperature increase up to 250 K leads to a j_{th} increase up to 5.0 kA/cm^2. The threshold voltage, V_{th}, amounts to 14 and 17 V at $T = 78$ K and 250 K, respectively.

The approximation of experimental values of j_{th} corresponding to various temperatures is based on the $j_{th}(T) = j_0\times\exp(T/T_0)$ expression and provides an evaluation of the characteristic temperature of threshold current stability, $T_0 = 94$ K, and the threshold current density at zero temperature, $j_0 = 2.1$ kA/cm^2. The intermodal distance between extensional modes, $\Delta\lambda$ is ~ 0.0053 μm. The mirror loss factor, $\alpha_m = \ln(R_{eff})/L$, is 7.3 cm^{-1}. Figure 2 demonstrates the generation spectra measured at various temperatures. Multimode laser generation has been demonstrated. With a temperature increase, a shift of the maximum intensity of the stimulated radiation spectrum to a long-wave region is observed, which is due to a decrease in the conduction band offset with a T increase.

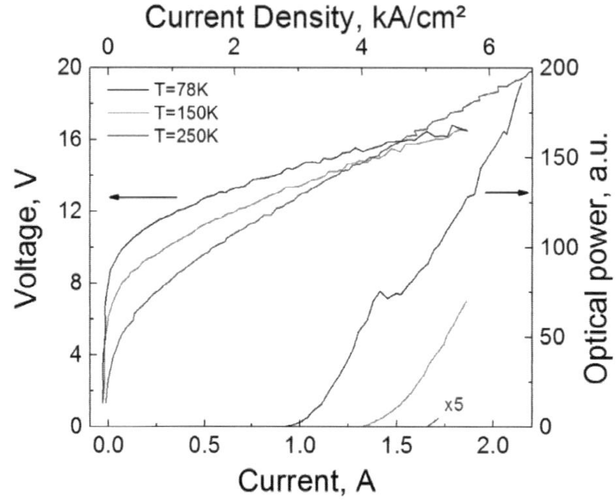

Figure 1. The current-voltage characteristics and power-voltage characteristics measured at various temperatures.

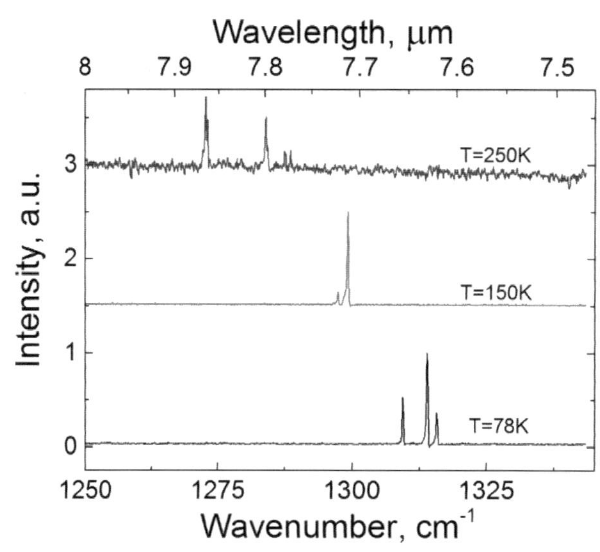

Figure 2. The stimulated radiation spectra of QCL samples measured at various temperatures.

Conclusions

Lasing of multistage quantum-cascade lasers in the 7-8 μm spectral range in the pulse generation mode has been demonstrated. The active region structure based on the two-phonon resonance scheme has been used. The heterostructure of the quantum-cascade laser based on a heteropair of $In_{0.53}Ga_{0.47}As/Al_{0.48}In_{0.52}As$ solid alloys was grown by molecular beam epitaxy and included 50 identical stages. A waveguide geometry with top cladding with full top metallization (surface-plasmon quantum-cascade lasers) has been used. The developed QCLs have demonstrated multimodal generation in the 7-8 μm spectral range in the pulse mode in the 78 - 250 K temperature range. The threshold current density amounted to ~ 2.8 kA/cm^2 at a temperature of 78 K. A temperature increase to 250 K causes a long-wave shift of the wavelength from 7.6 to 7.9 μm and a j_{th} increase to 5.0 kA/cm^2. The determined j_{th} values are comparable with the results presented previously for a similar structure of the active region of a QCL heterostructure grown by MOCVD [12]. Further optimization of the conditions for epitaxial growth of QCL heterostructures and the selection of new contact types to enhance the laser performance are required.

Acknowledgments
The research was partly supported by the Russian Foundation of Basic Research (grant № 16-29-03289) and the Ministry of Education and Science of the Russian Federation (state assignment №3.933.2017/4.6).

References
[1] Yu J S, Slivken S and Razeghi M 2010 *Semicond. Sci. Technol.* **25** 125015
[2] Hildenbrand J, Herbst J, Wöllenstein J and Lambrecht A 2009 *Proc. SPIE*, ed M Razeghi *et al* vol 7222 (San Jose: SPIE) p 72220B
[3] Michel A P M, Kapit J, Witinski M F and Blanchard R 2017 *Appl. Opt.* **56** E23
[4] Moreau V, Colombelli R, Perahia R, Painter O, Wilson L R and Krysa A B 2008 *Opt. Express* **16** 6387
[5] Bidaux Y, Terazzi R, Bismuto A, Gresch T, Blaser S, Muller A and Faist J 2015 *J. Appl. Phys.* **118** 093101
[6] Maulini R, Lyakh A, Tsekoun A and Patel C K N 2011 *Opt. Express* **19** 17203
[7] Troccoli M 2015 *IEEE J. Sel. Top. Quantum Electron.* **21** 61
[8] Garcia M, Vermersch F J, Marcadet X, Bansropun S, Carras M, Wilk A and Sirtori C 2006 *Proc. SPIE* vol 6133 (San Jose: SPIE) p 613304
[9] Babichev A V *et al* 2016 *Semiconductors* **50** 1299–3
[10] Babichev A V *et al* 2017 *Tech. Phys. Lett.* **43** 666–9
[11] Bousseksou A, Colombelli R, Babuty A, De Wilde Y, Chassagneux Y, Sirtori C, Patriarche G, Beaudoin G and Sagnes I 2009 *Opt. Express* **17** 9391
[12] Bahriz M, Moreau V, Palomo J, Colombelli R, Austin D A, Cockburn J W, Austin D A, Cockburn J W, Wilson L R, Krysa A B and Roberts J S 2006 *Appl. Phys. Lett.* **88** 181103b

Photoactive n-type oxide AZO-AgNP-AZO multilayer films for photovoltaic application

E V Shirshneva-Vaschenko, T G Liashenko, P S Shirshnev, V E Bougrov and A E Romanov

ITMO University, Kronverkskii pr. 49, St. Petersburg, 197101 Russia

Email: tgliashenko@corp.ifmo.ru

Abstract. We report on the simulation of the absorption cross-section of an aluminum-doped zinc oxide (AZO) - silver nanoparticles (AgNP)-AZO multilayer structure by the finite element method in Comsol Multiphysics. A key feature of our approach is consideration of two plasmon materials such as AgNP and highly doped AZO. The simulation is based on the Drude-Lorentz-Sommerfeld model. The absorption cross section spectra were obtained for structures with different thicknesses of AZO layers and different amounts of silver nanoparticles, and at a different incidence of illumination. As a result, several resonance peaks were detected in the absorption cross-section spectra. The enhancement of absorption in the visible and near-infrared regions of the spectrum is associated with plasmon oscillations of interacting silver nanoparticles and a plasmon resonance in an AZO nanolayer, respectively. The presence of several resonances in the visible region, as well as the resonance in the near-IR region, makes this material promising for photovoltaics.

1. Introduction

Worldwide trends show new developments in the search for alternative energy sources. Current efforts are focused on increasing the efficiency of photoelectric devices while reducing their cost. Multilayered oxide films with metal nanoparticles are promising functional components for photoelectric devices. In the last decade, the metal nanoparticles have come to be viewed as "activators" in photovoltaics, owing to their plasmon properties [1]. The excitation of nanoparticles at the plasmon frequency leads to an increase in their absorption and scattering cross sections. The frequency of the plasmon resonance is determined by the material of particles, their shape and the material of environment [2, 3]. At least three possibilities exist for the realization of metal nanoparticles in photovoltaic systems. First, the local near-field amplification caused by the excitation of plasmon oscillations results in the increasing generation of charge carriers [1]. Near-field effects were identified by an increase in the photocurrent in solar cells sensitized by the dye and in organic polymer-based solar cells [4-6]. The second approach is based on a possibility to redirect the radiation to an active layer of a solar cell by increasing the scattering cross section of metal nanoparticles at the plasmon resonance wavelength [7, 8]. The third approach suggests that the plasmon particles embedded in an environment of a semiconductor act as elements that accumulate light energy and trigger the mechanism of photoinduced charge separation. The first research in this direction was published in 1996 and pointed to the generation of electron-hole pairs in a TiO_2 semiconductor in contact with a plasmon metal nanoparticle [9]. Zhao and co-authors were the first to register a photocurrent generated by a TiO_2 electrode coated with gold and silver nanoparticles under the action of visible radiation. Later, a number of papers have confirmed the phenomenon of photoconductivity in such structures by detecting the generation, injection, and regeneration of "hot" electrons [10, 11].

In this work, we consider optical properties of a multilayer structure consisting of silver nanoparticles (AgNP) and AZO. AZO is of interest as a host material due to the fact that it is a transparent semiconductor that exhibits plasmon properties in the near–IR region [12].

2. The model

In this work, we use the experimentally measured parameters of the composite AZO-AgNP-AZO structures [13] as the initial data for the absorption cross-section simulation. The composite structure fabricated by the sol-gel technique consists of ZnO:Al (AZO) layers deposited by centrifugation on a quartz substrate with an intermediate layer of silver nanoparticles. AgNP were deposited from a silver colloidal solution.

The structure parameters for the simulation were determined based on the scanning electron microscopy data and the data on the thickness of the AZO layers checked by the interference method. The shape of all nanoparticles was assumed to be spherical with an average nanoparticle diameter of 40 nm and a distance between them of 10 nm. This approximation was sufficient to simulate the position of plasmon resonances in the structure. The dispersion in shape and size of nanoparticles in the real structure leads to broadening of the peaks in the absorption spectra.

Simulating optical properties of metal nanoparticles was based on the solution of the Maxwell equations by approximate methods. The spectral dependences of the dielectric permittivities of the material were analyzed in the framework of the Drude-Lorentz-Sommerfeld model. For noble metals (silver and gold), the real parts of the permittivity become negative in the visible range of the spectrum, which determines their optical properties. The quasistatic approximation of the Mie scattering theory allows considering a metal nanoparticle as an oscillating dipole.

The absorption cross section is determined by the following expression [14]:

$$\sigma_{abs} = k\,\mathrm{Im}\{\alpha(\lambda)\},$$

where $\alpha = 4\pi r^3 \dfrac{\varepsilon(\lambda) - \varepsilon_m(\lambda)}{\varepsilon(\lambda) + 2\varepsilon_m(\lambda)}$ is the polaribility in the case of spherical metal nanoparticles with a radius r, k is the wave vector, $\varepsilon(\lambda)$ is the dielectric permittivity of metal nanoparticles, and $\varepsilon_m(\lambda)$ is the permittivity of the surrounding medium.

The simulation was carried out by the finite element method (FEM) for the periodic structure shown in figure 1 in the RF module of Comsol Multiphysics software. A key feature of the model is the consideration of metal nanoparticles embedded in a medium of a heavily doped semiconductor that exhibits metal properties and demonstrates a plasmon resonance in the near-IR region (a low-loss material). Optical characteristics were calculated using Comsol Multiphysics and accounting for the spectral dependences of AZO permittivity [12].

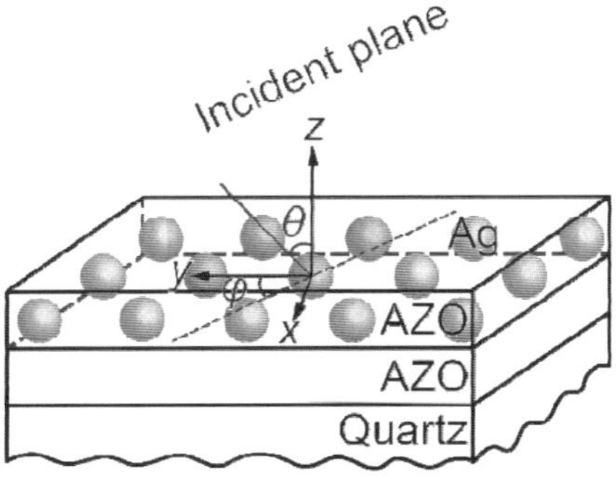

Figure 1. Schematic representation of the AZO-AgNP-AZO structure: φ is the azimuth angle and θ is the colatitude angle of incident light.

IOP Conf. Series: Journal of Physics: Conf. Series **993** (2018) 012032 doi:10.1088/1742-6596/993/1/012032

3. Results

In this work, the absorption cross section spectra (figure 2, 3), the electromagnetic field distribution (figure 4(a)), and the field amplification at resonant wavelengths (figure 4(b)) for various thicknesses of AZO layers and for various incident illumination of the AZO-AgNP-AZO structure in the wavelength range of 340-1600 nm were simulated. The simulation of the absorption cross section spectrum of a silver nanoparticle enclosed in a spherical shell of AZO was carried out (figure 5).

The optical properties of an assembly of metal nanoparticles are determined both by the properties of individual particles and by the electrodynamic interaction between them [15]. Multiple peaks appearing in the optical absorption cross section spectrum are related to the splitting of the resonance due to the interaction of nanoparticles with each other. Silver nanoparticles located in the AZO layer can be characterized by a distribution function of the eigenfrequency of collective excitations. This distribution is broader in comparison to the homogeneous width of the resonances, so that the width of the extinction spectrum of the entire ensemble of particles is precisely determined by the inhomogeneous distribution of the resonance frequencies [16]. Owing to the broadening of the spectra for the real structure, the peaks corresponding to the interaction of non-neighboring particles are absent in the absorption cross section spectra.

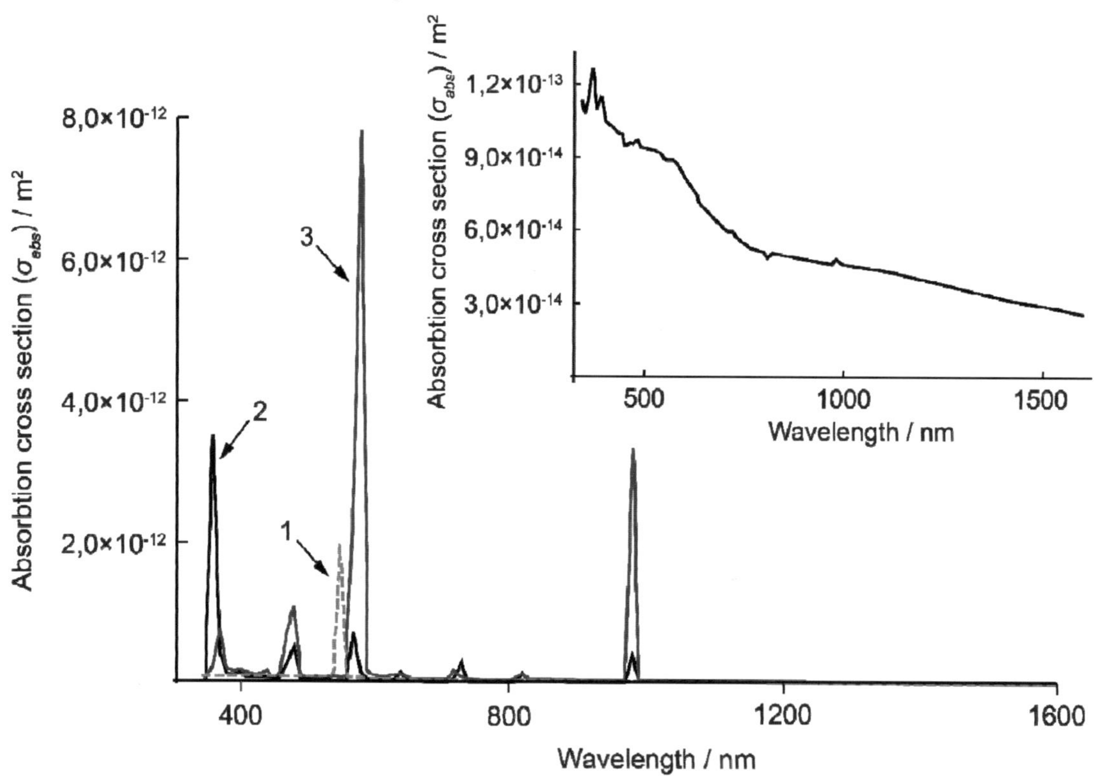

Figure 2. Absorption cross section spectra of the AZO-AgNP-AZO structure on a quartz substrate under normal illumination (1 - the structure consisting of one silver nanoparticle in AZO layers on a quartz substrate; 2 - the AZO-AgNP-AZO structure with an AZO layer thickness of 45 nm on a quartz substrate; 3 - the AZO-AgNP-AZO structure with an AZO layer thickness of 90 nm on a quartz substrate). The insert shows the absorption cross section for the AZO layer on a quartz

The number of silver nanoparticles interacting with each other in the layer and the thickness of the surrounding AZO layer vary depending on the angle of light incidence at the AZO-AgNP-AZO structure. Thus, the plasmon peak located in the near-IR region at the azimuthal angle ($2\pi/3$) is the consequence of the increasing the fraction of AZO in the environment.

IOP Conf. Series: Journal of Physics: Conf. Series **993** (2018) 012032 doi:10.1088/1742-6596/993/1/012032

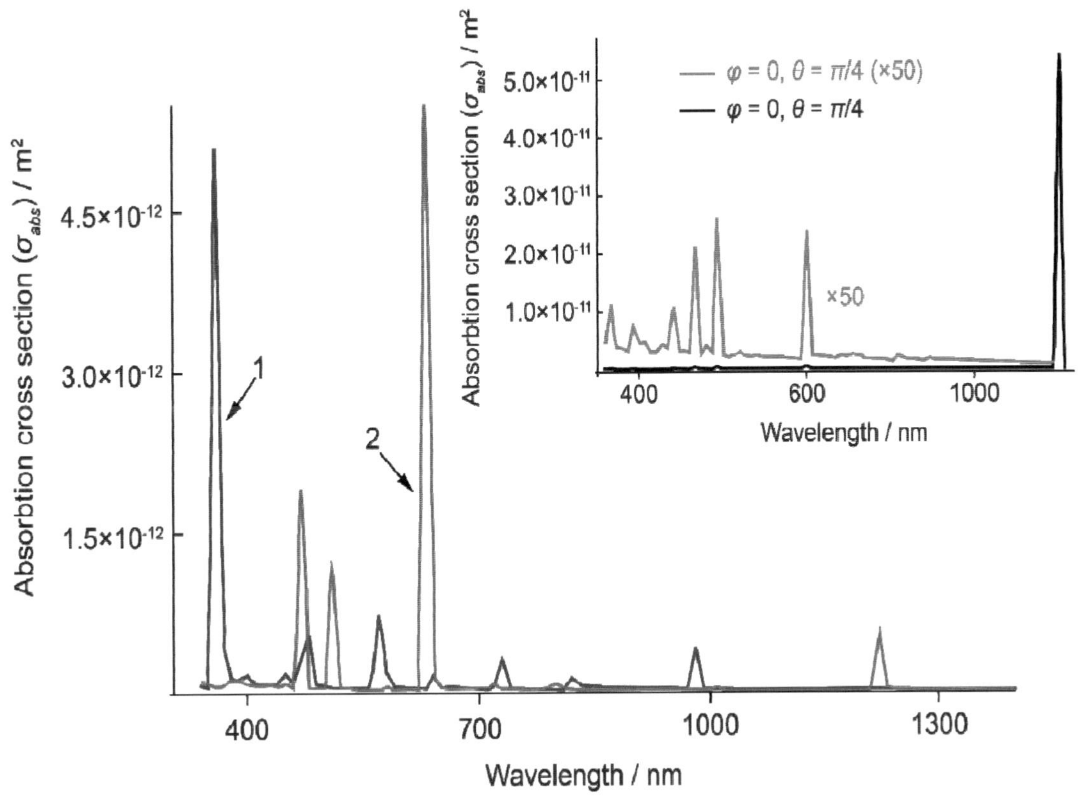

Figure 3. Absorption cross section spectra of the AZO-AgNP-AZO structure on a quartz substrate at different angles of light incidence (1 – $\varphi = 0$, $2\pi/3$, $\theta = 0$; 2 – $\varphi = 0$, $\theta = \pi/3$, $2\pi/3$). The insert shows the absorption cross section spectrum for the angle of incidence: $\varphi = 0$, $\pi/4$.

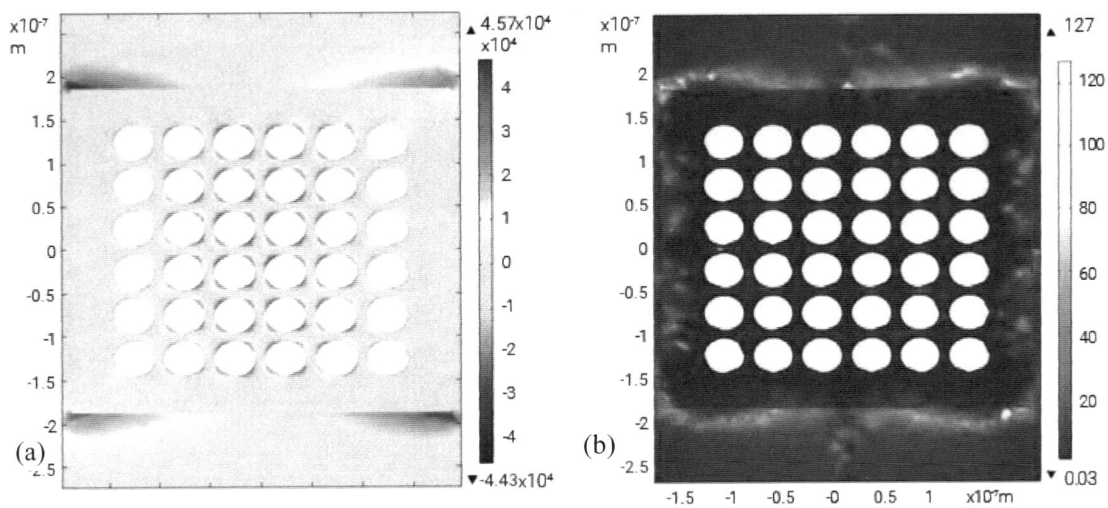

Figure 4. FEM simulation results for the AZO-AgNP-AZO structure under illumination of a 980 nm wavelength normal to the surface of the structure (a) – the electromagnetic field (E_x) in the AZO-AgNP-AZO structure; (b) – the field enhancement in the AZO-AgNP-AZO structure. The field near the nanoparticles is enhanced by up to 100 times.

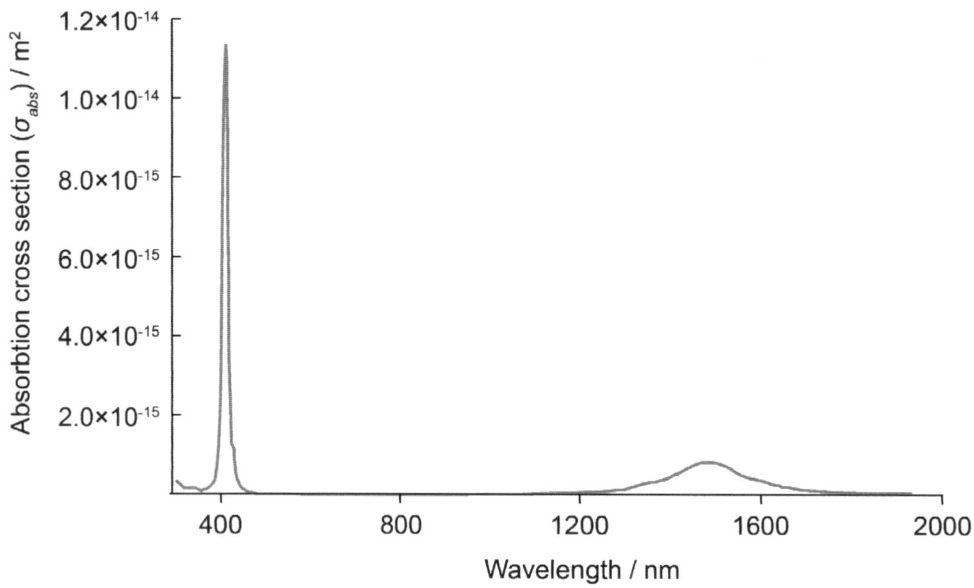

Figure 5. The absorption cross section spectrum of a silver nanoparticle (with a diameter of 40 nm) in an AZO shell (10 nm thick).

4. Conclusions

The absorption cross section spectra of AZO-AgNP-AZO structures were simulated at various angles of incidence light and at different thicknesses of AZO layers. The presence of peaks in the spectra indicates the excitation of resonances at the corresponding wavelengths. The analysis of the obtained spectra indicates that the optical properties of the structure correspond to the core-shell model (silver nanoparticle – AZO). The resonance in the near-IR region depends on the plasmon properties of AZO. The presence of plasmon resonances in the visible and near IR ranges indicates the possibility of formation of hot electrons in the structure and an increase in the photosensitivity of the material, which makes it promising for photovoltaics.

Acknowledgement

This work was supported by the Ministry of Education and Science of the Russian Federation in the framework of the Federal Target Program "Research and development in priority directions of the scientific-technological complex of Russia for 2014-2020", code 2017-14-576-0003, agreement № 14.575.21.0127 from September 26, 2017, unique ID RFMEFI57517X0127.

References

[1] Atwater H A and Polman A 2010 *Nature Materials* **9** 205
[2] Pelton M, Aizpurua J and Bryant G 2008 *Laser Photonics Reviews 2* 136-159
[3] Kelly K L, Coronado E, Zhao L L and Schatz G C 2003 *J. Phys. Chem. B* **107** 668-67
[4] Brown M D, Suteewong T, Kumar R S *et al* 2011 *Nano Letters* **11** 438-445
[5] Kawawaki T, Takahashi Y and Tatsuma T 2011 *Nanoscale* **3** 73-79
[6] Morfa A J, Rowlen K L, Reilly T H *et al* 2008 *Applied Physics Letters* **92** 013504
[7] Derkacs D, Lim S H, Matheu P *et al* 2006 *Applied Physics Letters* **89** 093103
[8] Nakayama K, Tanabe K and Atwater H A 2008 *Applied Physics Letters* **93** 121904
[9] Zhao G, Kozuka H and Yoko T 1996 *Thin Solid Films* **277** 147-154
[10] Tian Y and Tatsuma T 2005 *Journal of the American Chemical Society* **127** 7632–7637
[11] Y u K, Tian Y and Tatsuma T 2006 *Physical Chemistry* **8** 5417–5420
[12] Naik G V, Gururaj V, Shalaev V M and Boltasseva A 2013 *Advanced Materials* **25** 3264-3294

[13] Shirshneva-Vaschenko E V, Sosnin I M, Nuryev R K, Gladskikh I A, Liashenko T G, Bougrov V E and Romanov A E 2016 *Materials Physics and Mechanics* **29** 145-149

[14] Bohren, C F and Huffman D R 2008 *Absorption and scattering of light by small particles* (New York: John Wiley & Sons)

[15] Rechberger W, Hohenau A, Leitner A, Krenn J R, Lamprecht Band Aussenegg F R 2003 *Optics communications* **220** 137-141

[16] Kreibig U and Vollmer M 1995 Optical Properties of Metal Clusters *Theoretical considerations (Electronic Materials* vol 25*)* (Berlin: Springer-Verlag Berlin Heidelberg) chapter 2 pp 13-201

A study of the influence of forward bias pulse duration on the switching process of a 4H-SiC drift step recovery diode

A A Smirnov and S A Shevchenko

Saint Petersburg Electrotechnical University "LETI", Saint Petersburg 197376, Russia
E-mail: smirnov.eltech@gmail.com, sashevchenko.rte@gmail.com

Abstract. The results of simulation of the switching process in a silicon carbide drift step recovery diode with various times for injecting non-equilibrium carriers are presented. The inefficiency of a long injection time is demonstrated.

1. Introduction

Silicon-based drift step recovery diodes (DSRD) are widely and successfully implemented in high-voltage pulse generators as superfast opening switches. The main principle underlying a typical DSRD operation cycle is based on the superfast recovery effect that occurs in a *p-n* junction after injection of non-equilibrium charge carriers into the base region of the structure with a short forward current pulse and their subsequent removal with reverse current [1].

In [2, 3] the results of the design of silicon carbide DSRDs are presented. The designed diodes demonstrate outstanding performance, providing voltage rise velocities of 2.5 - 3.5 V/ps, which is unachievable for a silicon DSRD because its switching velocity is limited by the value of 1 V/ps. In [4, 5] the results of the design of the voltage pulse generators with 4H-SiC DSRDs implemented as superfast opening switches are demonstrated; they are capable of shaping voltage pulses with subnanosecond rise times.

However, despite the existence of experimental results and theoretical works [6-8], many questions relating to 4H-SiC DSRD operation remain unanswered. In order to improve the efficiency of these switches and to reach a voltage rise velocity of up to 10 V/ps that is theoretically achievable for 4H-SiC [1], it is necessary to conduct further studies of the electrical and thermal properties and, in particular, dynamic characteristics of these diodes. Thus, the objective of the present paper is to study the influence of forward bias pulse duration on charge losses in a silicon carbide drift step recovery diode.

2. Simulation

The study was conducted by simulation in Synopsys Sentaurus TCAD software in the drift-diffusion approach with the following effects accounted for: bandgap narrowing at high doping levels, avalanche generation in high fields, incomplete dopant ionization, which is a very specific effect for 4H-SiC, Shockley-Read-Hall and Auger recombination mechanisms, and various mobility models including mobility lowering in heavily doped regions and carrier velocity saturation at high fields.

The examined diode has a p^+–p–n^+ structure with epilayer thicknesses of 2, 18 and 350 um, respectively. The concentrations of acceptors (*Al*) and donors (*N*) in *p*-type and *n*-type layers are $2\cdot10^{19}$ cm^{-3}, $5\cdot10^{15}$ cm^{-3}, and 10^{19} cm^{-3}, respectively. The area of the top contact is 1 mm^2. The doping profile of the structure is presented in figure 1. Based on an approximate breakdown analysis involving solution of the Poisson equation with calculation of ionization integrals, the static breakdown voltage of the structure (and hence the reverse voltage the diode is to switch) is estimated to be 1800 V. The operation of the diode has been studied in a circuit containing a voltage source *V1*, a 4H-SiC-DSRD *D1* and a 50 Ω load resistor *R1*. This circuit is presented in figure 2.

A typical transient in the circuit the diode was studied in is presented in figure 3. During the time T_P, the diode is biased forward with a linearly rising voltage pulse, which is accompanied by injection of non-equilibrium carriers into the base region of the structure. After that, the voltage at the source changes its polarity and remains constant while the voltage drop at the diode is low as the injected carriers are extracted; the latter process lasts for time T_D. By the time the non-equilibrium carriers have been fully removed, the majority carriers are extracted from the base with saturated velocities, which goes along with a fast voltage rise at the diode and corresponds to superfast switching.

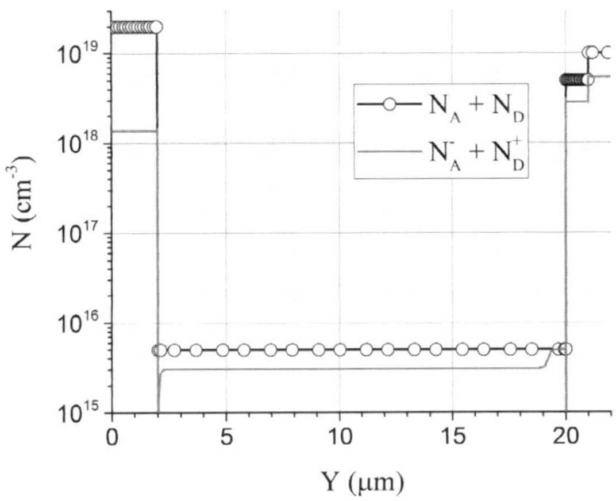

Figure 1. The doping profile of the drift step recovery diode (marked line) and ionized doping distribution (red line).

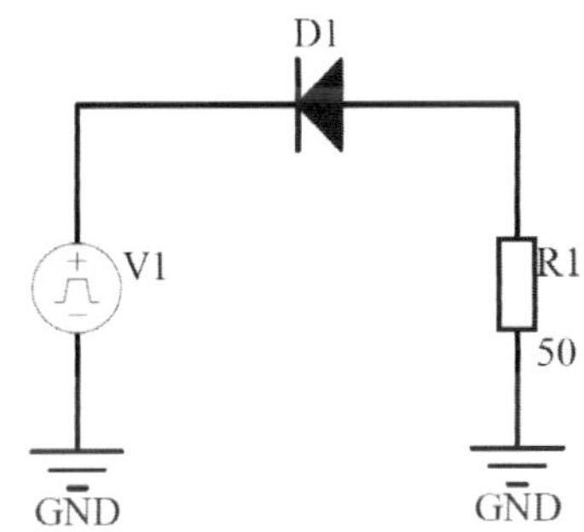

Figure 2. The circuit that the 4H-SiC ultrafast diode was examined in.

The results of several transient simulations described above with various values of T_P are presented in figure 4. In order to estimate the times T_D correctly, the endings of the forward bias pulses are shifted to the moment of time t = 0. Since the full charge that is injected into the structure and the extracted charge are, respectively:

$$Q_+ = \int_0^{TP} \frac{U}{R1}\, dt \approx \frac{1}{2} \cdot I_{D1} \cdot T_P; \qquad (1)$$

$$Q_- = \int_{TP}^{TD} \frac{U}{R1}\, dt \approx I_{D1} \cdot T_D, \qquad (2)$$

where I_{D1} is the current flowing through the diode, we may introduce the relation between Q_- and Q_+ as the percentage between the injected and extracted charge. So if Q_-/Q_+ is less than 100%, one can observe a loss of the injected charge. The quantified charge losses for various durations of the forward bias pulse T_P are summarized in table 1.

Table 1. Estimated values of charge losses at various durations of the forward bias pulse T_P.

T_P, ns	1	5	10	15	20	25	30	35	40	50	60	70	80	90	100
T_D, ns	0,5	1,8	3	3,9	4,5	5,2	5,8	6,3	6,8	7,7	8,6	9,3	9,9	11	11
Q_+, nC	18	90	180	270	360	450	540	630	720	900	1080	1260	1440	1620	1800
Q_-, nC	18	67	110	140	164	188	208	228	244	276	308	335	357	382	401
Q_-/Q_+, %	100	74	61	52	46	42	39	36	34	31	29	27	25	24	22

Figure 5 demonstrates that the concentrations of the injected non-equilibrium holes and electrons in the base region of the structure rise with the increase of the forward bias pulse duration. This evidences that the absolute amount of the charge stored in the diode's base region increases as well, and is limited by the maximum injection ability of p^+-p and n^+-p junctions [10].

It is also worth noting that the absolute value of the charge stored in the structure is relatively small compared to the injected charge at times T_P higher than 40 nanoseconds.

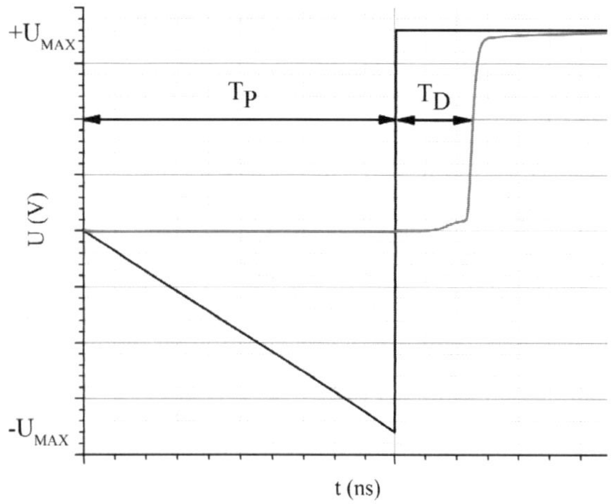

Figure 3. Typical transient in the circuit the diode has been studied in (black and red lines represent the voltages of the voltage source *V1* and the studied diode *D1*, respectively).

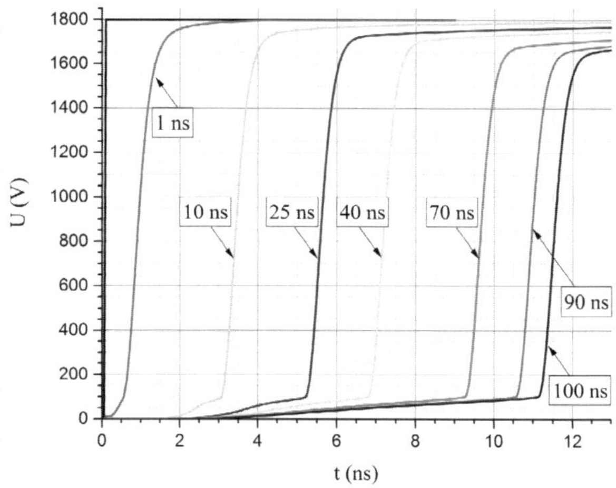

Figure 4. The switching processes of the studied diode at various durations of the forward current pulse.

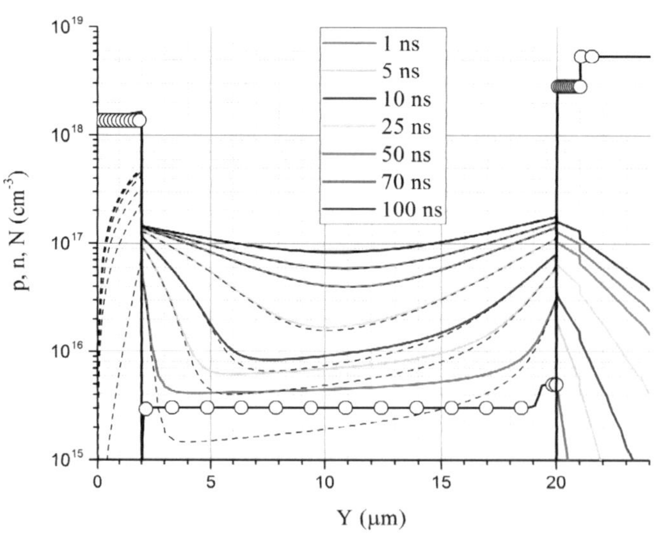

Figure 5. Distribution of holes (coloured lines) and electrons (dashed lines) at the end of the injection process (the marked line represents the ionized doping distribution).

3. Results and conclusion

It is shown that the losses of the injected charge increase with an increase of the duration T_P of the forward bias pulse injecting non-equilibrium carriers into the diode base region.

Based on the shapes of the dependences obtained in figure 4 and briefly stated in table 1, one can conclude that biasing the diode forward for a time longer than 40 ns is impractical due to the limited value of the charge that can be stored in the diode structure. Therefore, long injection times lead to vast loss of the injected charge. It can also be clearly seen that with an increase in T_P, the efficient part of the diode's switching process shrinks up to 10% at the end of it.

In figure 5, one can also observe that the electrons and holes are accumulated in the p^+ and n^+-regions, which undoubtedly shows that conservation of the injected charge is possible not only by the accumulation of non-equilibrium carriers in the base region, but also by the accumulation of minority carriers in passive heavily doped regions of the diode.

We assume that the loss of the injected charge is caused by recombination of minority charge carriers (electrons) at the ohmic contact to the p^+ emitter. Since the thickness of this region is much smaller than the electron diffusion length, a certain amount of electrons injected into a p^+ emitter diffuse towards the ohmic contact to this layer, where the recombination rate is infinite, so the loss of these electrons occurs. This mechanism of charge loss may only be enhanced by incomplete dopant ionization, which is a very common phenomenon for wide bandgap semiconductors such as 4H-SiC. Thus, incomplete dopants ionization may result in an increase in mobility of minority carriers in heavily doped regions and a consequent rise in their diffusion lengths.

According to the results presented in this paper, in order to improve the accumulation of electrons in a p^+ emitter and hence to reduce charge losses in the diode structure, an increase of doping ionization degree is of practical interest and will be the subject of our future research.

References

[1] Grekhov I V and Mesyats G A 2005 Phys.-Usp. **48** 703-712
[2] Afanasyev A V, Ivanov B V, Ilyin V A, Kardo-Sysoev A F, Luchinin V V and Serkov F B. 2012 *Proceedings of All-Russian Scientific and Technical Conference Microwave Electronics and Microelectronics* **2** 260-262
[3] Afanasyev A V, Ivanov B V, Ilyin V A, Kardo-Sysoev A F, Kuznetsova M A and Luchinin V V 2013 *Materials Science Forum* **740-742**, 1010-1013
[4] Afanasyev AV, Demin Y A, Ivanov BV, Ilyin VA, Kardo-Sysoev AF, Luchinin V V and Smirnov A A. 2015 *Proceedings of the Russian Universities: Radioelectronics* **3** 21.
[5] Afanasyev A V, Ivanov B V, Ilyin V A, Smirnov A A and Shevchenko S A 2016 *Proc. of 57th International Scientific Conference on Power and Electrical Engineering RTUCON* 2016 7763090
[6] Kyuregyan A S 2004 *Tech. Phys* **49** 720
[7] Ivanov P A and Grekhov I V 2015 *Tech. Phys* **60** 897
[8] Grekhov I V, Kyuregyan A S, Mnatsakanov T T and Yurkov S N 2003 *Semiconductors* **37** 1123
[9] Ilyin V A, Afanasyev A V, Ivanov B V, Kardo-Sysoev A F, Luchinin V V, Reshanov S A, Schöner A, Sergushichev K A and Smirnov A A 2016 *Materials Science Forum* **858** 786
[10] Herlet A 1968 *Sol. St. Electron.* **11** 717

Influence of interface layer preparation on the electrical and spectral characteristics of GaN/Si solar cells

K U Shugurov[1], A M Mozharov[1], G A Sapunov[1], V V Fedorov[1,2], A D Bolshakov[1,2] and I S Mukhin[1,2]

[1] Saint-Petersburg Academic University – Nanotechnology Research and Education Centre Russian Academy of Sciences, 8/3 Khlopina st., St. Petersburg 194021, Russian Federation

[2] ITMO University, 49 Kronverksky pr., St. Petersburg 197101, Russian Federation

E-mail: shugurov17@mail.ru

Abstract. Volt-ampere and spectral characteristics of GaN/Si solar cell samples differing in interface layer preparation are obtained and analyzed. External quantum efficiency curves are experimentally determined via excitation with a 532 nm incident radiation wavelength. It is demonstrated that interface preparation has a significant influence on photovoltaic characteristics of the studied samples.

1. Introduction

The amount of electricity being annually consumed constantly increases with time. Since the main energy resources are currently exhaustible, there is a severe need for the development of alternative energy sources, and solar energy conversion is known to be the most promising approach. The development of photovoltaic devices is usually realized through improvement of the characteristics and efficiency of photoconverters (solar cells). The most effective solar radiation converters at the moment are multijunction cells based on $A^{III}B^{V}$ [1] compounds, which have several advantages over other compounds due to a combination of crucial properties. Today, particular attention is focused on compounds of the $A^{III}N$ type. It is related to good electronic, structural and other properties of these materials [2].

First of all, most $A^{III}N$ compounds are wide-band semiconductors with high electron mobility [3], which allows to use them in solar cells (SCs) as a wide-gap window, and as a material for an antireflection coating to improve SC efficiency through light absorption enhancement. Secondly, Group III nitrides such as GaN are capable of operating at higher temperatures [3] compared to other compounds. As a consequence, this material can be used in the development of cost-effective photoconverters with solar concentrators. Thirdly, on an industrial scale, the devices based on $A^{III}N$ compounds on silicon substrates are economically more profitable than their nearest competitors.

The characteristics and efficiency of SCs are influenced by many factors, and among them interface layer passivating properties are important ones [4]. Characteristics of similar SCs can differ significantly due to distinctive interface preparation and, as a consequence, different surface recombination rates. For SCs based on indirect-band materials (silicon, germanium), recombination through surface states and volume recombination through impurity levels are the main limiting factors affecting the SC efficiency.

Our work is devoted to studying the electrical and spectral characteristics of SCs based on n-GaN / p-Si heterojunctions obtained with various types of interface layer preparation.

2. Experiment

Specially fabricated heterostructures based on arrays of n-GaN nanowires (NWs) grown on p-type silicon substrates using plasma-assisted molecular beam epitaxy were studied. The presence of an emitter layer in the form of an n-GaN NWs array is a distinctive feature of the SCs design. It is supposed to lead to an increase in the efficiency of silicon SCs and also to provide good antireflection properties without the use of expensive special multilayer antireflection coatings [5].

The synthesized heterostructures differed in the interface layer preparation process, which affects both the morphological properties of NWs (their diameter and density) and semiconductor properties of the heterointerface. There are three structure types considered in this paper. Structure 1 (sample 1): after thermal silicon oxide removal, the substrate was cooled down to a temperature of 810 °C (according to thermocouple measurements), and growth of NWs was initiated. Structure 2 (sample 2): after oxide removal, the hot substrate was treated with atomic nitrogen, cooled down to a temperature of 630 °C and exposed under gallium flow, and then the growth of NWs was initiated. Structure 3 (sample 3): before loading the substrate into a growth chamber, a GaO_x layer was deposited in a plasma-activated vapor deposition machine using the layer-by-layer deposition method, then the substrate was annealed at 850 °C under an atomic nitrogen flux, and NWs growth was initiated at a temperature of 800 °C.

At the next technological step, a back Al contact was fabricated on each of the synthesized GaN/Si heterostructures by thermal vacuum deposition of metals and further thermal annealing. Using various methods of post-growth treatment, a dielectric insulating layer was deposited over the emitter layer, while the tips of the NWs were left uncovered. Then a conductive transparent indium-tin oxide (ITO) coating was applied via magnetron sputtering. These heterostructures were used as the basis for fabricating solar photoconverters with a photoactive region in the form of a circular maze (2 mm in diameter).

Measurement of the current-voltage characteristics (*I - V* characteristics) was performed using a Keithley 2400 SourceMeter multimeter and a thermo-stabilized stage under illumination conditions of the AM 1.5G solar spectrum. The temperature was maintained at 25 °C during the measurements.

The spectral dependence of external quantum efficiency (EQE) was measured in the 350 – 1200 nm wavelength range using a M266 (Laser Solar) monochromator and a calibrated silicon photodiode with known spectral characteristics.

The study of EQE dependence on the incident radiation intensity was performed using a table with temperature stabilization over a wide intensity range from 10^{-2} to 10^{3} mW/cm^2. The radiation source was a laser with a wavelength of 532 nm. A laser beam was transmitted through an aperture of 1 mm in diameter. The power of the laser beam applied to the photoactive region of the SCs was adjusted by a set of optical filters and was measured by a calibrated laser power meter PM100D (THORLABS).

3. Experimental results and discussion

The measured *I - V* characteristics of the SC samples with different interface layers are shown in figure 1(a). A distinguishable influence of the interface treatment on the SCs main characteristics, namely open-circuit voltage (*Uoc*) and short-circuit current (*Isc*), was detected. Relatively low values of *Uoc* and *Isc* are presumably related to an intensive irradiative recombination process. By analyzing the difference between the light and dark *I - V* characteristics (figure 1 (b)) at large negative bias voltages, we can conclude that the dominant contribution to the recombination process is provided with surface recombination (on the substrate surface free of NWs and on the GaN/Si heterointerface).

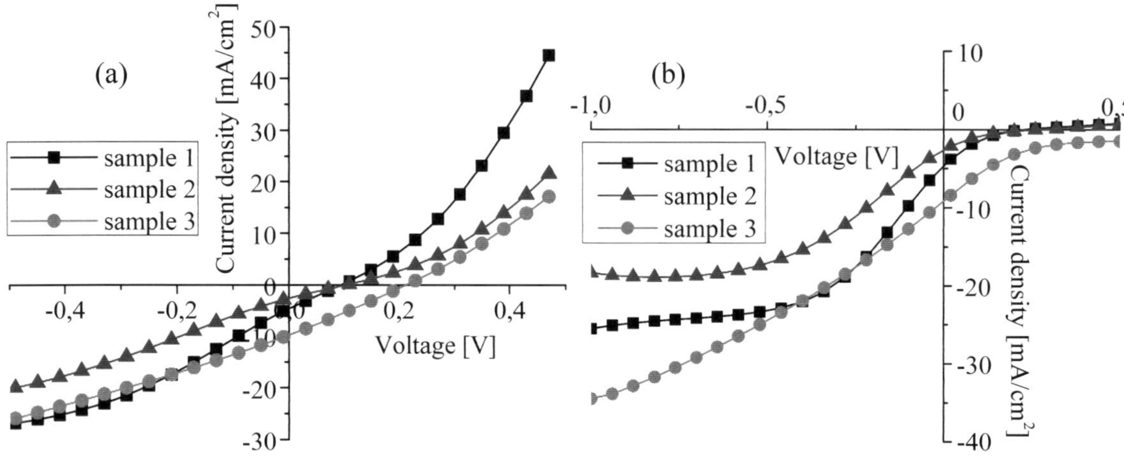

Figure 1. Light I - V characteristics (a) and the difference between light and dark I - V characteristics (b).

Figure 2 shows dependencies of the EQE spectra of the studied SCs. It is clear that the method of preparation of the GaN/Si heterointerface has a significant influence on the spectral characteristics of the solar cell.

EQE was calculated using the following formula:

$$EQE = \frac{hc}{\lambda e} \cdot \frac{I}{P}, \tag{1}$$

where h is the Planck constant, c is the speed of light in vacuum, λ is the wavelength of the incident radiation, e is the electron charge, P is the power of the incident radiation, and I is the current flowing through the solar cell under illumination.

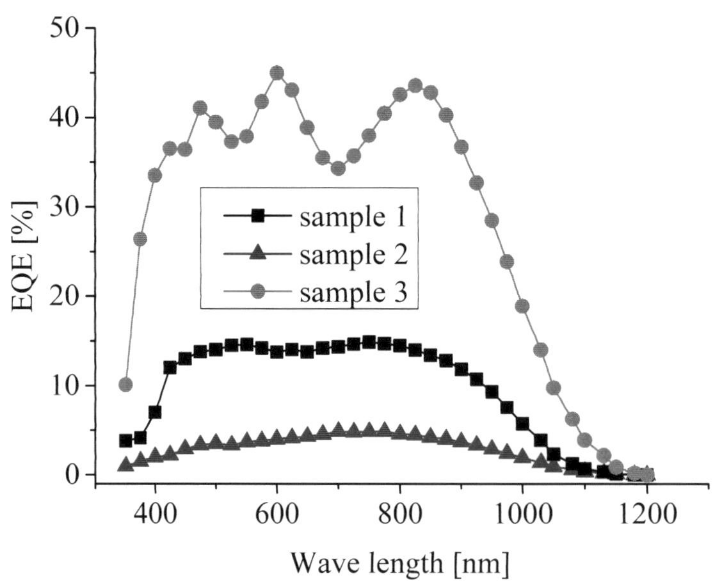

Figure 2. Experimental spectral characteristics.

We assume that the characteristic oscillations of the EQE spectrum in the range from 400 to 900 nm in sample 3 are related to the presence of an NWs array with an optimal morphology providing antireflective properties.

The experimental results on the calculation of the EQE dependence on the incident radiation intensity for the SCs are shown in figure 3.

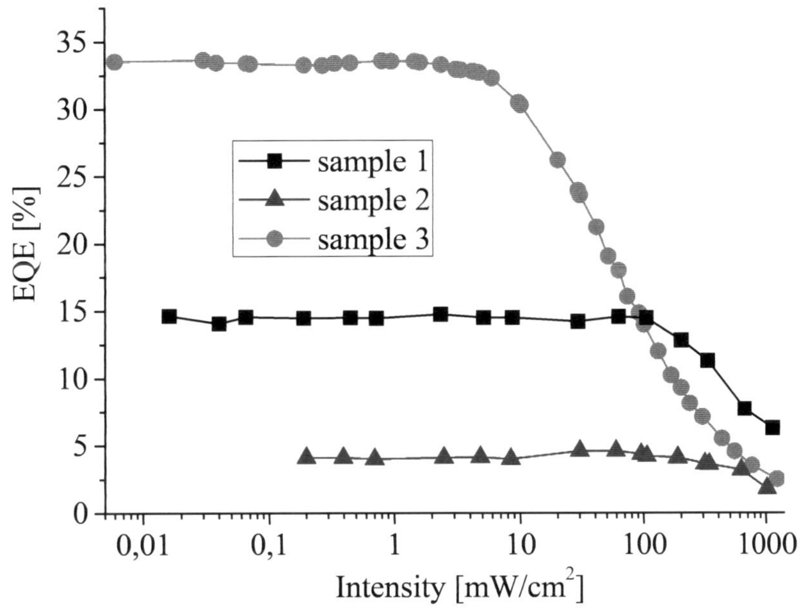

Figure 3. EQE experimental dependence on incident radiation intensity with a wavelength of 532 nm.

It was found that the EQE of sample 3 drops almost twofold from its maximum level at a light intensity of 100 mW/cm^2, which corresponds to the solar radiation intensity at the Earth's surface (see figure 3). This phenomenon relates to a high value of parasitic series resistance of the SCs. At low radiation intensities, the EQE does not change sufficiently, which is consistent with numerical modeling results not presented in this paper. We should also mention here that the rate of EQE decrease with the radiation intensity is subject to change by the surface passivation technique.

4. Summary

In our paper, we studied electrical and spectral characteristics of the SCs series based on GaN/Si heterostructures. We investigated the devices characterized by three different methods of the interface preparation. Volt-ampere characteristics of the samples were obtained using a solar spectrum simulator. EQE dependencies were calculated in the 350 – 1200 nm region of the incident light wavelength. We then carried out analysis of the obtained experimental data to compare the characteristics of the fabricated SCs. We also carried out experimental measurements of the EQE dependencies on the intensity of the 532 nm wavelength incident light. According to the results, the characteristics of GaN/Si SCs change sufficiently depending on the method of substrate treatment prior to GaN deposition.

Acknowledgments

This work was supported by the Russian Foundation for Basic Research (grants 16-32-00560, 15-02-06839), Council for grants of the Russian Federation President (grant MK-3632.2017.2). The authors acknowledge support from a mega-grant № 14.Y26.31.0015, grants 16.2593.2017/4.6, 16.2593.2017/8.9 of the Ministry of Education and Science of the Russian Federation and Leading universities of the Russian Federation grant 074-U01.

References

[1] Morozov I A and Gudovskikh A S 2014 *Semiconductors* **48** 475 – 80
[2] Ivantsov V A, Sukhoveev V A, Nikolaev V I, Nikitina I P and Dmitriev V A 1997 *Physics of the Solid State* **39** 858-60
[3] Sergeev S A, Mikhailov A I and Sergeeva B V 2014 *Memoirs of the Faculty of Physics* **2** 142301
[4] Bessolov V N and Lebedev M V 1998 *Semiconductors* **32** 1281-99
[5] Mozharov A M, Bolshakov A D, Cirlin G E and Mukhin I S 2015 *Phys. Status Solidi RRL* **9** 507-10

Simulation of charge transport in micro and nanoscale FETs with elements having different dielectric properties

A M Blokhin[1,2], E A Kruglova[3] and B V Semisalov[3]

[1] Sobolev Institute of Mathematics SB RAS, 630090, Novosibirsk, Russia
[2] Novosibirsk State Univercity, 630090, Novosibirsk, Russia
[3] Institute of Computational Technologies SB RAS, 630090, Novosibirsk, Russia

E-mail: ViBiS@ngs.ru

Abstract. The hydrodynamical model is used for description of the process of charge transport in semiconductors with a high rate of reliability. It is a set of nonlinear partial differential equations with small parameters and specific conditions at the boundaries of field effect transistors (FETs), which essentially complicates the process of finding its stationary solutions. To overcome these difficulties in the case of FETs with elements having different dielectric properties, a fast pseudospectral method has been developed. This method was used for advanced numerical simulation of charge transport in DG-MOSFET.

1. Introduction

The size of semiconductor electron devices is becoming increasingly smaller, up to micro and nanoscales, which leads to the complication of both fabrication methods and structures of modern field effect transistors (FETs). Therefore, for designing such devices, one needs to develop a theoretical basis that includes new reliable mathematical models and accurate numerical methods. Indeed, an adequate description of processes taking place in the semiconductor requires accounting for energy transport effects and impact ionization, as well as dielectric, doping and geometric properties of structural elements of FETs. Since the study of such models by analytical methods can be technically impossible, methods of predictive numerical simulation are often used. However, due to small parameters and complex boundary conditions, many of today's numerical methods turn out to be not sufficiently stable, accurate and fast. The aim of our work is to develop the mathematical model and method for solving these problems. As an example, we consider the problem of charge transport in DG-MOSFET (nano-size Double Gate-Metal Oxide Semiconductor Field Effect Transistor).

2. Hydrodynamical model based on the Maximum Entropy Principle and its applications

Here we shall focus on a hydrodynamical model based on the Maximum Entropy Principle (or MEP) describing the process of charge transport in semiconductors with a high degree of reliability. This model was proposed in [1,2], and it is a set of quasilinear non-stationary partial differential equations (PDEs) written in the form of conservation laws obtained from the system of moments of the Boltzmann transport equation. While constructing this model, the MEP was used for closing the system of moments. In [3], a specific procedure was applied for deriving the dimensionless form of the MEP model and for obtaining its resolving equations. In the 2D case, these equations are:

$$\begin{cases} R_t + \mathrm{div}(\vec{J}) = 0, \quad \vec{J}_t + \frac{2}{3}\nabla(RE) = R\vec{Q} + c_{11}\vec{J} + c_{12}\vec{I}, \\ (RE)_t + \mathrm{div}(\vec{I}) = (\vec{J},\vec{Q}) + cR\sigma, \quad \vec{I}_t + \frac{10}{9}\nabla(RE^2) = \frac{5}{3}RE\vec{Q} + c_{21}\vec{J} + c_{22}\vec{I}. \end{cases} \tag{1}$$

Here R is the electron density, E is the electron energy, $\vec{J} = R\vec{u}$, $\vec{I} = R\vec{q}$, $\vec{u} = (u^{(x)}, u^{(y)})$ is the vector of the electron velocity in a Cartesian coordinate system (x, y), $\vec{q} = (q^{(x)}, q^{(y)})$ is the energy flux, $\vec{Q} = \nabla\varphi = (\varphi_x, \varphi_y)$, $\varphi = \varphi(t,x,y)$ is the electric potential satisfying the Poisson equation:

$$\Delta\varphi = \varphi_{xx} + \varphi_{yy} = \beta(R - \rho); \tag{2}$$

$\rho = \rho(x, y)$ is the doping density (a given function in the interior of the semiconductor). In certain domains of semiconductor devices, the values of this function are less than in other domains by a factor of $10^5 - 10^{10}$. The coefficients $c_{11}, ..., c_{22}, c$ of system (1) are smooth functions of the energy E, detailed in [3], and $\beta > 0$ is a dimensionless constant (see [3]) that depends on the dielectric properties of the semiconductor and on other parameters and takes high values. Note that equations (1), (2) of the MEP model were studied both analytically and numerically in [3–5].

Below, the FETs containing several domains Ω_q, $q = 1,\ldots,Q$ with different dielectric properties are considered (in this case β is the piecewise constant, $\beta = \beta_q$ in Ω_q); S_q is the border separating adjacent domains Ω_q and Ω_{q+1}. The domain of FET is $\Lambda = \cup\Omega_q$. In each Ω_q, the boundary conditions for the electric potential should be imposed. They include the following:

1) *Voltage conditions*, i.e. $\varphi = B$ on a source of a device, $\varphi = G$ on a gate and $\varphi = D$ on a drain,
2) *Matching conditions* on S_q, which are:

$$\varphi|_{S_q^-} = \varphi|_{S_q^+}, \quad \frac{\partial\varphi}{\partial n_q}|_{S_q^-} = \mu_q \frac{\partial\varphi}{\partial n_q}|_{S_q^+}, \tag{3}$$

3) *Free-slip conditions* at all other points of a boundary of Λ. They are $(\alpha, \nabla\varphi) = 0$.

Here $\varphi|_{S_q^-}$ $\varphi|_{S_q^+}$ are the values of φ as its argument goes to S_q from Ω_q and from Ω_{q+1}; α, n_q are the vectors of outward normal to Λ and to S_q. The following parameters are dimensionless:

$$B = \ln\left(\frac{N^+}{N_i}\right), \quad D = \frac{e}{K_B T_0}V_D + B, \quad G = \frac{e}{K_B T_0}V - 10 + B; \quad \mu_q = \frac{\beta_q}{\beta_{q+1}}, \quad \beta_q = \frac{e^2 L^2 N^+}{\kappa_q K_B T_0} \tag{4}$$

and V_D, $V^\pm$, e, L, N^+, κ_q, K_B, T_0, N_i are dimensional parameters described in table 1.

Following [3–5], we recall that equations (1), (2) in the stationary case can be reduced to a system of nonlinear PDEs of the elliptic type:

$$\Delta\sigma = a_1|\nabla\sigma|^2 + a_2(\nabla\sigma,\vec{X}) + a_3(\nabla\sigma,\vec{Q}) + a_4(\vec{X},\vec{Q}) + a_5|\vec{Q}|^2 + bc\sigma, \tag{5}$$

$$\Delta\chi = -|\vec{X}|^2 + b_1|\nabla\sigma|^2 + b_2(\nabla\sigma,\vec{X}) + b_3(\nabla\sigma,\vec{Q}) + b_4(\vec{X},\vec{Q}) + b_5|\vec{Q}|^2 + \frac{\beta}{1+\sigma}(e^\chi - \rho) + nc\sigma, \tag{6}$$

$$(1/\beta)\Delta\varphi = (e^\chi - \rho) \tag{7}$$

and the components of the vectors $\vec{u}$, $\vec{q}$ can be obtained as follows:

$$\vec{u} = F(E)\{\vec{Q} - (1+\sigma)\vec{X} - F_0(E)\nabla\sigma\}, \quad \vec{q} = G(E)\{-\vec{Q} + (1+\sigma)\vec{X} + G_0(E)\nabla\sigma\}. \tag{8}$$

Here $\vec{Q}=(\varphi_x+D-B,\varphi_y)^T$, $\vec{X}=\nabla\chi$, $\chi=\ln R$, $|\nabla\sigma|^2=\sigma_x^2+\sigma_y^2$, $\sigma=2E/3-1$ etc., and the coefficients a_1-a_5, b_1-b_5, n, F_0, G_0 are nonlinear functions of E of a certain form (see [3]). We impose the conditions at the boundaries $\partial\Lambda, S_q$ for functions χ,σ as follows:

1) the dimensionless electron density R on $\partial\Lambda$ is equal to unity, therefore $\chi|_{\partial\Lambda}=0$,
2) the dimensionless electron energy E on $\partial\Lambda$ is equal to 3/2. On the borders S_q, the normal derivative of E is equal to zero, then $\sigma|_{\partial\Lambda}=0$, $\frac{\partial\sigma}{\partial n_q}|_{S_q}=0$ (see [3]),
3) charge carriers don't flow through S_q, therefore $(\vec{u},n_q)=0$ and the first of equations (8) yields

$$\frac{\partial\chi}{\partial n_q}|_{S_q}=\frac{(\varphi_x+D-B)n_{qx}+\varphi_y n_{qy}}{1+\sigma},\quad n_q=(n_{qx},n_{qy}).$$

3. Pseudospectral numerical method for finding distributions of φ,σ,χ in FET

Below we shall assume that Λ is a rectangular domain and Ω_q are its rectangular subdomains, i.e. S_q are parallel to the x or y axis. Without loss of generality, let them be parallel to the x axis. Since it can be easily checked by (4) that for real devices $1/\beta\sim10^{-4}-10^{-6}$, then (5) – (7) is a nonlinear coupled system with a small parameter at the highest order derivatives of φ. To solve it, we shall design a highly accurate and stable algorithm based on an iterative stabilization technique and interpolation polynomials with Chebyshev nodes; see [4]. It is briefly described here by the example of a test boundary value problem

$$\Delta\varphi=f(x,y,\varphi),\quad \Lambda=[0,\pi]\times[0,\pi], \tag{9}$$

where φ satisfies Neumann or Dirichlet conditions on $\partial\Lambda$ that are the *free-slip* or the *voltage conditions* (see above). After a linear change of variables, the domain Λ can be fit to any given rectangular domain. Introducing the additional time variable t, we derive the evolution equation:

$$\varphi_t=\Delta\varphi-f(x,y),\quad \varphi=\varphi(x,y,t),\quad t>0. \tag{10}$$

For (10), we take just the same boundary conditions and add the initial data to them at $t=0$. We seek a solution to (9) as the limit of solutions to (10) as $t\to\infty$. Introducing a grid with respect to the time variable t with the step τ and the nodes $t_n=\tau n$, $n=1,2,\ldots$, denoting $\varphi^{[n]}=\varphi^{[n]}(x)=\varphi(x,y,t_n)$ and approximating the derivative φ_t by the difference quotient $(\varphi^{[n]}-\varphi^{[n-1]})/\tau$, one obtains the formula:

$$\varphi^{[n]}-\tau\Delta\varphi^{[n]}=\varphi^{[n-1]}-\tau f(x,y,\varphi^{[n-1]})=\tilde{f}(x,y,\varphi^{[n-1]}). \tag{11}$$

In accordance with [4], if Neumann conditions are imposed at the boundary $x=0,\pi$, then the solution to (11) on the segment $x\in[0,\pi]$ is approximated by the following interpolation polynomial obtained from the polynomial with Chebyshev nodes by a simple transformation of a variable:

$$\mathcal{P}_1(x,\varphi^{[n]})=\frac{1}{N}\sum_{j=1}^{N}(-1)^{j-1}\frac{\sin x_j\cos(Nx)}{\cos x-\cos x_j}\varphi_j^{[n]}(y)+f_1(x,y),\quad x_j=\frac{(2j-1)\pi}{2N},\quad j=1,\ldots,N. \tag{12}$$

If at $x=0,\pi$ Dirichlet conditions are imposed, then the following approximation of φ is applied:

$$\mathcal{P}_2(x,\varphi^{[n]})=\frac{1}{N+1}\sum_{j=1}^{N}(-1)^{j-1}\frac{\sin x_j\sin((N+1)x)}{\cos x-\cos x_j}\varphi_j^{[n]}(y)+f_2(x,y),\quad x_j=\frac{\pi j}{N+1},\quad j=1,\ldots,N, \tag{13}$$

where x_j are the interpolation nodes, $\varphi_j^{[n]}(y)=\varphi^{[n]}(x_j,y)$ and $f_{1,2}$ are the functions of a simple form satisfying the given boundary conditions. Denoting by μ one of the derivatives $\mu\in\{x,xx\}$ and setting $(\varphi_j^{[n]}(y))_\mu=\varphi_\mu^{[n]}(x_j,y)$, one forms the vectors: $\vec{\Phi}(y)=(\varphi_j^{[n]}(y))_N$, $\vec{\Phi}_\mu(y)=((\varphi_j^{[n]}(y))_\mu)_N$, $\vec{\mathcal{F}}(y)=\tilde{f}(x_j,y,\varphi_j^{[n-1]})$. Using l'Hospital rule for each of (12), (13), one can find the limit expressions of its first order and second order derivatives as $x\to x_i$, which are linear combinations of $\varphi_j^{[n]}(y)$ for each i. Collecting the coefficients of these combinations into matrices A and B, one can finally obtain the pseudospectral matrix approximations of the derivatives:

$$\vec{\Phi}_x(y)\approx A\vec{\Phi}(y),\quad \vec{\Phi}_{xx}(y)\approx B\vec{\Phi}(y). \tag{14}$$

Substituting (12) or (13) into (11) and using (14), one obtains the system of second-order equations

$$\vec{\Phi}''(y)=\mathcal{B}\vec{\Phi}(y)+\vec{\mathcal{F}}(y)/\tau, \tag{15}$$

where a dash means differentiation with respect to y, $\mathcal{B}=((I_N-B)/\tau)$ with I_N being the identity matrix of size N. It is worth noting that the matrices B are symmetric in both cases of Neumann and Dirichlet conditions (see [4]), which allows one to use stable and fast methods for inverting $\mathcal{B}$. Approximations (12), (13) provide high accuracy only if φ has a high order of smoothness, therefore (3) means that the formulas similar to (12), (13) cannot be applied along the y axis. Thus, for approximating $\vec{\Phi}(y)$ an interpolation cubic C^2-spline is used (see [3,4]). Using the continuity of spline derivatives, one obtains the following difference schemes in each Ω_q, $q=1,\ldots,Q$ from (15):

$$\left\{I-\frac{h^2}{6}\mathcal{B}\right\}\vec{\mathcal{R}}_{m-1}^q-2\left\{I+\frac{h^2}{3}\mathcal{B}\right\}\vec{\mathcal{R}}_m^q+\left\{I-\frac{h^2}{6}\mathcal{B}\right\}\vec{\mathcal{R}}_{m+1}^q=\frac{h^2}{6}\{\vec{\mathcal{F}}_{m-1}+4\vec{\mathcal{F}}_m+\vec{\mathcal{F}}_{m+1}\}, \tag{16}$$

where $\vec{\mathcal{R}}_m^q=\vec{\Phi}(y_m)$ in Ω_q and $\vec{\mathcal{F}}_m=\vec{\Phi}(y_m)$ $y_m=mh_q$, $m=0,\ldots,M_q$, $M_qh_q=\pi$. The schemes (16) should be supplemented with the conditions at the boundary of Λ described in the first section and with matching conditions (3). To solve (16), we propose to use a matrix Thomas algorithm, which requires $O(N^3)\sum_{q=1}^{Q}8M_q$. It is worth noting that a similar algorithm was used in [5] for simulation of charge transport in a single domain of MOSFET. Here, it is significantly modified in order to perform the computations for an arbitrary number of subdomains Ω_q with different conditions (*voltage or free-slip*) at their boundaries. To this end, the following new methods were developed.

Firstly, the method of coordination of interpolation nodes x_j of (12), (13) was designed for the case when subdomains have different types of conditions at the boundaries $x=0,\pi$. The idea is to approximate the values of $\varphi_j^{[n]}(y)$ in the nodes x_j of (12) using the linear combination of its values in four nearby nodes of (13). Thus, the space grid of (13) becomes common for all subdomains and (14) remains valid, but the expressions for elements of A and B become much more cumbersome.

Secondly, the Thomas algorithm has been significantly changed in order to correctly specify its sweep coefficients on the borders S_q. The forward and backward sweeps of our algorithm have Q stages (each stage corresponds to a subdomain). On the current stage of forward sweep, the coefficients are resolved using the coefficients from the previous stage and the *matching conditions*. Similarly, the solution on the current stage of backward sweep is obtained using the solution from the previous one.

Finally, all these methods have been implemented using the Object Pascal environment in a programm package for solving the problems with an arbitrary number of subdomains Q, arbitrary geometrical and physical parameters, and boundary conditions.

4. Application of the proposed method to simulation of charge transport in DG-MOSFET

A detailed description of DG-MOSFET is given in [6]. A special feature of this silicon transistor is the presence of two subdomains fabricated of silicon oxide. The dielectric properties of silicon and silicon oxide differ by three times (see table 1). The DG-MOSFET in dimensionless variables x and y is sketched in figure 1. Here $b = 10\,\text{nm}$ (0.35) is the length of gates, $l_y^-, l_y^+ = 1-2\,\text{nm}$ $(1/30 - 1/15)$ are thicknesses of silicon oxide layers, $l_x = 1.5 - 4.5\,\text{nm}$ (0.05–0.15) characterizes the distribution of ρ in the domain Ω_2, and $a = 4-8\,\text{nm}$ $(2/15 - 4/15)$ is the thickness of the domain Ω_2. Here the dimensionless values are given in brackets. It should be noted that there is no charge transport in silicon oxide layers $\Omega_{1,3}$ and, therefore, the electric potentials in them satisfy the Laplace equation $\Delta\varphi = 0$. Table 1 contains all physical parameters necessary to perform the simulation.

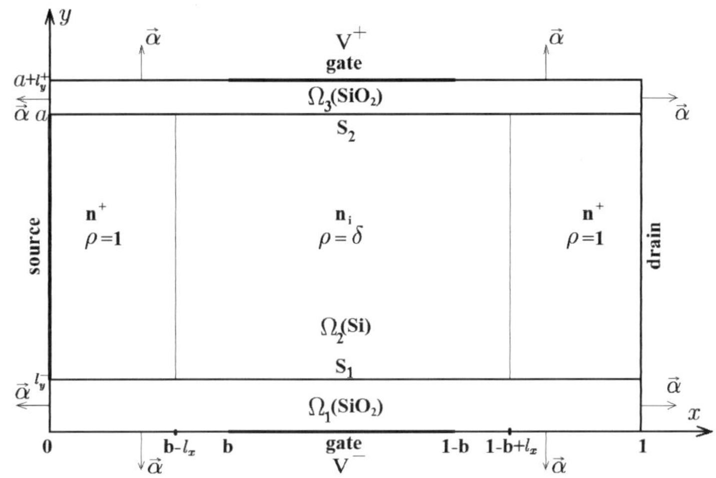

Figure 1. Schematic sketch of a 2D silicon DG-MOSFET.

Table 1. Values of physical parameters.

Variable	Description	Value
e	Electron charge	1.6×10^{-19} C
L	Characteristic linear size	30 nm (3×10^{-8} m)
N^+	Doping density in the domains n^+	2×10^{26} m^{-3}
T_0	Temperature of lattice	300° K
K_B	Boltzmann constant	1.38×10^{-23} J / K
κ (Si)	Dielectric constant of silicon	1.03545×10^{-10} C / Volt m
$\kappa(SiO_2)$	Dielectric constant of SiO_2	$1/3\,\kappa$(Si)
N_i	Internal electron concentration in the domain n_i	10^{16} m^{-3}
$V^\pm$, V_D	Bias voltages	0 –2 Volt

Using the proposed algorithm, we got the distribution of (a) the electric potential (in Volts), (b) the dimensionless electron energy (see figure 2) and the current-voltage curves (see figure 3).

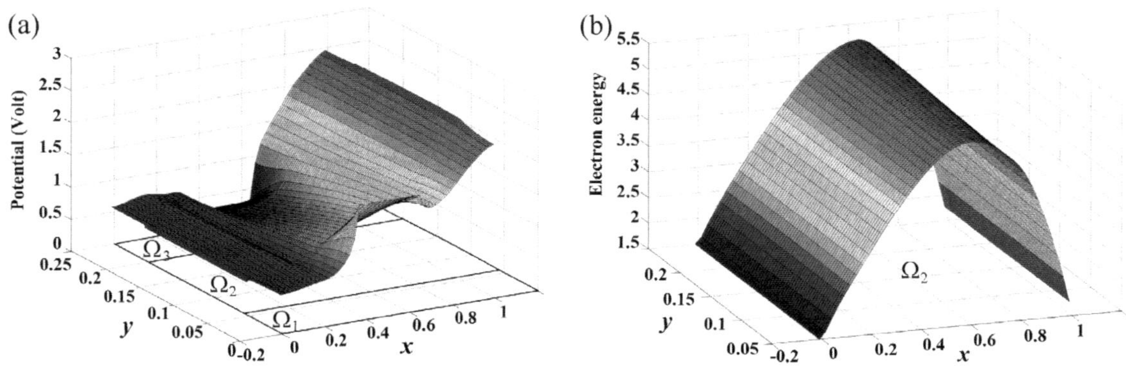

Figure 2. Solutions obtained for $V_D = 2$ Volt, $V^+ = 0.01$ Volt, $V^- = 1.5$ Volt $\delta = 5\times10^{-10}$, $l_y^+ = 0.03334$, $l_y^- = 0.06$, $l_x = 0.15$, $a = 0.2$.

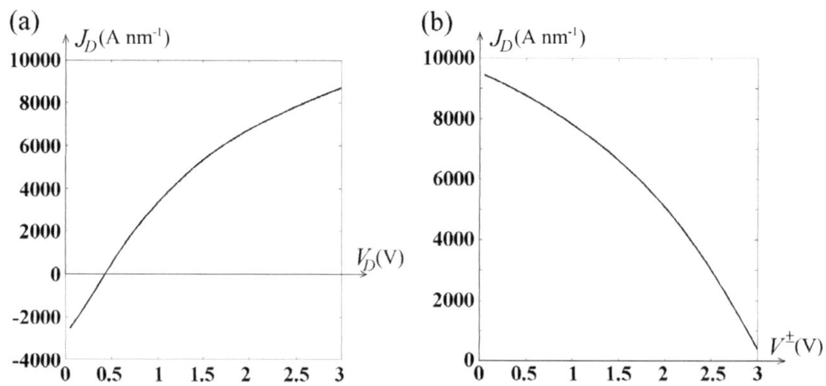

Figure 3. Current-voltage curves for DG-MOSFET obtained for $l_y^+ = l_y^- = 1$ nm, $a = 4$ nm, $l_x = 3$ nm: (a) the dependence of the electrical current J_D flowing through the drain on the drain voltage V_D for $V^{\pm} = 0.538$ Volt; (b) the dependence of J_D on the gate voltage $V^{\pm}$ for $V_D = 3$ Volt.

5. Conclusions

Finally, we pay attention to an interesting fact that was discovered while drawing the current-voltage curves. Analyzing the behavior of the curve in figure 3(a), we can see that for DG-MOSFET the regimes with "blockage" are possible, i.e., when the electric current begins to flow from the drain into the device. There are other important effects that have been discovered by analyzing the solutions.

The first is a small potential well along the line $x = b - l_x$. It is caused by an extreme doping density difference on this line ($\delta = N_i / N^+ \sim 10^{-10}$; see figure 1 and table 1). This effect has been found due to the high stability of the interpolations (12), (13) and a special smoothing of doping density.

The second effect is a break of the electric potential on the lines $S_{1,2}$ caused by an essential difference of dielectric properties of materials. It significantly affects all the physical processes, including the current.

The third effect is that by specifying different bias voltages on two gates, one can manipulate all the parameters of charge transport in very wide ranges, making their profiles more or less symmetric. The quantitative data for these manipulations can be obtained by the designed algorithm.

Acknowledgments
This research was performed with the financial support of the Russian Science Foundation (grant agreement № 17-71-10135), except the derivation of equations (1), (2), (4) – (6), which was made by A.M. Blokhin.

References
[1] Anile A M and Romano V 1999 *Cont. Mech. Thermodyn.* **11** 307
[2] Romano V 2000 *Cont. Mech. Thermodyn.* **12** 31
[3] Blokhin A M, Bushmanov R S, Rudometova A S and Romano V 2006 *Nonlinear Analysis* **65** 1018
[4] Blokhin A M and Ibragimova A S 2009 *SIAM J. Sci. Comp.* **31(3)** 2015
[5] Blokhin A M and Semisalov B V 2011 *J. Phys.: Conf. Ser.* **291** 012016
[6] Jiang H and Cai W 2010 *J. Comp. Phys.* **229** 4461

IOP Conf. Series: Journal of Physics: Conf. Series **993** (2018) 012036 doi:10.1088/1742-6596/993/1/012036

New connecting elements for cascade photoelectric converters based on InP

A E Marichev[1, 2], B V Pushnyi[1,2], R V Levin[1,2], N M Lebedeva[1,2], N D Prasolov[1] and E V Kontrosh[1]

[1] Ioffe Institute, Polytechnicheskaya 26, St. Petersburg 194021, Russia
[2] Submicron Heterostructures for Microelectronics Research and Engineering Center of the RAS, Polytechnicheskaya 26, St. Petersburg 194021, Russia

E-mail: segregate1@yandex.ru

Abstract. In this paper, we report on the initial studies of connecting elements for cascade photodetectors. The heterostructures used in this work are based on InP. As a connecting element, it is proposed to use nanocrystalline inclusions instead of the tunnel junction. GaP nanocrystals are most suitable for this purpose because this material does not cause absorption of the incident radiation.

1. Introduction

Single-stage *p–n* or *p–i–n* structures can be used for conversion of low-power radiation. To increase the converted power of photovoltaic converters (PhVC) (concentrator solar cells), it is preferable to use cascade photodetectors. A *p++–n++* tunnel junction is traditionally used as a connecting element of photoelectric converter cascades. However, with a significant increase in the incident power, the generated photocurrent may exceed the peak current of the tunnel junctions, which leads to an increase in the resistance of the structure. As a result, the efficiency of radiation conversion decreases [1].

It was proposed to employ a connecting element by introducing an aggregation of crystalline inclusions into the bulk charge area (BCA) of the *p+–n+* connecting junctions between the cascades. In this case, tunnel junctions can be completely excluded and the mechanism of ohmic current flow over conductivity channels in the crystalline inclusion layer in a *p–n* junction space charge region SCR can be facilitated.

In [3], the properties of tunnel and photovoltaic GaSb *p–n* junctions with crystalline Si inclusions introduced into their BCA were studied for the first time. An analysis of the dark current–voltage characteristics showed a significant decrease in ohmic losses in such structures.

In this study, a solid solution based on InP was used, as it is suitable for converting the solar spectrum in the range of 0.95 to 1.2 μm. GaP was used for aggregation of nanocrystalline inclusions, because this material does not cause absorption of incident radiation, and because the large difference between the InP and GaP lattice constants (~ 7%) means that an aggregation of nanocrystalline inclusions forms, rather than a continuous layer (Stranski–Krastanov growth).

2. Experimental details

The material for developing conductivity channels in the BCA should satisfy the following conditions: (1) low absorption of optical radiation converted by underlying cascades of a narrower gap; and (2) the crystal material should form only isolated crystals larger than the BCA thickness at the *p–n* junction

IOP Conf. Series: Journal of Physics: Conf. Series **993** (2018) 012036 doi:10.1088/1742-6596/993/1/012036

boundary, rather than a continuous layer. For photoactive InP *p–n* junctions, GaP satisfies these conditions. It has a wider band gap than indium phosphide, hence, low light absorption in the InP photosensitivity area.

All the InP-based structures were grown by metalorganic chemical vapor deposition (MOCVD) using an AIX-200 setup operating at a reduced pressure. Trimethylindium (TMIn), Triethylgallium (TEGa) and phosphine (PH3) were used as In, Ga and P sources. The growth temperature was 600°C, while the reactor pressure was 100 mbar. *n*-InP(Te) (100) wafers misoriented by 4° towards (111) with an electron concentration of (1–5) $\times 10^{18}$ cm^{-3} were used as substrates.

Technological experiments on GaP crystal growth on an InP substrate surface were performed in order to analyze the size and shape of GaP inclusions. It became possible to vary the size of GaP inclusions by varying the triethylgallium flow, providing low ohmic losses in the connecting *p–n* junctions. The following conditions of crystal growth were chosen as optimal for the study: the temperature T_g=600°C and time t_g= 1 min.

Figures 1–3 show micrographs for the GaP crystals in the BCA, obtained using an atomic force microscope. The data presented show that GaP crystals 60–95 nm high and 78–92 nm wide are formed under the chosen growth conditions. Table 1 presents the results of research on GaP nanocrystals shown in figures 1–3.

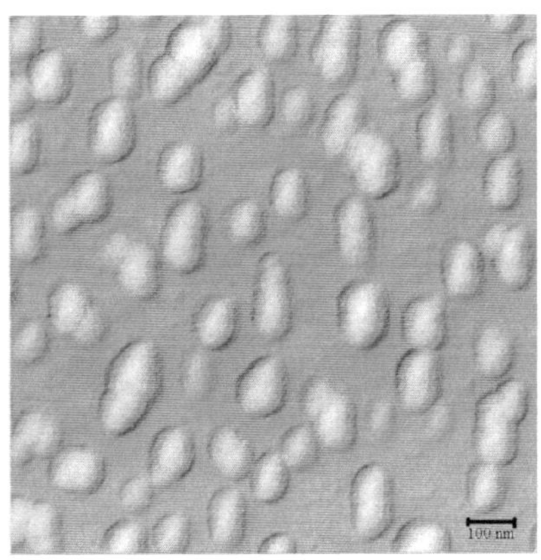

Figure 1. Photo of GaP nanocrystals grown on the surface of an n-InP substrate. Growth parameters: t_g = 30 seconds, V/III = 300.

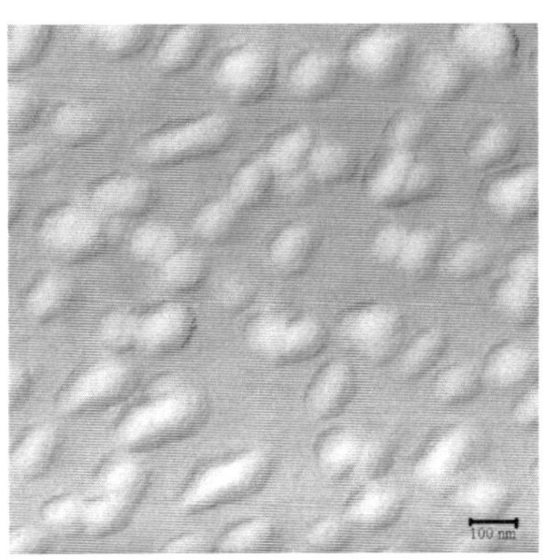

Figure 2. Photo of GaP nanocrystals grown on the surface of an *n*-InP substrate. Growth parameters: t_g = 30 seconds, V/III = 590.

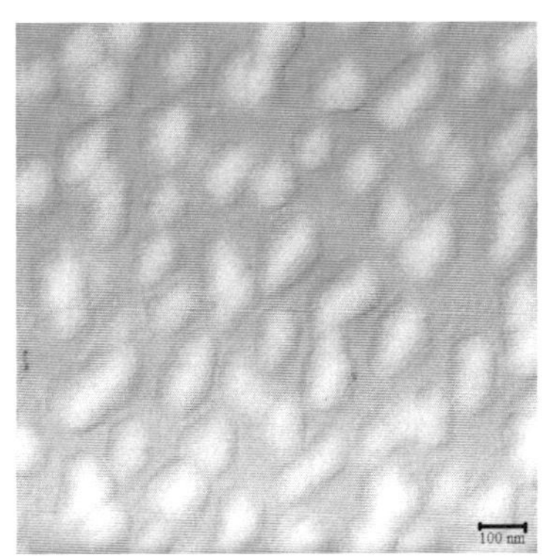

Figure 3. Photo of GaP nanocrystals grown on the surface of an *n*-InP substrate. Growth parameters: t_g = 45 seconds, V/III = 590.

Table 1. The results of research on GaP nanocrystals

Sample No	Number	Average height (nm)	Average width (nm)
1	72	60	86
2	62	95	78
3	69	80	92

When growing the multi-junction cells, it is important to retain the qualitative characteristics of the photoactive junctions grown over the layers containing crystalline GaP inclusions. For this study, the structure presented in figure 4 was grown.

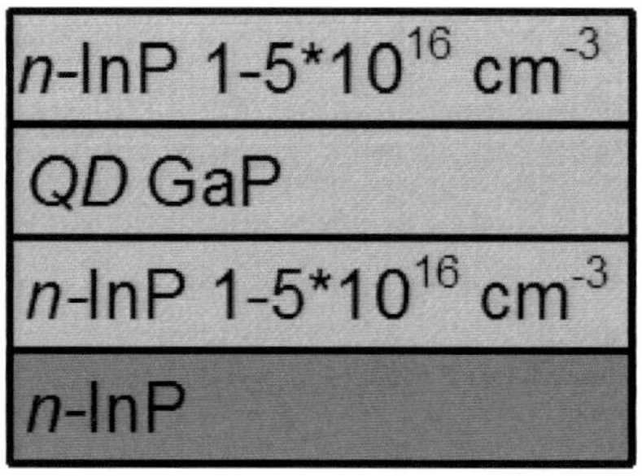

Figure 4. A sample of an *n-n* transition with GaP crystallites.

3. Results and discussion

The study was carried out by sample photoluminescence at 77 K in the wavelength range of 600-1500 nm.

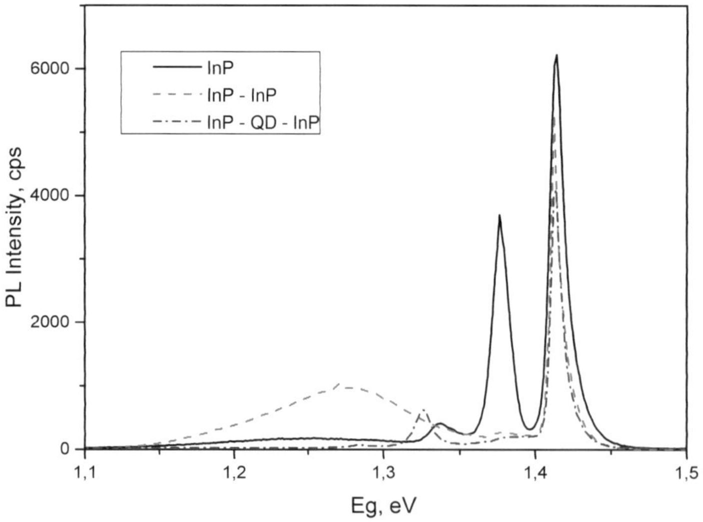

Figure 5. The photoluminescence spectrum ($T = 77$ K) of the samples under study.

Figure 5 shows the photoluminescence spectra of the samples under study. Thus, for InP layers with and without QDs, the intensity of peaks with an energy of 1.41 eV (interband transitions in InP) differs by less than 20%, which may indicate that there are practically no defects arising from growth of epitaxial layers on a layer of GaP nanocrystals.

When a layer of GaP nanocrystals is introduced into InP layers, which is, in fact, equivalent to introduction of defects, the quality of the obtained layers does not deteriorate, as evidenced by the change in the photoluminescence half-width of 1 meV lying within the measurement error.

4. Conclusions

In this paper, we report on the initial stages of research on new connecting elements based on nanocrystalline inclusions for use in monolithic multi-transient solar cells and photoconverters of powerful optical radiation based on InP. GaP nanocrystals were grown as conductivity channels in the IPF, with dimensions of 60–80 nm in height and 86–92 nm in width. At the same time, it is shown that the quality of epitaxial, intentionally non-doped InP layers grown on GaP nanocrystals does not deteriorate.

Acknowledgements

The authors are grateful to A.S. Vlasov for assistance in research and discussion of the results. This work was supported by the Russian Science Foundation (Grant Agreement №17-79-30035).

References

[1] Alferov Zh I, Andreev V M and Rumyantsev V D 2008 *Semiconductors* **8** 937
[2] Kalinovsky V S, Levin R V, Pushniy B V, Mizerov M N, Rumyantsev V D and Andreev V M 2013 *Semiconductors* **12** 1677

IOP Conf. Series: Journal of Physics: Conf. Series **993** (2018) 012037 doi:10.1088/1742-6596/993/1/012037

Ultraviolet light-emitting diodes and photodiodes grown by plasma-assisted molecular beam epitaxy

D V Nechaev[1], E A Evropeytsev[1], A N Semenov[1], S I Troshkov[1], V I Egorkin[2], V E Zemlyakov[2], M V Rzheutski[3], E V Lutsenko[3], A A Toropov[1], S V Ivanov[1] and V N Jmerik[1]

[1] Ioffe Institute, St. Petersburg, 194021 Russia
[2] National Research University of Electronic Technology - MIET, Zelenograd, Moscow, 124498 Russia
[3] Stepanov Institute of General Physics, Minsk, 220072 Belarus

E-mail: nechayev@mail.ioffe.ru

Abstract. We demonstrate a Schottky ultraviolet photodiode (UV-PD) and a UV light-emitting diode (UV-LED) based on AlGaN heterostructures grown by plasma-assisted molecular beam epitaxy. The spectral responsivity of the Schottky UV-PD is 3 mA/W at 271 nm at zero bias and decreases by two orders of magnitude for the spectral range of longer wavelengths (> 300 nm). The sub-monolayer digital alloying technique was used for growing the AlGaN quantum wells in UV-LEDs emitting a single electroluminescence peak at 273 nm.

1. Introduction

Ultraviolet photodiodes and light-emitting diodes (UV-PD and LEDs, respectively) are used in many civilian and military areas, the former including water and air purification, medicine, curing and printing, etc. For the latter, the development of advanced non-line-of-sight covert UV-communications, solar-blind detectors of UV-radiation, and sensors for chemical and bioagents are of great importance.

At present, most AlGaN UV-PD and UV-LED heterostructures with an operating wavelength $\lambda < 300$ nm are grown by using both metal-organic chemical vapor deposition (MOCVD) and plasma-assisted molecular-beam epitaxy (PA MBE) [1-7]. The maximum values of the output power and external quantum efficiency (EQE) as large as 150 mW and ~5%, respectively, have been achieved in the UV-LEDs grown on bulk AlN substrates [8]. Unfortunately, UV-LEDs grown on commercially available *c*-sapphire substrates possess an output power of several mW and the EQE of a few percent, which are related to a high density of non-radiative recombination centers in the active area. Previously, only threading dislocations (TDs) were supposed to be the origin of the non-radiative centers. However, a rather modest improvement of the EQE for the devices grown on TD-free AlN substrates led to a conclusion about the influence of other non-radiative centers [9]. Most probably, different point defects in AlGaN heterostuctures can also restrict the EQE of UV-LEDs, and their influence is actively studied now [10]. Therefore, PA MBE operating in the ultra-high vacuum growth environment could be considered as a very attractive growth technique for reducing point defects. Moreover, this technology allows one to grow the AlGaN quantum wells (QWs) in the active area with a sub-monolayer control of the thickness, sharp interfaces, and various composition profiles for the improvement of carrier localization in QWs.

In addition, fabrication of UV-LEDs with a high Al content is hindered by the low doping efficiency of *p-* type layers, high ohmic contact resistance and electron leakage through an electron blocking layer (EBL). The latter leads to a decrease in the EQE and the appearance of a second long-wavelength emission peak originating from a *p-* (Al,Ga)N cladding layer [1].

In our previous works, UV-LEDs emitting at $\lambda = 320$ nm have been grown by PA MBE on standard *c*-Al_2O_3 substrates by using a new sub-monolayer digital alloying (SDA) technique to form $Al_xGa_{1-x}N/Al_yGa_{1-y}N$ QWs [4]. Recently, we have also demonstrated the ability of PA MBE to fabricate solar-blind p-i-n photodiodes with polarization-doped *p-* emitters, which exhibit a maximum photoresponse of 35 mA/W at $\lambda = 271$ nm [11].

In this paper, we demonstrate the capabilities of PA MBE in fabrication of sub-300-nm UV-LEDs with SDA AlGaN QWs, as well as initial results on sub-300-nm Schottky PDs.

2. Experiment

The (Al,Ga)N-based heterostructures were grown on *c*-Al_2O_3 substrates by a PA MBE setup Riber21T equipped with a plasma source offering a flux of activated nitrogen of 0.6 ML/s at 220 W. Firstly, 65-nm-thick AlN nucleation layers were grown by migration enhanced epitaxy on annealed and nitridated substrates at a substrate temperature $T_S = 780°C$, as described earlier [12]. Sketches of the UV-PD and UV-LED structures are shown in figure 1. Both devices have the same 2 -µm-thick AlN buffer layer grown at slightly Al-rich growth conditions ($F_{Al} / F_{N*} = 1.05$ at $T_S = 780°C$) by the metal-modulation epitaxy technique [13]. During the growth of the $Al_{0.75}Ga_{0.25}N$:Si (UV-LED) and $Al_{0.55}Ga_{0.45}N$:Si (UV-PD) layers with an electron concentration of 10^{18} cm^{-3}, a periodic change (with a period of about 10 min) of stoichiometric growth conditions from $F_{III} / F_{N*} = 2$ to 1.6 was used, which led to a corresponding variation of the surface morphology from 2D to 3D. This was done to prevent the Ga-droplets accumulation on the surface and to achieve a relatively smooth and continuous surface morphology [14]. As a result, we could avoid the formation of short-circuiting due to the excess metal accumulated along the extended defects. The results of our previous calculations of the optimal design for AlGaN-based QWs for UV-LEDs were used to determine the nominal parameters of the QWs with an $Al_{0.7}Ga_{0.3}N$ barrier layer and three QWs having a thickness of 1.5 nm [15]. The QWs were fabricated by the SDA technique using ten 0.25-ML-thick GaN insertions spaced by 0.32-ML-thick AlGaN barriers (figure 1(b)). An electron blocking $Al_xGa_{1-x}N$ and *p-* doped AlGaN:Mg ([Mg]~10^{18} cm^{-3}) layers were grown with a gradient $x = 0.90$-0.85 and $x = 0.85$-0.30, respectively, to obtain polarization-enhanced *p-* doping therein [11]. The UV-LED structure was capped by a 20-nm-thick GaN:Mg layer. Figure 2 shows a cross-section SEM image of the UV-LED heterostructure.

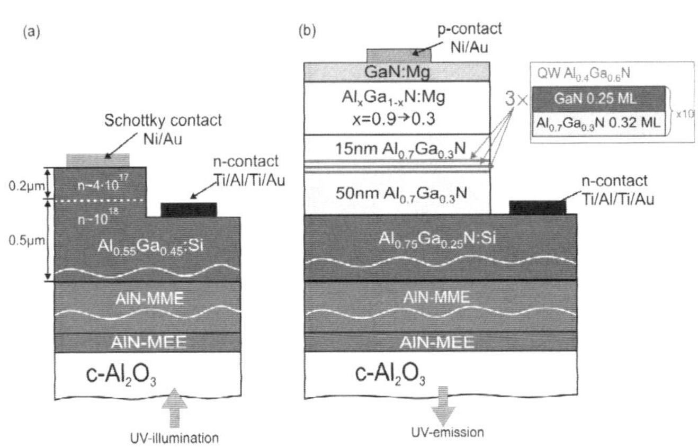

Figure 1. Schematic design of (a) UV-PD and (b) UV-LED structures

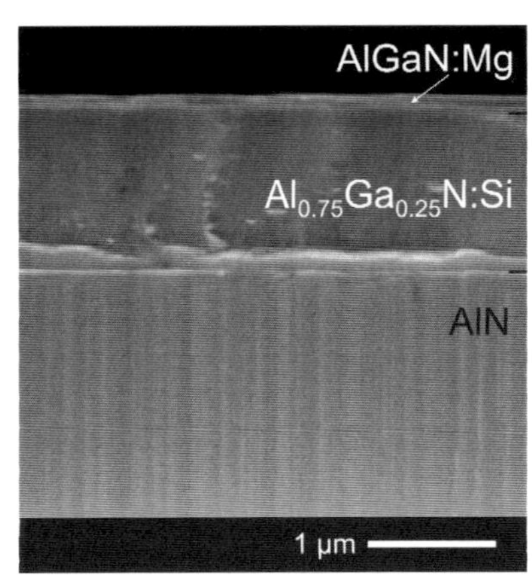

Figure 2. Cross-section SEM image of the UV-LED heterostructure

The 100 × 100 μm^2 mesa type devices were fabricated by accessing the bottom *n*- AlGaN layers using CCl_4 plasma reactive ion etching. The Ti/Al/Ti/Au and Ni/Au metallizations annealed at 850°C and 450°C were used for *n*- and *p*- ohmic contacts for UV-LEDs, respectively. The former was also used for an *n*- type ohmic contact in UV-PDs, whereas the latter provided a Schottky contact. UV-PD possesses single Schottky and ohmic *n*- contact pads with areas of 100 × 100 μm^2 and 200 × 200 μm^2, respectively. Each UV-LED had a single 200 × 200 μm^2 *n*- type contact and a varied number of *p*- contact pads (1 × 1, 2 × 2, or 3 × 3) which were merged.

Scanning electron (SEM) and optical (OM) microscopies were used for post growth control of UV-LED and UV-PD devices. Current-voltage (I-V) characteristics were measured at room temperature by point contacts by using a Keysight B1505A setup. Electroluminescence (EL) spectra of the UV-LEDs were measured from the substrate side at a cw drive current up to 60 mA. The spectral photoresponse of UV-PDs was measured for the back-illuminated geometry.

3. Results

All devices grown on 2-μm-thick AlN/*c*-Al_2O_3 buffer layers and ~1-μm-thick AlGaN:Si layers demonstrate the relatively smooth and crack-free surface morphology. Nevertheless, on the ring area located near the edge of a 2-inch-diameter wafer, rare metallic microdroplets with a density of 10^4 - 10^5 cm^{-2} are formed due to a lower substrate temperature there as compared to the wafer center. The precise control of excessive metal across the whole wafer surface grown at varied stoichiometric conditions should be developed to provide droplet-free growth.

Figure 3 shows I-V and spectral responsivity characteristics of the Schottky UV-PD. The reverse breakdown voltage of the diode is of 18 V, and the dark current is as low as 0.4 μA at -5 V. The non-exponential reverse I-V curve is apparently related to leakage currents exceeding the thermionic emission limited by the Schottky barrier. The typical value of the ideality factor for the Schottky PD is around 2. The maximum spectral responsivity of 3 mA/W at U_{bias} = 0 V was measured at a wavelength of 271 nm, as shown in figure 3(b). This relatively low value (by one order of magnitude lower than that demonstrated by us for a *p-i-n* UV-PD [11]) was presumably caused by the back-illuminated scheme of the measurements and insufficient transparency of the thick $Al_{0.55}Ga_{0.45}N$ buffer layer with the same Al-content as in the active area of the device under the Schottky contact. To overcome this issue, we plan to develop the devices with either a semi-transparent Schottky contact or the increased Al-content in the buffer layers.

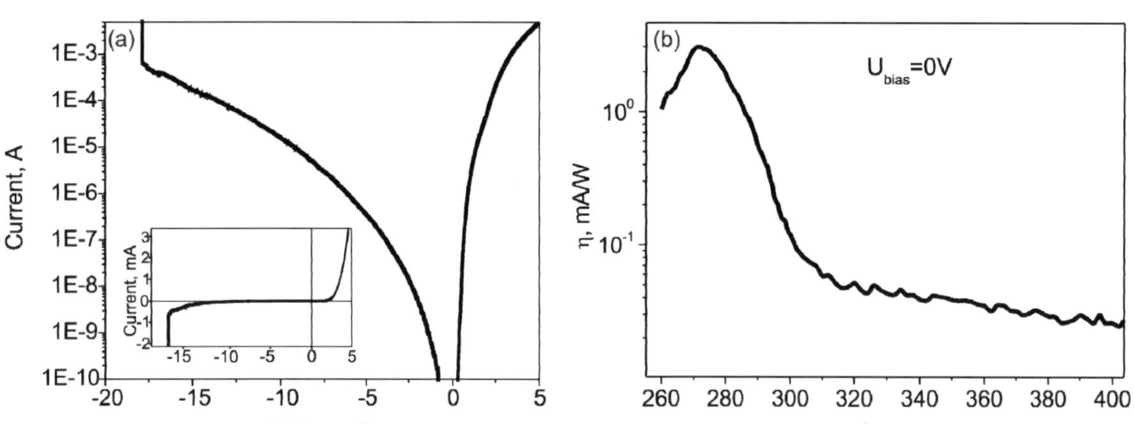

Figure 3. I-V characteristic (a) and spectral responsivity at U_{bias} = 0 V (b) of the Schottky UV-PD.

Figure 4(a) exhibits I-V characteristics of the UV-LEDs with different areas of a *p*- contact. The devices turn on at ~3 V with a series resistance reduction from 1000 to 130 Ohm with an increase in the number of p- contact pads from 1× to 9× (100×100 μm^2), respectively. It is worth noting that the analogous first UV-LEDs grown using the MOCVD technology demonstrated a smaller series resistance of 20-60 Ohm for devices with a smaller contact area [1,2]. The dependence of the series resistance of UV-LEDs on the p-contact area allows us to conclude that its relatively high value is more likely related to poor conductivity of the polarization-doped p- layer than to a high n- contact resistance. Thus, the method of polarization-doping of AlGaN:Mg layers should be optimized for use in UV-LEDs, or another approach to p- type doping should be applied in further experiments. One should also stress that UV-LEDs demonstrate a relatively low turn-on voltage of < 4 V that indicates sufficient leakage currents. The ideality factor of the UV-LEDs is as high as 6.

Despite the moderate I-V characteristics, the UV-LEDs demonstrate a distinct EL even for the simplest measurement scheme through the substrate backside and with a top GaN:Mg layer and a Ni/Au p- contact adsorbing UV-radiation. Figure 4(b) shows the EL spectra measured at different drive currents for the UV-LED with the largest p- contact area. These spectra reveal dominant peaks at a wavelength of 270-275 nm with a FWHM value of around 30 nm. In addition, a second emission peak of much lower intensity can be recognized in the spectra at 305 nm. This is apparently related to a leakage of electrons through the EBL, followed by their recombination in p- AlGaN:Mg cladding layers with the Al-content lowering towards the top contact. It should be noted that our first UV-LEDs with a non-optimized design of the top layers demonstrated the EL spectra in which these long wavelength peaks prevailed (not shown), while the QW-related emission was negligible.

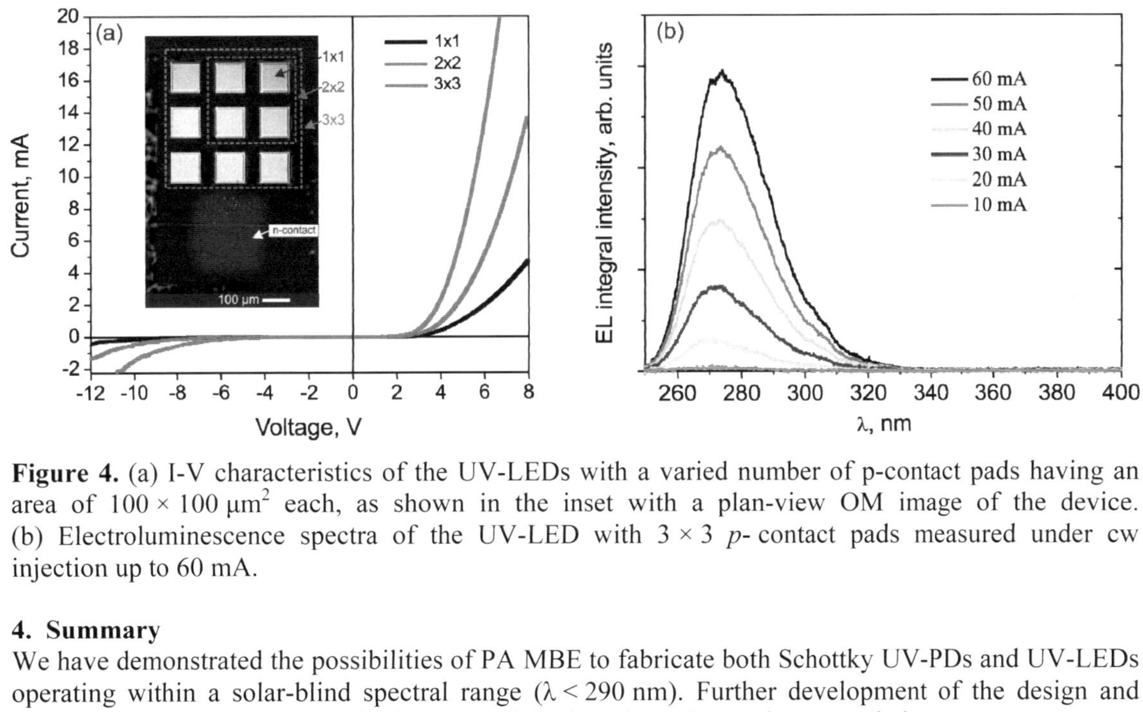

Figure 4. (a) I-V characteristics of the UV-LEDs with a varied number of p-contact pads having an area of 100×100 μm^2 each, as shown in the inset with a plan-view OM image of the device. (b) Electroluminescence spectra of the UV-LED with 3×3 p- contact pads measured under cw injection up to 60 mA.

4. Summary

We have demonstrated the possibilities of PA MBE to fabricate both Schottky UV-PDs and UV-LEDs operating within a solar-blind spectral range ($\lambda < 290$ nm). Further development of the design and technologies is necessary for both types of devices in order to improve their output parameters: namely, to increase the photoresponsivity of the Schottky UV-PD above the achieved value of 3 mA/W and to achieve a higher output optical power for UV-LEDs.

Acknowledgments
This work was supported by the Russian Foundation for Basic Research (#15-02-05206) and the Belarusian Republican Foundation for Fundamental Research (F15SO-052). D.V.N. acknowledges the support of the Grant of the President of the Russian Federation. E.V.L. acknowledges the support of the Grant of the President of the Republic of Belarus.

IOP Conf. Series: Journal of Physics: Conf. Series **993** (2018) 012037 doi:10.1088/1742-6596/993/1/012037

References

[1] Adivarahan V *et al* 2002 *Appl. Phys. Lett.* **81** 3666

[2] Yasan A, McClintock R, Mayes K, Darvish S R, Kung P and Razeghi M 2002 *Appl. Phys. Lett.* **81** 801

[3] Sampath A V, Garret G A, Collins C J, Sarney, Readiger E D, Newman P G, Shen H and Wraback M 2006 *J. Electron. Materials* **35** 641

[4] Jmerik V N *et al* 2008 *Semiconductors* **42** 1420

[5] Liao Y, Thomidis C, Kao C-K and Moustakas T D 2011 *Appl. Phys. Lett.* **98** 081110

[6] Kim B-K, Kim J-K, Park S-J, Lee H-B, Lee J-H, Rue G-H, Lee M-B, Lee Y-H, Lee J-H and Hahm S-H 2003 *Phys. Stat. Sol. C* **0** 2309

[7] Jianga H and Egawa T 2007 *Appl. Phys. Lett.* **90** 121121

[8] Inoue S I, Tamari N and Taniguchi M 2017 *Appl. Phys. Lett.* **110** 141106

[9] Kneissl M and Rass J 2016 *III-Nitride Ultraviolet Emitters: Technology and Applications* (Springer Series in Materials Science) vol 227 (New York: Springer)

[10] Reddy P *et al* 2016 *J. Appl. Phys.* **120** 185704

[11] Kuznetsova N V *et al* 2016 *Technical Physics Letters* **42** 635

[12] Nechaev D V, Aseev P A, Jmerik V N, Brunkov P N, Kuznetsova Ya V, Sitnikova A A, Ratnikov V V and Ivanov S V 2013 *J. Crys. Growth* **378** 319

[13] Jmerik V N, Mizerov A M, Nechaev D V, Aseev P A, Sitnikova A A, Troshkov S I, Kop'ev P S and Ivanov S V 2012 *J. Cryst. Growth* **354** 188

[14] Nechaev D V, Brunkov P N, Troshkov S I, Jmerik V N and Ivanov S V 2015 *J. Crys. Growth* **425** 9

[15] Toropov A A, Shevchenko E A, Shubina T V, Jmerik V N, Nechaev D V, Evropeitsev E A, Kaibyshev V K, Pozina G, Rouvimov S and Ivanov S V 2016 *J. Electron. Materials* **46** 3888

IOP Conf. Series: Journal of Physics: Conf. Series **993** (2018) 012038 doi:10.1088/1742-6596/993/1/012038

ECV profiling of GaAs and GaN HEMT heterostructures

G Yakovlev and V Zubkov
Department of micro- and nanoelectronics, St. Petersburg State Electrotechnical University "LETI", 5 Prof. Popov str., St. Petersburg 197376, Russia

E-mail: gy@etu.ru

Abstract. AlGaAs/InGaAs/GaAs and AlGaN/GaN HEMT heterostructures were investigated by means of electrochemical capacitance-voltage technique. A set of test structures were fabricated using various doping techniques: standard doping, δ-doping GaAs pHEMT and non-doping GaN HEMT. The concentration profiles of free charge carriers across the samples were experimentally obtained. The QW filling was analyzed and compared for different mechanisms of emitter doping and 2DEG origins.

1. Introduction

HEMT transistors are the fastest active elements of microelectronics and nanoelectronics. The new record results associated with the improved technology and the use of new materials, such as GaN and InP, are published regularly [1]. The range of application areas for such devices is constantly expanding beyond broadband communication and special purpose devices [2]. For example, the 2DEG channel located near the surface was found to be very sensitive to chemical adsorption. This led to the development of various HEMT-based sensors with high selectivity and sensitivity [3]. In particular, as stated in [4], "Au gated AlGaN/GaN HEMTs functionalized in the gate region with label-free 3' thiol modified oligonucleotides serves as a binding layer to the AlGaN surface, which can detect the hybridization of matched target DNAs. X-ray photoelectron spectroscopy (XPS) shows immobilization of thiol-modified DNA that is covalently bonded with gold in the gated region. The drain source current shows a clear decrease of 115 μA, as this matched target DNA is introduced to the probe DNA on the surface, showing the promise of DNA sequence detection for biological sensing". A similar principle is used for detecting different proteins but in this case the current changes by 4 μA [4]. The use of Sc_2O_3 in the gate region for ungated HEMTs exhibits a linear change in current by value of 37μA/pH between pH 3–10 with a resolution of <0.1 pH. Such HEMTs-based pH sensors show stable operation over the entire pH range. The results indicate that HEMTs may be successfully used for monitoring the pH solution changes between 7 and 8, the range of high interest for human blood testing [3].

Semiconductor heterostructures of modern HEMT devices are a set of epitaxial layers of heterogeneously doped materials of various composition. The improvement of such devices with the purpose of increasing operating frequencies and currents steadily leads to high requirements for reproduction accuracy of the sequence, composition, and doping level of the layers. In addition, to satisfy certain device requirements, it is necessary to create sharp impurity profiles, including the use of δ-doping. For HEMTs, it is also extremely important to achieve a balance between the emitter impurity concentration, which determines the 2DEG concentration in the channel, and the undoped layer thickness, which determines the mobility of the charge carriers and device speed [5]. Otherwise, there is a probability of shunting the channel. All this requires careful monitoring of the parameters of heterostructures at various stages of the technological process. From this point of view, it is especially important to control the depth distribution of both the impurity and free charge carriers.

The electrochemical capacitance-voltage (ECV) profiling technique allows to obtain the concentration-depth distribution profiles of the impurity and free charge carriers. The ECV is included in SEMI standards and successfully used for the study of various semiconductor heterostructures. The measurement of homogeneously-doped semiconductor structures by the ECV technique is well known and represents a routine task. At the same time, ECV measurements of modern multilayer HEMT structures, including those with δ-layers, require careful selection of etching parameters and an operating measurement point [6, 7].

The main purpose of this work was to study the features of the spatial distribution of free charge carriers in GaAs and GaN HEMT heterostructures, depending on the type of doping and the 2DEG channel formation mechanism.

2. Samples and experimental technique

In this work, we investigated a set of three types of HEMT heterostructures grown by MBE (figure 1). Samples 1 and 2 are single-doped GaAs pHEMT heterostructures with emitters of different types: standard (bulk) and δ-one. The asymmetrical $Al_{0.22}Ga_{0.78}As/In_{0.22}Ga_{0.78}As/GaAs$ quantum well (QW) of these samples had a 14 nm width and was located at depths of 39 nm and 31.5 nm, respectively. In this case, sample 1 had a sufficiently thick emitter layer. Sample 3 had a standard layer configuration for GaN HEMT device structures. In such structures, a QW is formed due to the band offset at the AlGaN/GaN interface, while the 2DEG channel is located on the side of the GaN layer. The surface emitter layer of AlGaN had a 22 nm thickness. There were very thin (1–2 nm) GaN and AlN layers grown on both sides of the emitter layer for improving the structural properties and roughness. In particular, the presence of the AlN layer leads to sufficient improvement of the GaN/AlGaN interface quality and a noticeable increase in 2DEG mobility, respectively.

Sample №1	*Sample №2*	*Sample №3*
GaAs cap layer	GaAs cap layer	
$Al_{0.7}Ga_{0.3}As$ stop layer	$Al_{0.7}Ga_{0.3}As$ stop layer	GaN cap layer
GaAs buffer	GaAs buffer	$Al_{0.33}Ga_{0.67}N$ barrier
$Al_{0.22}Ga_{0.78}As$ emitter	δ-$Al_{0.22}Ga_{0.78}As$ emitter	AlN interlayer
$In_{0.22}Ga_{0.78}As$ QW	$In_{0.22}Ga_{0.78}As$ QW	GaN channel
GaAs buffer	GaAs buffer	GaN buffer
GaAs S/I substrate	GaAs S/I substrate	Al_2O_3 substrate

Figure 1. The sequence of layers in the investigated samples.

The measurements were performed at room temperature using an ECVPro profiler (Nanometrics). For GaAs samples, a 0.1M Tiron solution was chosen as the electrolyte [8], while for GaN we used 0.2M H_2SO_4 [9]. The area of the electrolytic rectifying contact was 0.1 cm^2. The etching current was maintained at a level of 0.5 mA/cm^2. During ECV profiling, the samples were etched gradually with a 1 nm step, while etching of GaN HEMT was performed in a pulsed mode [10]. The frequency of an ac test signal for arsenide and nitride structures was 300 Hz and 100 Hz, respectively. In some of the experiments, an Agilent E4980A LCR meter was used [11].

The structure of HEMT consists of many layers of different materials, electrochemical etching of which occurs at different rates. Therefore, in the ECV etching process, it is important to check the amount of etched material at different stages of the measurement. The etching depth and the surface quality were verified by atomic force microscopy (Solver NEXT AFM).

3. The experimental results and discussion

In this part of the paper, the results of the measured depth distributions of free charge carriers are presented.

3.1. GaAs pHEMT

Figure 2 shows the apparent depth distribution of free charge carriers in sample 1 with bulk QW doping, measured at several measurement points. Note that an apparent profile of free charge carriers is a result of CV derivation, in contrast to the real profile obtained under equilibrium conditions (without applied bias) [12]. As could be seen from the figure, the obtained ECV data show high reproducibility. There are two concentration peaks on the measured profile. The first peak located at a depth of 30 nm corresponds to the highly doped (emitter) region, while the second one at a depth of 44 nm originates from the QW. The concentration of free charge carriers in the emitter layer is $1.5 \cdot 10^{18}$ cm^{-3}, in the QW $2 \cdot 10^{18}$ cm^{-3}, which corresponds to a 2DEG concentration of $1.6 \cdot 10^{12}$ cm^{-2}. The results correlate well with Hall measurements, and the position of the peaks corresponds to the regions of charge carrier localization according to the structure specification. The observed slight mismatch (less than 5 nm) with respect to the position of the peak associated with the quantum well is explained by a QW skew [13]. It is important to note that the peak corresponding to the emitter layer in such structures could be registered only by the ECV technique. According to the literature, conventional CV gives only one peak (attributed to QW) in the concentration profile [14]. The presence of two peaks is explained by a smaller Schottky barrier in case of an electrolytic rectifying contact [15].

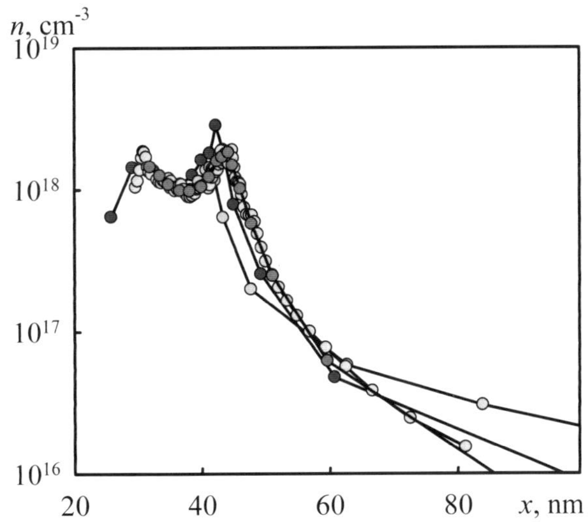

Figure 2. Apparent profiles of free charge carriers in sample 1 (bulk-doped GaAs pHEMT), measured at 4 different measurement points.

Figure 3 shows the depth distribution profile of free charge carriers in sample 2 with δ-doping, measured at several points. As above, there are two peaks in the concentration profile corresponding, in this case, to the δ-layer (29 nm) and the quantum well (37 nm). The peak concentration of free charge carriers in the δ-layer and the QW are $3 \cdot 10^{18}$ cm^{-3} and $2 \cdot 10^{18}$ cm^{-3} ($1.6 \cdot 10^{12}$ cm^{-2}), respectively.

Comparing the observed profiles of free charge carriers for samples 1 and 2, it could be seen that there is no obvious advantage in using a δ-doped emitter in terms of 2DEG concentration for the considered layer configuration and doping levels. Hall measurements of these samples also show the same concentration of 2DEG in a channel for both doping techniques. However, a smaller thickness of upper layers in the case of δ-layer structure design allows to obtain a greater value of the steep subthreshold slope characteristic, and therefore better controllability for the final device at the same values of operating current and power. This also helps to reduce the cutoff bias.

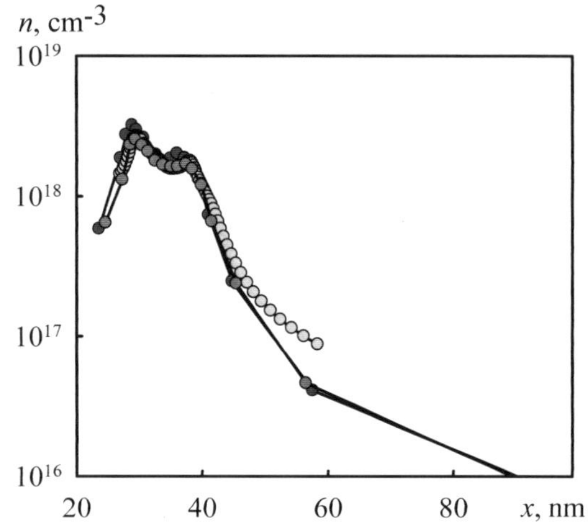

Figure 3. Apparent profiles of free charge carriers in sample 2 (δ-doped GaAs pHEMT), measured at 3 different measurement points.

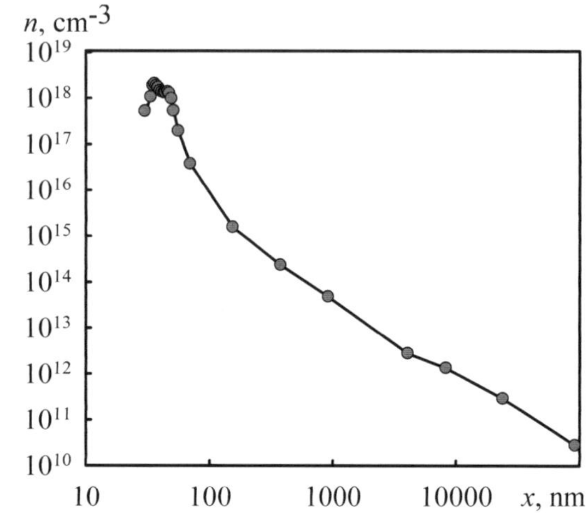

Figure 4. Apparent profile of free charge carriers in sample 2 (δ-doped GaAs pHEMT), measured in a wide depth range.

The measured concentration profile of free charge carriers in sample 2 over a wider depth range (up to 100 μm) is shown in figure 4. As could be seen from the figure, after the QW response, the observed concentration value monotonically decreases according to the Debye law (investigated in detail in [16]) to a semi-insulating substrate with a concentration of ~ 10^{10} cm^{-3}. Note that this is the lowest concentration of free charge carriers in a semiconductor ever measured using the CV technique.

3.2. GaN HEMT

Figure 5 shows the depth distribution profile of free charge carriers in sample 3, measured at several points. This sample is a GaN HEMT structure and typically does not need the presence of additional doping areas. Therefore, unlike the previous samples, we observe only one peak in the concentration profile. Moreover, because of the polarization effects and a larger band offset at the interface compared to GaAs, the peak on the concentration profile has a significantly larger amplitude. So, a maximum of $1.2 \cdot 10^{20}$ cm^{-3} ($2 \cdot 10^{13}$ cm^{-2}) is located at a depth of 22 nm. Thus, even without the use of doping, GaN HEMT allows achieving 2DEG densities of more than 10^{13} cm^{-2}.

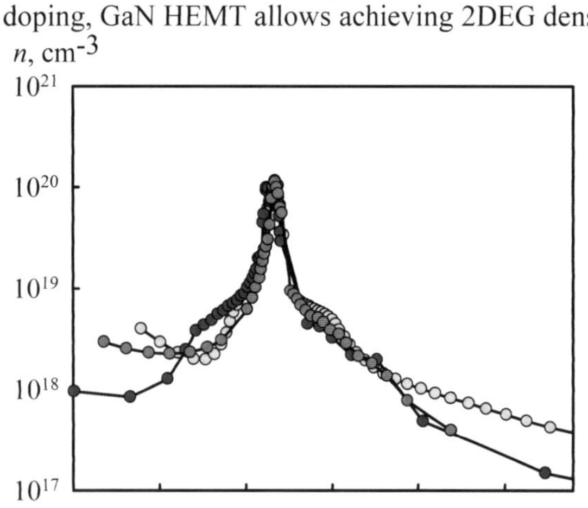

Figure 5. Apparent profile of free charge carriers in sample 3 (GaN HEMT).

A strategy that is similar to the GaAs HEMT approach is used to increase the subthreshold slope characteristic of GaN HEMTs. It means the reduction of the thickness of upper (emitter) layers (in our case, a AlGaN layer). With decreasing thickness of these layers, the magnitude of the induced electric field also decreases, which results in a decrease in the number of charge carriers in the channel. In this case, in order to keep the same device performance, it is necessary to compensate the lost amount of the charge (due to the thickness decreasing). This problem is solved by additional Si-doping of AlGaN. Thus, the general principles of operation and the approaches to performance optimization are similar for GaAs and GaN HEMT, with the differences due solely to the nature of the 2DEG channel formation. The measured concentrations of free charge carriers in the studied samples are compared in table 1.

Table 1. The comparison of obtained concentrations of free charge carriers in GaAs and GaN HEMT structures

Sample number	Emitter carrier concentration in maximum, cm^{-3}	QW carrier concentration in maximum, cm^{-3}	2DEG density, cm^{-2}
1	$1.5\cdot10^{18}$	$2\cdot10^{18}$	$1.6\cdot10^{12}$
2	$3\cdot10^{18}$	$2\cdot10^{18}$	$1.6\cdot10^{12}$
3	-	$1.2\cdot10^{18}$	$2\cdot10^{13}$

4. Conclusion

In this work, a set of GaAs and GaN heterostructures were investigated using the electrochemical CV technique. The apparent depth distribution profiles for free charge carriers were obtained. The position of the emitter layer and the 2DEG concentration in the quantum well were estimated. The obtained results show high repeatability and correlate well with Hall measurements.

The comparison of the apparent profiles for free charge carriers in GaAs pHEMT structures with different types of emitter layer doping (bulk and δ-) showed identical QW filling. Despite the apparent absence of δ-doping advantages, reduced thickness of the upper layers allows to obtain a greater value of the steep subthreshold slope characteristic, and therefore the ability of better control of the final device with the same operating current and power. This also helps to reduce the cutoff bias. For one of the structures, the record minimum value of the free charge carrier concentration ever measured by the CV technique (10^{10} cm^{-3}) was obtained.

A HEMT structure typically does not need the presence of additional doping areas. However, due to the polarization effects and a larger band offset at the interface compared to GaAs, the 2DEG density in the channel ($2\cdot10^{13}$ cm^{-2}) is an order of magnitude higher than that in GaAs pHEMT structures.

Acknowledgments
The research was supported by FASIE (№12063GU/2017 from 14.07.2017).

References

[1] Aadit M, Kirtania S, Afrin F, Alam K and Khosru Q 2017 *Different Types of Field-Effect Transistors – Theory and Applications* (Rijeka: InTech) ed M M Pejovic and M M Pejovic chapter 3 pp 45-64
[2] Xiao-Guang H, De-Gang Z and De-Sheng J 2015 *Chin. Phys. B* **24** 067301
[3] Chu B, Kang B, Wang H, Chang C, Tseng Y, Goh A, Sciullo A, Wu W, Lin J, Gila B and Pearton S 2009 *Proc. of SPIE - The International Society for Optical Engineering* (San Jose: SPIE) p 72162A
[4] Kang B 2005 *Fabrication and Characterization of Compound Semiconductor Sensors for Pressure, Gas, Chemical, And Biomaterial Sensing* (Doctoral dissertation, University of Florida)
[5] Daoudi M, Dhifallah I, Ouerghi A and Chtourou R 2012 *Superlattices and Microstructures* **51** 497

[6] Dudin A, Kogan I, Yakovlev G, Mironova M, Schukov I, Frolov D, Zubkov V and Glinskii G 2017 *Proc. of 2017 11th Int. Workshop on the Electromagnetic Compatibility of Integrated Circuits (EMCCompo)* (St. Petersburg: IEEE) p 108
[7] Yakovlev G, Frolov D and Zubkov V 2016 *J. Phys. Conf. Ser.* **690** 012015
[8] Shestakova L, Yakovlev G and Zubkov V 2017 *J. Phys. Conf. Ser.* **816** 012022
[9] Knübel A, Polyakov V, Kirste L and Aidam R 2010 *Appl. Phys. Lett.* **96** 082106
[10] Wolff T, Rapp M and Rotter T 2004 *Phys. Stat. Sol. A* **201** 2067
[11] Frolov D and Zubkov V 2016 *Semicond. Sci. Technol.* **31** 125013
[12] Petrovskaya A and Zubkov V 2011 *Bull. Russ. Acad. Sci. Phys.* **75** 1416
[13] Vasilevskii I, Galiev G, Klimov E, Mokerov V, Shirokov S, Imamov R and Subbotin I 2008 *Semiconductors* **42** 1084
[14] Tarasova E *et al* 2016 *Semiconductors* **50** 1599
[15] Dudin A, Mironova M, Yakovlev G, Frolov D, Kogan I, Schukov I, Zubkov V and Glinskii G 2017 *Applied Physics* **3** 78
[16] Yakovlev G, Zubkov V, Frolov D, Zubkova A, Sterlyadkin O and Sorokin S 2016 *Semiconductors* **50** 320

Study and optimization of heterojunction silicon solar cells

N A Chuchvaga[1], D V Zhilina[2], S R Zhantuarov[1], S Zh Tokmoldin[1], E I Terukov[2,3] and N S Tokmoldin[1]

[1] Institute of Physics and Technology, Satbayev University, Ibragimova 11, Almaty 050032, Kazakhstan
[2] R&D Center TFTE, Polytechnicheskaya 28, St. Petersburg 194064, Russia
[3] Ioffe Institute, Polytechnicheskaya 26, St. Petersburg 194021, Russia

E-mail: nikolay.chuchvaga@gmail.com

Abstract. This paper reports on the optimization of fabrication parameters of a heterojunction silicon solar cell using computer simulations and comparison of the simulation data with experimental results. The optimization of the photovoltaic devices was performed with respect to the thicknesses of amorphous silicon layers. As a result, a reasonable correlation between the simulation data and experimental results is observed within the acceptable margin of error with respect to heterojunction silicon solar cells with efficiency exceeding 20%.

1. Introduction

Heterojunction silicon solar cells, also known as HIT (heterojunction with intrinsic thin layer) cells, currently demonstrate efficiencies exceeding 26% [1-3]. Their advantages over diffusion-based technologies include, among others, low manufacturing temperatures below 200°C (compared to 900-1000°*C*) and a weaker dependence of power output on operating temperature [4]. Despite impressive recent achievements, there is still potential for further improvement of the heterojunction silicon technology both in terms of reaching higher efficiency and reducing the cost. The HIT cells have a rather complex structure representing a p-i-n-i-n^+ junction, where an n-type monocrystalline silicon wafer is "sandwiched" between intrinsic and doped layers of amorphous silicon (a-Si). The multilayered structure of the cell brings significant complications to optimization of its performance in laboratory due to the requirement to consider several parameters at once. Computer simulation is an important tool in the study of solar cells that enables their optimization in the absence of significant resources with respect to a number of parameters [5-8]. This approach is employed herein to optimize fabrication parameters of a heterojunction silicon solar cell and compare the obtained data with experimental results.

2. Experiment

Simulation of HIT solar cells was performed using the AFORS-HET simulation tool. The AFORS-HET software is intended for simulation of heterojunction solar cells based on silicon and amorphous silicon and features a rich database with parameters of these materials. The underlying principles of AFORS-HET have been reported elsewhere [9]. The built solar cell models used material parameters both integrated within the software and measured separately on fabricated samples. Some of the key parameters of monocrystalline and amorphous silicon layers used in the simulation are given in tables 1 and 2. Device fabrication was performed using commercial 170 μm-thick n-type monocrystalline silicon wafers subjected to proprietary processes of texturing and oxidation. Oxide layer removal was performed immediately prior to the amorphous layer deposition in a dilute aqueous solution of hydrogen fluoride. This was followed by plasma-enhanced chemical vapour deposition (PECVD) of intrinsic (*i*) and doped (*p*/*n*) a-Si layers on both sides of the wafer to obtain a p-i-n-i-n^+ structure, which was capped with indium-tin oxide (ITO) by means of magnetron sputtering. Finally, the contact grid was printed using a commercial Ag-based metallization paste. Measurements of the cell I-V characteristics were made using a PV Measurements IVL-16 setup.

Table 1. Parameters of monocrystalline *n*-Si simulated using the AFORS-HET tool.

Dielectric constant	11.9	Layer density	2.328 g·cm^{-3}
Electron affinity	4.05 eV	Doping concentration, donors	2.15×10^{17} cm^{-3}
Effective conduction band density	2.846×10^{19} cm^{-3}	Doping concentration, acceptors	0 cm^{-3}
Effective valence band density	2.685×10^{19} cm^{-3}		
Band gap E_g	1.124 eV	Electron/hole thermal velocity	10^7 cm/s
Effective electron mobility	1111 cm^2/Vs	Effective hole mobility	421.6 cm^2/Vs
Auger electron recombination coefficient	2.2×10^{-32} cm^6/s	Auger hole recombination coefficient	9.9×10^{-33} cm^6/s

Table 2. Parameters of amorphous *p*/*n*-Si simulated using the AFORS-HET tool.

Dielectric constant	11.9	Effective electron mobility	20 cm^2/Vs
Electron affinity	3.9 eV	Effective hole mobility	5 cm^2/Vs
Effective conduction/valence band density	10^{20} cm^{-3}	Doping concentration acceptors/donors	10^{20} cm^{-3}
Band gap E_g	1.72 eV		

3. Results

The results presented in this paper represent both the data obtained from computer simulations and the measurements on fabricated devices. Further, we describe the use of HIT cell simulations using the AFORS-HET tool to optimize the device structure with respect to a number of parameters. This is followed by comparison of the simulation data with the results of the structure optimization performed on fabricated devices. Finally, the device model is adjusted based on comparison of the simulation data and experimental results.

The structure of a HIT cell is given in figure 1. Separation of the electron-hole pairs generated within the device proceeds by means of a built-in electric field between an *n*-type crystalline silicon wafer and a *p*-type amorphous silicon layer. These two layers are separated by a thin intrinsic amorphous layer employed for passivation of surface states in a crystalline silicon substrate. The rear side of the crystalline wafer is also coated with an intrinsic amorphous layer, followed by an *n*-type amorphous back-surface field (BSF) layer intended to reduce the minority carrier current in this region. This structure is "sandwiched" from both sides by a transparent conducting oxide and metallic contacts. The initial parameters of the simulated solar cell structure are presented in table 3.

The simulated initial device performance was quite poor. It was characterized by the following parameters: short-circuit current J_{SC} = 29.25 mA·cm^{-2}, open-circuit voltage V_{OC} = 683.6 mV, and fill factor FF = 56.3%, resulting in a power conversion efficiency (PCE) of 11.26%.

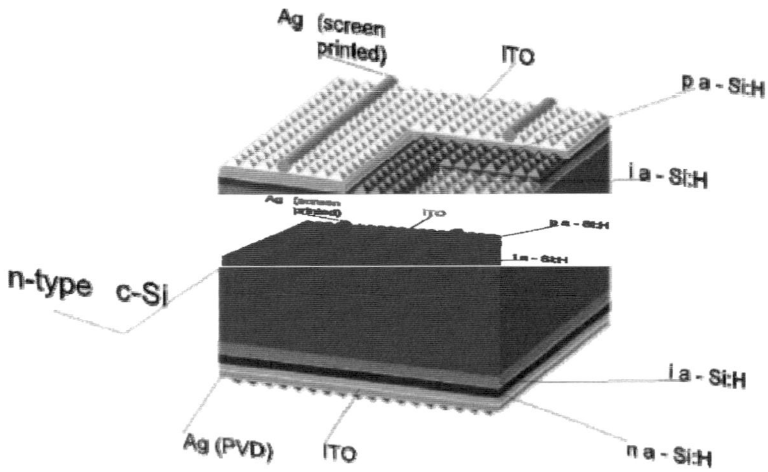

Figure 1. Schematic representation of the heterojunction silicon (HIT) solar cell structure.

Table 3. Initial parameters of the HIT solar cells simulated using the AFORS-HET tool.

$d_{p\text{-a-Si}}$[a]	10 nm	$N_{a,p\text{-a-Si}}$[b]	7.47×10^{19} cm^{-3}
$d_{n\text{-c-Si}}$	300 μm	$N_{d,n\text{-c-Si}}$[c]	1.5×10^{16} cm^{-3}
$d_{n\text{-a-Si}}$	5 nm	$N_{d,n\text{-a-Si}}$	1×10^{20} cm^{-3}
$d_{i\text{-a-Si, front}}$	10 nm	$d_{i\text{-a-Si, rear}}$	3 nm

[a] '*d*' denotes the layer thickness.
[b] 'N_a' denotes the doping concentration of acceptors.
[c] 'N_d' denotes the doping concentration of donors.

Optimization of the device structure was performed by varying the following parameters in the sequence of their diminishing impact on cell efficiency (identified separately in a work due for publication):

- Doping of the front emitter layer,
- Thickness of the front intrinsic layer,
- Doping of the silicon wafer,
- Thickness of the front emitter layer,
- Thickness of the silicon wafer,
- Doping of the back-surface field (BSF) layer at the rear,
- Thickness of the back-surface field (BSF) layer at the rear,
- Thickness of the rear intrinsic layer.

Taking into consideration the arguments presented elsewhere [10], the intrinsic layer thickness was set to be 5 nm, which is related to the dependence of the minority carrier lifetime within the silicon wafer on the thickness of the passivating *i*-a-Si layer. As a result, the following output characteristics of the optimized cell were obtained: $J_{SC} = 33.60$ mA·cm^{-2}, $V_{OC} = 735.2$ mV, $FF = 84.95\%$, and $PCE = 21.02\%$. The respective cell fabrication parameters are given in table 4.

Figure 2 compares the results of computer simulations with the experimental data obtained on fabricated devices with varying thickness of the front *i*-a-Si layer. The intrinsic layer thickness is expected to affect the device performance in two ways: firstly, due to the above mentioned passivation requiring thicker layers to obtain lower recombination and a higher V_{OC}, and secondly, due to the increase in series resistance with growing thickness. The optimal *i*-a-Si layer thickness, as suggested by the experimental results, is 7 nm. Qualitative differences between the simulation and the experiment may be explained by the fact that the simulation did not account for the significance of passivation of crystalline silicon surface states. Therefore, increasing thickness of the intrinsic amorphous layer in this case is only expected to result in a higher series resistance. In reality, the requirement for surface passivation leads to a minimum intrinsic layer thickness of 5-7 nm, based, for example, on the dependence of the minority carrier lifetime in a crystalline silicon wafer on the passivation layer thickness presented earlier [10].

Table 4. Optimized parameters of HIT solar cells simulated using the AFORS-HET tool.

$d_{p\text{-a-Si}}$	7 nm	$N_{a,p\text{-a-Si}}$	1×10^{20} cm^{-3}
$d_{n\text{-c-Si}}$	260 μm	$N_{d,n\text{-c-Si}}$	2×10^{17} cm^{-3}
$d_{n\text{-a-Si}}$	20 nm	$N_{d,n\text{-a-Si}}$	1×10^{20} cm^{-3}
$d_{i\text{-a-Si, front}}$	5-7 nm	$d_{i\text{-a-Si, rear}}$	5 nm

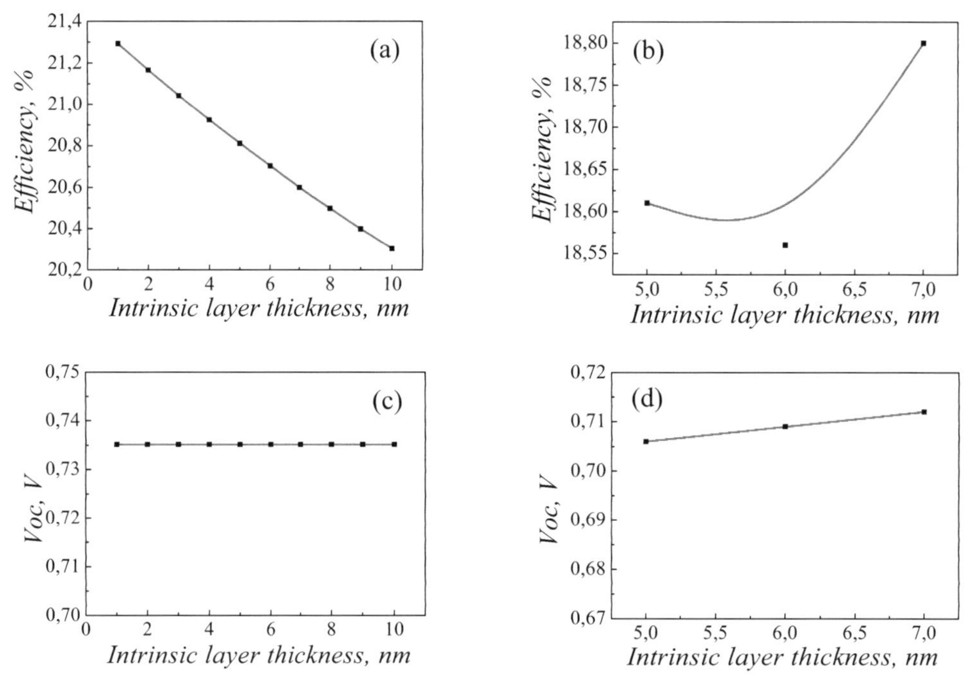

Figure 2. Variation in heterojunction silicon solar cell performance with front *i*-a-Si layer thickness: (a), (c) – simulation; (b), (d) – experimental results.

Figure 3 considers the change in device performance with thickness of the *p*-type emitter layer. Preliminary simulation results showed that the optimal emitter thickness was 1 nm (or as low as possible), which is explained by the tendency to minimize the scattering of charge carriers and the resistance of cell series. Separately, based on the calculations using the Anderson model, the minimal thickness of the amorphous silicon layer was evaluated with the assumption that it must exceed the width of the space charge region [11]. The calculations have shown that the internal electric field propagates into the amorphous silicon layer to a distance of 3 nm at the 'amorphous silicon-crystalline silicon' interface and to a distance of 4 nm at the 'amorphous silicon-ITO' interface. In total, the minimal *p*-type emitter layer thickness is expected to be at least 7 nm. This argument has been taken into account to specify the minimum *p*-type layer thickness in the final optimized model of the solar cell. However, the simulation results in figure 3 do not appear to account for these considerations, as evidenced by the data corresponding to a *p*-type layer thickness below 7 nm. The experimental results in figure 3 suggest that the optimum *p*-type layer thickness is 10 nm, agreeing rather well with the theory.

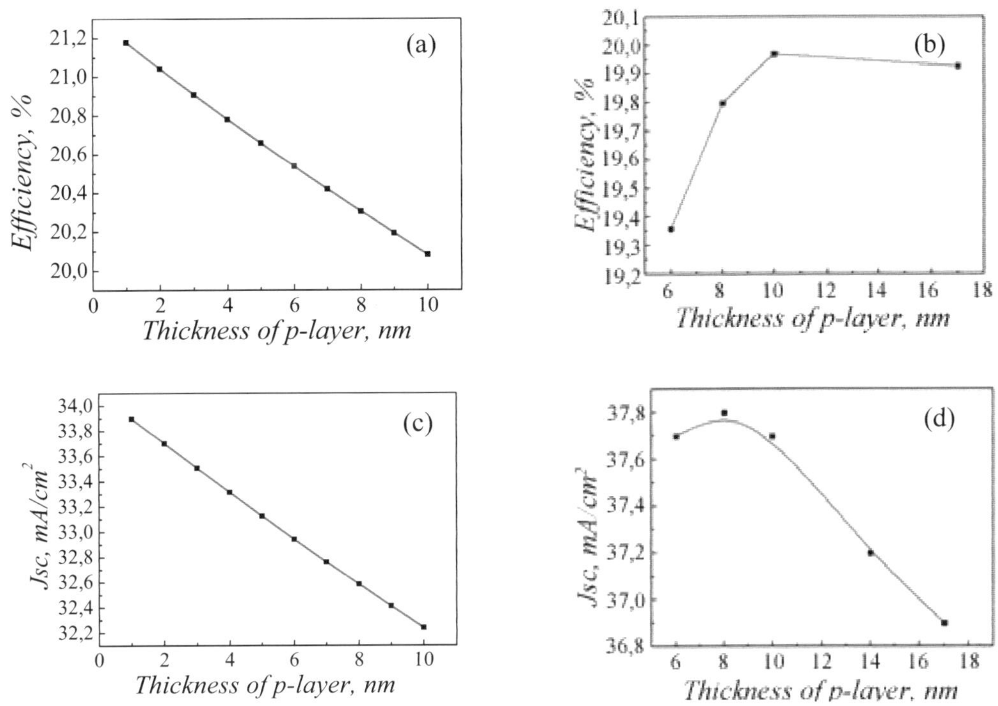

Figure 3. Variation in heterojunction silicon solar cell performance with *p*-type emitter layer thickness: (a), (c) – simulation; (b), (d) – experimental results.

In contrast to the front *p*-type emitter layer acting as a component of a semiconductor junction and an optical filter, the n^+-type layer is located at the rear of the device, so the influence of its thickness on solar cell output characteristics is not expected to be significant (figure 4). At the same time, it is evident that, on the one hand, the n^+-type layer thickness must be sufficient to enable formation of the back surface field, and, on the other hand, reduction in thickness is consistent with minimization of series resistance. In terms of computer simulations, variation in thicknesses of the rear *i*-type and n^+-type a-Si layers in the range from 1 to 50 nm did not result in any changes in the device output characteristics.

Based on the laboratory optimization of the HIT cell structure, optimal parameters of a HIT-structured solar cell were elucidated for the given wafer quality, and an optimized solar cell was fabricated with the following output characteristics: $J_{SC} = 37.51$ mA·cm^{-2}; $V_{OC} = 713$ mV; $FF = 77.18\%$; $PCE = 20.64\%$. The simulated values for comparison were as follows: $J_{SC} = 30.33$ mA·cm^{-2}; $V_{OC} = 781.6$ mV; $FF = 85.44\%$; $PCE = 20.26\%$. Respective device parameters are given in table 5.

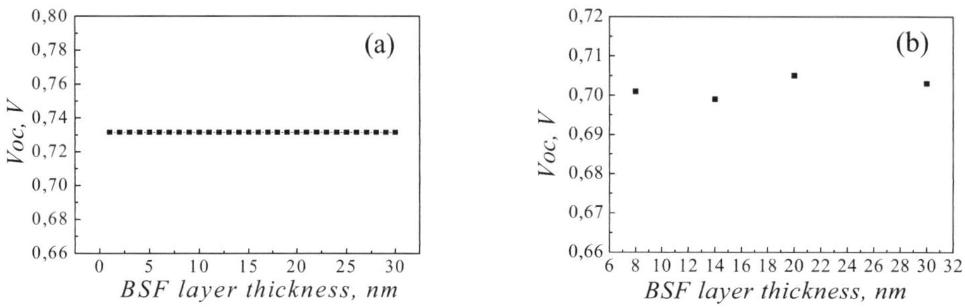

Figure 4. Variation in open-circuit voltage of a heterojunction silicon solar cell with n^+-type rear layer thickness: (a) – simulation; (b) – experimental results.

Table 5. Final optimized parameters of HIT solar cells.

$d_{p\text{-a-Si}}$	10 nm	$N_{a,p\text{-a-Si}}$	3.59×10^{20} cm^{-3}
$d_{n\text{-c-Si}}$	170 μm	$N_{d,n\text{-c-Si}}$	2.15×10^{17} cm^{-3}
$d_{n\text{-a-Si}}$	20 nm	$N_{d,n\text{-a-Si}}$	1×10^{20} cm^{-3}
$d_{i\text{-a-Si, front}}$	7 nm	$d_{i\text{-a-Si, rear}}$	9 nm

Here, the rear *i*-a-Si layer thickness was increased from 7 to 9 nm in order to ensure an efficient surface defect passivation.

4. Conclusion

This paper reports on the efforts of using computer simulation results to optimize the device structure and performance of a heterojunction silicon solar cell. Optimization of the cell performance was done with respect to thicknesses and doping concentrations of the functional layers. In addition, HIT cell fabrication was performed in laboratory to compare the simulation data and experimental results. Reasonable matching of simulation and experimental results is observed, however, further studies are required to account for material properties in the search for higher device efficiencies.

Acknowledgments

We are grateful to the Science Committee of the Republic of Kazakhstan Ministry of Education and Science (Grant No. AP05133645/GF) and the Newton – al-Farabi Institutional Links program of the British Council and JSC Science Fund of Kazakhstan (Grant No. 172723657).

References

[1] Masuko K, Shigematsu M, Hashiguchi T, Fujishima D, Kai M, Yoshimura N, Yamaguchi T, Ichihashi Y, Mishima T, Matsubara N and Yamanishi T 2014 *IEEE Journal of Photovoltaics* **4** 1433-1435

[2] Green M A, Emery K, Hishikawa Y, Warta W, Dunlop E D 2015 *Progress in photovoltaics: research and applications* **23** 1–9

[3] Tokmoldin N S, Chuchvaga N A, Verbitsky V N, Terukov E I, Titov A S, Tokmoldin S Zh and Zholdybayev K S 2017 *Technical Physics* **62** 1877–81

[4] Tanaka M, Taguchi M, Matsuyama T, Sawada T, Tsuda S, Nakano S, Hanafusa H and Kuwano Y 1992 *Japanese Journal of Applied Physics* **31** 3518

IOP Conf. Series: Journal of Physics: Conf. Series **993** (2018) 012039 doi:10.1088/1742-6596/993/1/012039

[5] Stangl R and Leendertz C *Physics and Technology of Amorphous-Crystalline Heterostructure Silicon Solar Cells* (Berlin Heidelberg: Springer) pp 445–58

[6] Zhao L, Zhou C L, Li H L, Diao H W and Wang W J 2008 *Solar Energy Materials and Solar Cells* **92** 673–81

[7] Yang L, He F, Zheng L, Zhang Z, Ma P and Liu S F 2016 *Proc. China Semiconductor Technology Int. Conf. (CSTIC), Shanghai* (IEEE)

[8] Varache R, Leendertz C, Gueunier-Farret M E, Haschke J, Muñoz D and Korte L 2015 *Solar Energy Materials and Solar Cells* **141** 14–23

[9] Froitzheim A, Stangl R, Elstner L, Kriegel M and Fuhs W 2003 *Proceedings of 3rd World Conference, IEEE* **1** 279–82

[10] Herasimenka S Y 2013 *Large Area Ultrapassivated Silicon Solar Cells Using Heterojunction Carrier Collectors* (Arizona State University, USA)

[11] Ilyin V I, Musikhin S F and Shik A Ya 2000 *Varionic semiconductors and heterostructures, in Russian* (Saint Petersburg: Nauka) pp 21–44

Internal friction in particulate composites of $(x)Mn_{0.4}Zn_{0.6}Fe_2O_4$–$(1\text{-}x)PbZr_{0.53}Ti_{0.47}O_3$ in the vicinity of the structural phase transition temperatures

A V Kalgin and S A Gridnev

Voronezh State Technical University, 14 Moskovsky prospect, Voronezh 394026, Russia

E-mail: kalgin_alexandr@mail.ru

Abstract. The internal friction in particulate ceramic composites of $(x)Mn_{0.4}Zn_{0.6}Fe_2O_4$–$(1\text{-}x)PbZr_{0.53}Ti_{0.47}O_3$ (x = 0, 0.1, 0.2, 0.3, 0.4, and 0.6) in the vicinity of the phase transition temperatures was studied. We observed the influence of the composite composition on the exponent that characterizes a temperature dependence of the internal friction near the ferroelectric Curie point. The reason for this influence is shown to be the doping of the $PbZr_{0.53}Ti_{0.}$4703 ferroelectric phase with atoms of the $Mn_{0.4}Zn_{0.6}Fe_2O_4$ ferrite phase that occurs during high-temperature sintering of composite samples.

1. Introduction

In modern solid-state physics, special attention is paid to systems with structural phase transitions (PT) of different natures because in these systems the anomalies of some physical parameters (heat capacity, dielectric constant, magnetic susceptibility, etc.) are observed in the vicinity of the Curie temperatures T_C. These anomalies are usually discussed in terms of the Landau thermodynamic theory and different microscopic models as well.

In most cases, at a temperature $T \rightarrow T_C$, some theories predict a simple power dependence of physical parameters on temperature in the form $A(T - T_C)^p$, where A is a temperature-independent factor and p is the exponent obtained from thermal, dielectric, magnetic and other measurements.

The main task of theoretical and experimental research is to determine the values of exponents and to reveal regularities and reasons for their variations in systems of different natures. The solution of this task makes it possible to reveal the limits for various theoretical approaches and to predict general features of the physical parameter variation for systems at a temperature $T \rightarrow T_C$.

Until now, the effect of the composite structure and composition on the exponent value has been studied in simple and complex systems using mainly measurements of the dielectric permittivity and magnetic susceptibility [1, 2]. At the same time, this influence has rarely been studied by the internal friction method characterized by the high structural sensitivity that is often unattainable by other methods.

Therefore, this study seeks to investigate the effect of the composition of particulate composites of $(x)Mn_{0.4}Zn_{0.6}Fe_2O_4$ –$(1\text{-}x)PbZr_{0.53}Ti_{0.47}O_3$ [hereafter, (x)MZF–$(1\text{-}x)$PZT] on the exponents characterizing the temperature dependences of the internal friction Q^{-1} near the magnetic and ferroelectric Curie points.

The magnetoelectric composites were chosen due to their multifunctional properties and potential applications based on the magnetoelectric effect in these materials (the possibility of mutual conversion of magnetic and electric fields). Besides, there exists a unique correlation between the ME response and internal friction Q^{-1} in composites [3].

2. Experiment

Samples of particulate magnetoelectric (x)MZF–(1-x)PZT composites, where x = 0, 0.1, 0.2, 0.3, 0.4, and 0.6, were prepared using the conventional ceramic technology. Mixtures of powders of the $Mn_{0.4}Zn_{0.6}Fe_2O_4$ (MZF) ferrite with the Curie point T_K = 473 K and $PbZr_{0.53}Ti_{0.47}O_3$ (PZT) ferroelectric with T_K = 593 K taken in stoichiometric ratio were sintered in air for 5 h at a temperature that was varied from 1373 to 1463 K as the MZF content in the mixture was increased from 0.1 to 0.6 mass fraction.

The prepared samples contained only initial phases (the ferrite phase with the spinel structure and the ferroelectric phase with the perovskite structure). This was confirmed by micrographs obtained using a Quanta 2003D scanning electron microscope and by X-ray diffraction patterns obtained using a DRON-3 diffractometer with FeK_α-radiation and a Ni-filter [4]. Samples were 12 x 2 x 2 mm^3 in size.

We measured the internal friction Q^{-1} in composites depending on temperature using an inverted torsion pendulum. The temperature in the measuring chamber was controlled and measured by a chromel-alumel thermocouple; the measurement error was less than ± 0.5 K.

3. Results and discussion

The measurement data for Q^{-1} in composites near the structural phase transition temperatures are shown in figure 1.

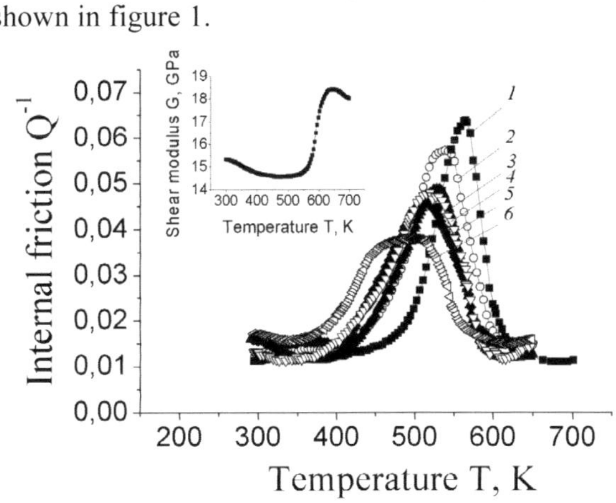

Figure 1. Temperature dependences of internal friction at a frequency of 30 Hz and a heating rate of 2 K/min for (x)MZF–(1-x)PZT composite samples with various x, mass fraction: *1* – 0, *2* – 0.1, *3* – 0.2, *4* – 0.3, *5* – 0.4, and *6* – 0.6. The inset shows the temperature dependence of the composite shear modulus for the sample with x = 0.

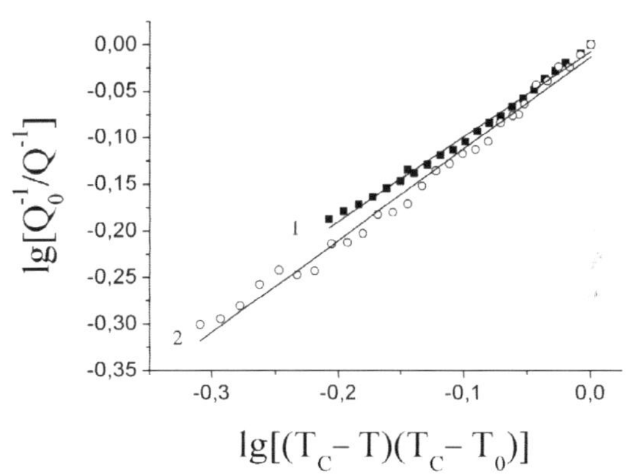

Figure 2. Dependences of $lg[Q_0^{-1}/Q^{-1}]$ on $lg[(T_C-T)/(T_C-T_0)]$ for (x)MZF–(1-x)PZT composite samples with various x, mass fraction: *1* – 0.6 and *2* – 0.2. Q_0^{-1} is the value of Q^{-1} at the temperature T_0.

The $Q^{-1}(T)$ dependences have a maximum at temperatures of 565, 538, 528, 519, 493, and 514 K for the samples with $x = 0$, 0.1, 0.2, 0.3, 0.4, and 0.6, respectively. The observed maximum can be attributed to the first-order structural ferroelectric phase transition (FPT). It is accompanied by a characteristic step-like increase in the composite shear modulus G with temperature. As an example, the $G(T)$ dependence for the composite sample with $x = 0$ is presented in the insert in figure 1. Both $Q^{-1}(T)$ and $G(T)$ anomalies are close to the temperature of the tetragonal-cubic FPT in PZT and far from the temperature of the magnetic PT in MZF.

It should be noted that no Q^{-1} anomalies were observed near the magnetic PT temperature (see figure 1. This can be explained by the substantial smallness (by approximately an order of magnitude) of a Q^{-1} maximum for pure MZF as compared to Q^{-1} maxima for all composite samples.

The nature of the Q^{-1} maximum in PZT can be related to the fluctuation mechanism of losses at the first-order phase transition [5]. According to this mechanism, PT occurs by the fluctuation nucleation of a new phase and subsequent growth of the nuclei of a supercritical size, i.e., because of the phase boundary motion through a system of point defects leading to the internal friction. In the case of small amplitudes of sample oscillations, the expression for the $Q^{-1}(T)$ maximum is written as:

$$Q_{\mathrm{m}}^{-1} = \frac{Ga^2V}{kT_{\mathrm{m}}} \cdot \frac{\dot{m}}{\omega}, \tag{1}$$

where G is the shear modulus; a is the spontaneous strain at T_C; V is the volume of a new-phase nucleus; $\dot{m}$ is the phase transformation rate; k is the Boltzmann constant; T_{m} is the temperature of the Q^{-1} maximum; and ω is the frequency.

As it follows from equation (1), the internal friction Q_m^{-1} should be directly proportional to the heating rate ($\dot{m}$) and inversely proportional to the frequency of a sample oscillation (ω) under other equal conditions. It was this dependence that was found in the investigation of the internal friction dependence on the heating rate in composites of (x)MZF–(1-x)PZT near T_{C} [6].

Figure 1 shows that the maximum of Q^{-1} decreases in height, broadens, and shifts to low temperatures with x at the same frequency. It appears natural to relate the decrease in the maximum height to the increase in the number of point defects that pinning domain walls moving under the mechanical stress in PZT. The increase in the number of point defects leads to larger fluctuations of the composition of PZT in the composite and to a broader distribution of the local Curie temperature over the bulk of PZT. Consequently, the Q^{-1} maximum becomes more diffuse. The shift of the Q^{-1} maximum to lower temperatures, which was revealed in our experiment, is associated with the substitution of Ti atoms (with the ionic radius $R_{\mathrm{Ti}} = 0.064$ nm) by Fe atoms (with the ionic radius $R_{\mathrm{Fe}} = 0.067$ nm $> R_{\mathrm{Ti}}$) [7]. Because of the decrease in the chemical pressure at such substitution, the internal field acting on ferroelectric active ions in unit PZT cells decreases. Consequently, the disordering of the polar state of PZT will occur at a lower temperature.

Temperature dependences of Q^{-1} near T_{C} were supposed to be approximated by the following power function:

$$\lim_{T \to T_{\mathrm{K}}} Q^{-1} = f(\omega, T) \cdot \left| \frac{T - T_{\mathrm{K}}}{T_{\mathrm{K}}} \right|^{-p}, \tag{2}$$

where p is the exponent that is equal to unity for the mechanism studied in detail in [6].

It was interesting to relate the value of p to the structure and composition of composites. For this purpose, we determined the exponent p from temperature dependences of Q^{-1} in (x)MZF– (1-x)PZT composites and revealed a dependence of p on the composite composition. To find the parameter p, the $Q^{-1}(T)$ dependences (figure 1) were plotted in coordinates of $lg[Q_0^{-1}/Q^{-1}]$ on $lg[(T_{\mathrm{K}} - T)/(T_{\mathrm{K}} - T_0)]$ (figure 2). The experimental points were revealed to fall well on a straight line and the slope gives

p = 0.99 over a temperature interval of 51 < $T_K - T$ < 106 K for the composites with x = 0.6 and p = 0.91 over a temperature interval of 70 < $T_K - T$ < 113 K for the composites with x = 0.2.

One can see that the temperature dependences of Q^{-1} for the composites studied are quite well approximated by equation (2), where p takes values from 0.91 to 1.06 depending on the composite composition (figure 3).

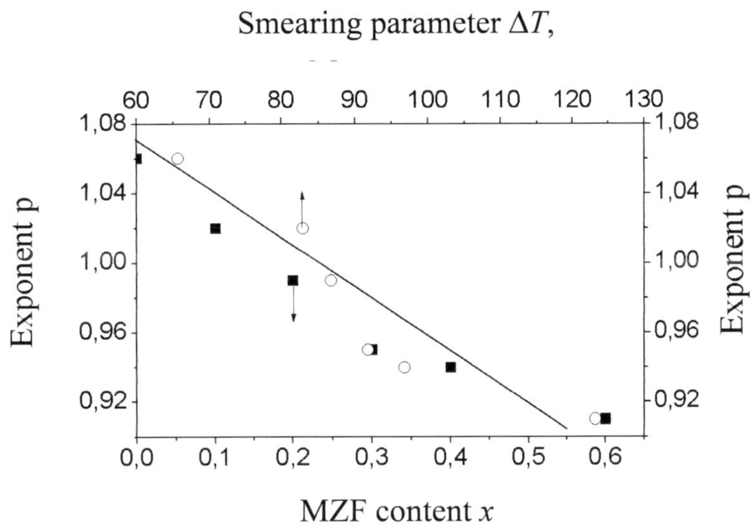

Figure 3. Dependences $p(x)$ and $p(\Delta T)$ for (x)MZF–$(1-x)$PZT composite samples.

Consequently, the parameter p reflects the composition variations occurring in the ferroelectric phase of the composite quite well. According to the theory [5], Q^{-1} is equal to zero in the paraelectric phase. In fact, the $Q^{-1}(T)$ dependence is observed in a certain temperature range above T_C in the paraelectric phase that can be related to the smearing of FPT owing to composition fluctuations and thermal fluctuations of the order parameter.

Since p decreases with x (figure 3) and FPT becomes more smeared with x (figure 1), it is natural to expect that there should be a clear correlation between p and the smearing parameter of the ferroelectric phase transition ΔT. The estimation of the width of the $Q^{-1}(T)$ maxima, ΔT, at their half-height for various composite samples confirms this conclusion. The relationship between p and ΔT is appeared to be inversely proportional (figure 3).

Since the cause of the increase in the degree of the FPT smearing with x in the composite was revealed, and a relation between p and ΔT was found, we came to conclude that the reason for the decrease in p with x is the increase in the content of doping atoms in the ferroelectric phase, which are diffusionally removed from the magnetic phase during high-temperature sintering of composite samples. Thus, the exponent p can be used as an indicator of the defectiveness degree for the ferroelectric phase in composites. We have not seen this approach mentioned before in the literature and believe that it offers certain advantages in material science.

4. Conclusions

Samples of particulate magnetoelectric composites of (x)MZF–$(1-x)$PZT (x = 0, 0.1, 0.2, 0.3, 0.4, and 0.6) were prepared using ceramic technology. The prepared samples were used to measure Q^{-1} over a temperature range of ferroelectric and magnetic structural phase transitions. A pronounced Q^{-1} maximum was found in temperature dependences of Q^{-1} that is due to the ferroelectric phase transition. The maximum behavior agrees rather well with conclusions of the thermodynamic theory that predicts the increase in the Q^{-1} maximum height with the heating rate. The temperature dependence of Q^{-1} near T_C

was approximated by a power function. The exponent p reflecting the smearing degree decreases with x in samples of (x)MZF–(1-x)PZT. The decrease in the parameter p with x is attributed to the substitution of PZT lattice atoms by MZF lattice atoms arising during high-temperature sintering of composite samples.

Acknowledgments
The work was supported by Russian Foundation for Basic Research (grant 16-02-00072 and grant 16-42-360412) and Russian Science Foundation (project 17-72-20105).

References

[1] Stanley H E 1971 *Introduction to phase transitions and critical phenomena* (Oxford: Clarendon Press) p 333
[2] Liu W C, Mak C L, Wong K H, Lo C Y, Or S W, Zhou W, Hauser A, Yang F Y and Sooryakumar R 2008 *J. Phys. D: Appl. Phys.* **41** 125402
[3] Kalgin A V, Gridnev S A and Grigorjev E S 2016 *Bull. Russ. Acad. Sci: Phys.* **80(9)** 1097
[4] Gridnev S A, Kalgin A V, Amirov A A and Kamilov I K 2010 *Ferroelectrics* **397** 142
[5] Gridnev S A and Darinskii B M 1978 *Phys. Status Solidi A* **47** 379
[6] Kalgin A V, Gridnev S A and Karaeva O A 2015 *Phys. Solid State* **57(11)** 2235
[7] Gridnev S A and Kalgin A V 2010 *Phys. Status Solidi B* **247(7)** 1769

Institute of Physics
Dirac House, Temple Back
Bristol BS1 6BE UK

CURRAN ASSOCIATES INC.
proceedings.com

ISSN: 1742-6588
ISBN 978-1-5108-6164-0